T0348373

PROGRESS IN

Molecular Biology and Translational Science

Volume 111

PROGRESS IN

Molecular Biology and Translational Science

Genetics of Stem Cells, Part A

edited by

Yaoliang Tang

Associate Professor
Division of Cardiovascular Disease
Internal Medicine College of Medicine, University of Cincinnati
Cincinnati, OH, USA

Volume 111

AMSTERDAM • BOSTON • HEIDELBERG • LONDON
NEW YORK • OXFORD • PARIS • SAN DIEGO
SAN FRANCISCO • SINGAPORE • SYDNEY • TOKYO
Academic Press is an imprint of Elsevier

Academic Press is an imprint of Elsevier
525 B Street, Suite 1900, San Diego, CA 92101-4495, USA
225 Wyman Street, Waltham, MA 02451, USA
The Boulevard, Langford Lane, Kidlington, Oxford, OX51GB, UK
32, Jamestown Road, London NW1 7BY, UK
Radarweg 29, PO Box 211, 1000 AE Amsterdam, The Netherlands

First edition 2012

Library of Congress Cataloging-in-Publication Data
A catalog record for this book is available from the Library of Congress

British Library Cataloguing in Publication Data
A catalogue record for this book is available from the British Library

ISBN: 978-0-12-398459-3
ISSN: 1877-1173

For information on all Academic Press publications
visit our website at store.elsevier.com

Printed and bound by CPI Group (UK) Ltd, Croydon, CR0 4YY

Transferred to digital print 2012

Contents

Induction of Somatic Cell Reprogramming Using the MicroRNA miR-302 83

Karen Kelley and Shi-Lung Lin

From Ontogenesis to Regeneration: Learning how to Instruct Adult Cardiac Progenitor Cells 109

Isotta Chimenti, Elvira Forte, Francesco Angelini, Alessandro Giacomello, and Elisa Messina

Roles of MicroRNAs and Myocardial Cell Differentiation 139

Tomohide Takaya, Hitoo Nishi, Takahiro Horie, Koh Ono, and Koji Hasegawa

Wnt Signaling and Cardiac Differentiation 153

Michael P. Flaherty, Timothy J. Kamerzell, and
Buddhadeb Dawn

Cross Talk Between the Notch Signaling and Noncoding RNA on the Fate of Stem Cells 175

Yaoliang Tang, Yingjie Wang, Lijuan Chen, Neal Weintraub,
and Yaohua Pan

Myocardial Regeneration: The Role of Progenitor Cells Derived from Bone Marrow and Heart 195

Xiaohong Wang, Arthur H.L. From, and Jianyi Zhang

Role of GATA-4 in Differentiation and Survival of Bone Marrow Mesenchymal Stem Cells 217

Meifeng Xu, Ronald W. Millard, and Muhammad Ashraf

Progenitor Cell Mobilization and Recruitment: SDF-1, CXCR4, α4-integrin, and c-kit 243

Min Cheng and Gangjian Qin

Genetically Manipulated Progenitor/Stem Cells Restore Function to the Infarcted Heart Via the SDF-1α/CXCR4 Signaling Pathway......................... 265

Yigang Wang and Kristin Luther

Genetic Modification of Stem Cells for Cardiac, Diabetic, and Hemophilia Transplantation Therapies ... 285

M. Ian Phillips and Yaoliang Tang

Role of Heat Shock Proteins in Stem Cell Behavior 305

Guo-Chang Fan

Preconditioning Approach in Stem Cell Therapy for the Treatment of Infarcted Heart. 323

Khawaja Husnain Haider and Muhammad Ashraf

Contributors

Numbers in parentheses indicate the pages on which the authors' contributions begin.

Francesco Angelini, Department of Medical Surgical Sciences and Biotechnologies, "Sapienza" University of Rome, Latina, Italy (109)

Muhammad Ashraf, Department of Pathology and Laboratory Medicine, University of Cincinnati College of Medicine, Cincinnati, Ohio, USA (217, 323)

Lijuan Chen, Division of Cardiovascular Disease, Internal Medicine, University of Cincinnati, Cincinnati, Ohio, USA (175)

Min Cheng, Department of Cardiology, Union Hospital, Tongji Medical College, Huazhong University of Science and Technology, Wuhan, Hubei, PR China (243)

Isotta Chimenti, Department of Medical Surgical Sciences and Biotechnologies, "Sapienza" University of Rome, Latina, Italy (109)

Buddhadeb Dawn, Division of Cardiovascular Diseases, University of Kansas Medical Center, Kansas City, Kansas, USA (153)

Guo-Chang Fan, Department of Pharmacology and Cell Biophysics, University of Cincinnati College of Medicine, Cincinnati, Ohio, USA (305)

Michael P. Flaherty, Division of Cardiovascular Medicine, University of Louisville School of Medicine, Louisville, Kentucky, USA (153)

Elvira Forte, Department of Molecular Medicine, Pasteur Institute, Cenci-Bolognetti Foundation, "Sapienza" University of Rome, Rome, Italy (109)

Arthur H.L. From, Department of Medicine, University of Minnesota Medical School, Minneapolis, Minnesota, USA (195)

Alessandro Giacomello, Department of Molecular Medicine, Pasteur Institute, Cenci-Bolognetti Foundation, "Sapienza" University of Rome, Rome, Italy (109)

Khawaja Husnain Haider, Department of Pathology and Laboratory Medicine, University of Cincinnati, Cincinnati, Ohio, USA (323)

Ji Woong Han, Department of Medicine, Division of Cardiology, Emory University School of Medicine, Atlanta, Georgia, USA (1)

Koji Hasegawa, Division of Translational Research, National Hospital Organization Kyoto Medical Center, Kyoto, Japan (139)

Takahiro Horie, Department of Cardiovascular Medicine, Graduate School of Medicine, Kyoto University, Kyoto, Japan (139)

Eneda Hoxha, Feinberg Cardiovascular Research Institute, Feinberg School of Medicine, Northwestern University, Chicago, Illinois, USA (27)

Timothy J. Kamerzell, Division of Cardiovascular Diseases, University of Kansas Medical Center, Kansas City, Kansas, USA (153)

Karen Kelley, Division of Regenerative Medicine, WJWU & LYNN Institute for Stem Cell Research, Santa Fe Springs, California, USA (83)

Raj Kishore, Feinberg Cardiovascular Research Institute, Feinberg School of Medicine, Northwestern University, Chicago, Illinois, USA (27)

Shi-Lung Lin, Division of Regenerative Medicine, WJWU & LYNN Institute for Stem Cell Research, Santa Fe Springs, California, USA (83)

Kristin Luther, Department of Pathology and Laboratory Medicine, College of Medicine, University of Cincinnati, Cincinnati, Ohio, USA (265)

Elisa Messina, Department of Molecular Medicine, Pasteur Institute, Cenci-Bolognetti Foundation, "Sapienza" University of Rome, Rome, Italy (109)

Ronald W. Millard, Pharmacology and Cell Biophysics, University of Cincinnati Medical Center, Cincinnati, Ohio, USA (217)

Hitoo Nishi, Department of Cardiovascular Medicine, Graduate School of Medicine, Kyoto University, Kyoto, Japan (139)

Koh Ono, Department of Cardiovascular Medicine, Graduate School of Medicine, Kyoto University, Kyoto, Japan (139)

Yaohua Pan, Division of Cardiovascular Disease, Internal Medicine, University of Cincinnati, Cincinnati, Ohio, USA (175)

M. Ian Phillips, Keck Graduate Institute, Claremont, California, USA (285)

Gangjian Qin, Feinberg Cardiovascular Research Institute, Department of Medicine – Cardiology, Northwestern University Feinberg School of Medicine, Chicago, Illinois, USA (243)

Johnson Rajasingh, Cardiovascular Research Institute, Division of Cardiovascular Diseases, Department of Internal Medicine, University of Kansas Medical Center, Kansas City, Kansas, USA (51)

Young-Doug Sohn, Department of Medicine, Division of Cardiology, Emory University School of Medicine, Atlanta, Georgia, USA (1)

Tomohide Takaya, Division of Translational Research, National Hospital Organization Kyoto Medical Center, Kyoto, Japan (139)

Yaoliang Tang, Division of Cardiovascular Disease, Internal Medicine, University of Cincinnati, Cincinnati, Ohio, USA (175, 285)

Xiaohong Wang, Department of Medicine, University of Minnesota Medical School, Minneapolis, Minnesota, USA (195)

Yigang Wang, Department of Pathology and Laboratory Medicine, College of Medicine, University of Cincinnati, Cincinnati, Ohio, USA (265)

Yingjie Wang, Division of Cardiovascular Disease, Internal Medicine, University of Cincinnati, Cincinnati, Ohio, USA (175)

Neal Weintraub, Division of Cardiovascular Disease, Internal Medicine, University of Cincinnati, Cincinnati, Ohio, USA (175)

Meifeng Xu, Department of Pathology and Laboratory Medicine, University of Cincinnati College of Medicine, Cincinnati, Ohio, USA (217)

Young-sup Yoon, Department of Medicine, Division of Cardiology, Emory University School of Medicine, Atlanta, Georgia, USA (1)

Jianyi Zhang, Department of Medicine, University of Minnesota Medical School, Minneapolis, Minnesota, USA (195)

Preface

The book *Genetics of Stem Cells* is a compilation of expert reviews in the stem cell and development field. This is a very rewarding experience for me to work with talented colleagues. As a volume editor, I want to acknowledge the publisher team: Sarah Latham, Mary Ann Zimmerman, Mohamed Ashiq Ibrahim Nazeer, Lisa Tickner, and Michael Conn; they are the unsung heroes of this book.

The pluripotent embryonic stem cells (ESC) have limitations in immunological incompatibility and the ethical problem associated with destroying a human embryo. To circumvent these limitations of ESC, scientists have approached many alternatives such as somatic cell nuclear transfer, cell fusion, or direct reprogramming. With recent advancements, adult somatic cells can now be reverted into induced pluripotent stem cells (iPS). In the first and the third chapters, Dr. Yoon, Dr. Rajasingh, and collaborators describe the current knowledge about the somatic cell reprogramming by means of integrating and nonviral methods. In the second chapter, Dr. Kishore and collaborators highlight the epigenetic mechanisms of induced pluripotent and review the protocols of cardiovascular lineage differentiation of iPS cells. Dr. Lin's lab first reported miR-302-induced iPSCs (mirPSCs). In the fourth chapter, he introduces his experience in the mirPSCs and related epigenetic mechanisms.

Urodeles has extraordinary regenerative capacity of heart, however, such ability has been largely lost in mammalian heart. In the fifth chapter, Dr. Messina and collaborators extensively review the regeneration and its mechanisms starting with the lesson learned from lower vertebrates and then focus on recent advancements and novel insights concerning regeneration in the adult mammalian heart, including the discovery of resident cardiac progenitor cells. The studies of microRNA-mediated gene regulatory network are hot now. In the sixth chapter, Dr. Hasegawa and collaborators review the role of microRNA-mediated myocardial cell differentiation. Wnt and Notch signaling pathways play a variety of important roles in stem cell differentiation and survival. In the seventh chapter, Dr. Dawn and collaborators provide an overview of canonical and noncanonical Wnt signaling and highlight the role of Wnt11 in cardiac differentiation. Recently, microRNA has been reported to cross-talk with Notch signaling pathway in stem/progenitor cell differentiation. In the eighth chapter, Dr. Tang and collaborators review the cross-talk between Notch signaling pathway and microRNA in determining cell fate.

Both bone marrow stem cells and cardiac progenitor cells have been used in clinical trial recently. In the ninth chapter, Dr. Zhang and collaborators discuss heart repair using bone marrow-derived cells or heart-derived cells. The major barrier for clinical stem cell therapy is the severely poor survival of donor cells in ischemic myocardium because over 90% of stem cells are lost within 1 week after cell transplantation. Enhancing the survival of engrafted stem cell survival in host tissue is important for successful progenitor cell therapy for patients. In the tenth chapter, Dr. Xu and collaborators review the role of GATA4, a cardiac early transcriptional factor, in improving bone marrow cell capacity for heart repair. In the eleventh chapter, Dr. Qin and collaborators review the c-kit marker, SDF-1/CXCR4 signaling, and integrin on progenitor cell homing and retention. In the twelfth chapter, Dr. Wang and collaborators introduce their experience in using genetically manipulated cell path for repairing infarcted heart. In the fourteenth chapter, Dr. Fan provide an overview of heat shock proteins, a multifaceted gene involved in stem cell survival, proliferation, and differentiation ageing. In the fifteenth chapter, Dr. Haider and Dr. Ashraf introduce their experience in developing and optimizing the protocols to enhance donor stem cell survival posttransplantation with special focus on preconditioning approach.

Like the famous Trojan horse, the gene modified cell has to gain entrance inside the host's walls and survive to deliver its transgene products. Using cellular, molecular, and gene manipulation techniques, the transplanted cell can be protected in a hostile environment from immune rejection, inflammation, hypoxia, and apoptosis. In the thirteenth chapter, Dr. Phillips and Dr. Tang discuss methods to deliver and construct gene cassettes with viral and nonviral delivery, siRNA, and conditional Cre/Lox P. The constitutive overexpression of therapeutic gene has a risk of unwanted side effects; to overcome this problem, Dr. Phillips and Dr. Tang introduce a novel tissue-specific, hypoxia, or glucose-inducible vigilant vector system for treating heart disease, diabetes, and hemophilia.

YAOLIANG TANG

Generation of Induced Pluripotent Stem Cells from Somatic Cells

YOUNG-DOUG SOHN,[1] JI WOONG
HAN,[1] AND YOUNG-SUP YOON

*Department of Medicine, Division of
Cardiology, Emory University School of
Medicine, Atlanta, Georgia, USA*

The technology for generation of induced pluripotent stem cell (iPSC) from somatic cells emerged to circumvent the ethical and immunological limitations of embryonic stem cell (ESC). The recent progress of iPSC technology offers an unprecedented tool for regenerative medicine; however, integrating viral-driven iPSCs prohibits clinical applications by their genetic alterations and tumorigenicity. Various approaches including nonintegrating, nonviral, and nongenetic methods have been developed for generating clinically compatible iPSCs. In addition, approaches for using more clinically convenient or compatible source cells replacing fibroblasts have been actively pursued. While iPSC and ESC closely resemble in genomic, cell biologic, and phenotypic characteristics, these two pluripotent stem cells are not identical in terms of

[1]These two authors contributed equally to this manuscript.

Progress in Molecular Biology
and Translational Science, Vol. 111
http://dx.doi.org/10.1016/B978-0-12-398459-3.00001-0

1

differentiation capacity and epigenetic features. In this chapter, we deal with the current techniques of generating iPSCs and their various characteristics.

The pluripotency and self-renewal properties of embryonic stem cells (ESCs) are valuable resources utilized for basic research and clinical applications; however, ESCs have limitations such as the ethical problem associated with destroying a human embryo and immunological incompatibility. To circumvent these limitations of ESC, scientists have considered many alternatives such as somatic cell nuclear transfer (SCNT),[1] cell fusion[2], or direct reprogramming.[3,4] With recent advancements, adult somatic cells can now be generated into pluripotent stem cells.

Since the successful generation of induced pluripotent stem cell (iPSC) from adult somatic cells by Yamanaka,[5,6] various methods have been developed for iPSC generation; the primary method is to deliver the ESC-specific transcription factor to the adult somatic cell. Even though the iPSCs were generated effectively by using an integrative virus system, they cannot be used for clinical purposes because of genetic alterations and tumorigenecity. Therefore, various approaches such as nonintegrating vector and non-DNA-based methods including proteins, RNAs, or small molecules have been attempted.

Generation of pluripotent stem cells from adult cells is an artificial manipulation that may not generate cells identical to naturally occurring PSCs. However, some features of iPSC generation may parallel the natural genetic processes that arise during embryonic development, including the reprogramming of the gamete pronuclei at fertilization under the influence of factors in the oocyte. Although remarkable effort has been put into generating patient-specific immunocompatible stem cells, capable methods were not successful until iPSC technology with defined transcription factors was developed.[5,6] In this chapter, we describe the identification, technologies, and advancements regarding iPSCs over the past 5 years.

I. Generation of iPSCs

IPSCs are a type of PSCs that are artificially derived from nonpluripotent cells, typically from an adult somatic cell, by inducing a "forced" expression of ESC-specific genes. iPSCs are similar to natural PSCs, such as ESCs, in many respects, such as the expression of pluripotency-related genes, DNA methylation patterns, self-renewal, embryoid body formation, teratoma formation, and viable chimera formation, but the full extent of their relation to natural PSCs is still being assessed. ESCs are derived from the inner cell mass (ICM) of preimplantation embryos, but alternative approaches such as SCNT, cell fusion, or direct reprogramming[1-4] are now available, which allow the generation of PSC lines directly from differentiated adult somatic tissue.

Based on the success and limitations of previous nuclear transfer and cell culture experiments, scientists began to experiment with more direct manipulation of genetic information of cells to create developmental plasticity in differentiated mature cells. The first report of generating iPSCs from somatic cells was achieved by ectopic overexpression of pluripotency-associated transcription factors.[5,6] Takahashi and Yamanaka introduced a mini-library of 24 candidate reprogramming factors, known as pluripotency-associated genes, which were known to be expressed in ESCs. The genes were introduced into mouse embryonic fibroblasts (MEFs), which carried a fusion of the β-galactosidase and neomycin-resistance genes expressed from the *Fbx15* locus.[7] When MEFs were infected with all 24 genes and cultured on feeder cell layers in an ES medium in the presence of G418, drug-resistant colonies emerged that had ESC-like proliferation, gene expression, and morphology. To narrow down the factors that are essential for reprogramming, all combinations of the 24 factors were attempted until four factors, namely, *Oct3/4*, *Sox2*, *Klf4*, and *c-Myc*, were identified. The resultant cells, which showed pluripotent features indistinguishable from those of ESCs, were indicated as iPSCs (Fig. 1). These four essential factors are often referred to as "Yamanaka factors" in recognition of the inventor of this method, Shinya Yamanaka.

After this discovery, several groups have improved upon the original reprogramming method. One group combined *Oct3/4* and *Sox2* with *Lin28* and *Nanog* to generate human iPSCs.[8] Remarkably, the same four factors identified in the murine system were able to confer pluripotency in primate cells, indicating that the fundamental transcriptional network governing pluripotency is preserved across species. Several groups have shown that the *c-Myc* gene is dispensable for reprogramming,[9-12] although efficiency was then quite low. Recently, Yamanaka and colleagues showed that *c-Myc* can be replaced with *L-Myc*, another Myc family member, for generation of human iPSCs, resulting in even higher efficiency. As *c-Myc* reactivation can trigger tumorigenicity of iPSC derivatives, this study is important.

Although most iPSCs were generated from skin fibroblasts using these reprogramming transcription factors, expression of the same reprogramming factors also appears to initiate a sequence of stochastic events that eventually lead to the generation of iPSCs in a variety of other differentiated cells, such as neural cells,[13-16] keratinocytes,[17] melanocytes,[18] adipose-derived cells,[16,19] amniotic cells,[20-22] pancreatic cells,[23] and blood cells[24-27] (Fig. 1). The forced expression of the four genes is required only temporarily, at the initiation stage of reprogramming, and can then be mostly silenced when endogenous pluripotency-related genes are turned on.[28]

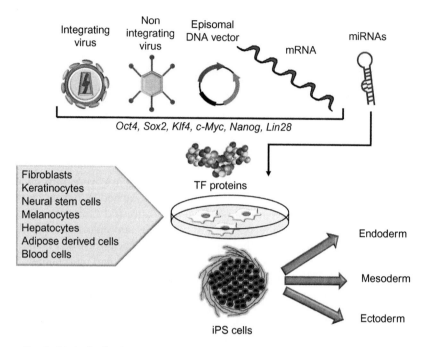

FIG. 1. Methods of iPSC generation and somatic cell sources. iPSCs were originally derived from embryonic fibroblasts in mice and skin fibroblasts in humans. Fibroblasts have been most widely used for reprogramming. Thereafter, other cells including neural stem cells, keratinocytes, melanocytes, adipose-tissue-derived cells, and various blood cells have been successfully used for achieving higher efficiency or enhancing accessibility. There has also been advancement in the delivery vectors for reprogramming factors, *Oct4, Sox2, Klf4, c-Myc, Nanog,* and *Lin28.* Initial studies used DNA-integrating viral vectors such as retroviruses and lentiviruses. Later studies demonstrated the usefulness of a nonintegrating viral vector, adenovirus, and nonviral plasmids for reprogramming. More recently, recombinant proteins, mRNAs and miRNAs, have been shown to reprogram somatic cells into iPSCs, suggesting safer and potentially clinically applicable systems for deriving iPSCs. (See Color Insert.)

II. Methods of Delivering Transcription Factors into Cells

A. Integrating Viral Vectors

Several types of integrating viral vectors, including retrovirus[6,29,30] and lentivirus,[23,31,32] have been used in iPSC generation. Retroviral vectors were the first type of vectors used to create iPSCs, and the site of viral integration has been closely studied. Takahashi and Yamanaka noted approximately 20 retroviral integration sites (RISs) per iPSC clone in their initial report.[6] In a recent report comparing the reprogramming of MEFs to that of murine hepatocytes or gastric epithelial cells, Aoi and colleagues examined the number of RISs for each of the four retroviruses by Southern blot. They detected 1–9 RISs in MEF-derived iPS clones and 1–4 RISs in gastric epithelium or hepatocyte-derived clones, suggesting that tissues of epithelial origin might be more readily reprogrammed. The integration sites were random and did not show common viral integration sites.[33] Several RISs found by Aoi had previously been identified by retroviral-mediated tumor induction in mice, so the safety of iPSCs is still questionable.[34,35] Bioinformatics analysis revealed no enrichment of any specific gene function, gene network, or canonical pathway by retroviral insertions.[36] Retroviruses have a propensity to integrate near transcription start sites, and may be more likely to cause malignant transformations.

Lentiviral vectors have a hypothetical safety advantage over retroviral vectors because they lack the propensity to integrate near transcription start sites.[37] To date, no insertion site analysis has been conducted and thus the biological relevance of vector differences remains theoretical. An additional advantage of lentivirus is its ability to transcribe large genetic packages, and two recent publications detailed the use of polycistronic lentiviral vectors that delivered the four reprogramming factors in a single construct, instead of the four separate vectors, each carrying one previously used gene.[38,39] Both papers demonstrate the derivation of iPSC clones from a single vector integration, which may minimize mutagenesis caused by viral insertion as well as increase efficiency.

B. Nonintegrating Viral Vectors

In the first report on germline competent iPSCs, ~20% of chimeric mice had tumors most likely caused by reactivation of the integrated *c-Myc* proviral transgene in the host genome.[29] Another study showed cancer-related mortality in 18 of 36 (50%) iPSC chimeric mice.[12] Insertional mutagenesis due to the integration of viral vectors into critical sites of the host genome, leading to malignant transformation, has been observed in preclinical and clinical gene therapy trials.[40–42] Because of these limitations and safety concerns, alternative methods of iPSC generation have been sought that focused on eliminating integration of retroviral and lentiviral vectors from the reprogramming procedure. Accordingly, the potential for using nonintegrating vectors was explored. As it is known that only transient expression of the four original factors are required, Stadtfeld and colleagues used a transiently active and nonintegrating adenoviral vector and succeeded in generating iPSC lines.[43] These viruses could contribute to the formation of teratomas and chimeric mice, but were unable to pass through the germline.[43] By contrast, Okita and colleagues were unable to obtain murine hepatocyte iPS clones when the four reprogramming factors were introduced by adenovirus alone and required additional transfections of *Oct3/4* and *Klf4* or *Oct3/4* and *Sox2* by the retrovirus.[44] For unknown reasons, some cell types may be amenable to the safer adenoviral vector transduction, while other cell types cannot be made into iPSCs using this technique. The biologic basis of this cell-specific plasticity is a fundamental but unanswered question in stem cell biology.

C. Nonviral Reprogramming

While the partial success in generating iPSCs with adenovirus suggests that this safer vector may be useful, the widespread use of any viral vector technology in human application is likely impossible. Because of the great potential of iPSCs, investigators have turned to nonviral vector systems for the generation of iPSCs. Yamanaka and colleagues used a polycistronic expression plasmid containing the *Oct3/4*, *Sox2*, and *Klf4* cDNAs linked by the foot-and-mouth disease virus 2A self-cleaving peptide.[44] When this construct, which lacks viral genetic material, was repeatedly transfected into MEFs together with a separate *c-Myc* cDNA expression vector over a 1-week time period (on days 1, 3, 5, and 7), 1–29 *Oct4*-positive iPSC colonies emerged from 1×10^6 cells in 7 out of 10 independent experiments, while from the same number of retrovirus-infected cells, 100 *Oct4*-positive iPSC colonies were routinely obtained. In 6 of 10 experiments, no evidence of plasmid integration into the host genome was detected by polymerase chain reaction (PCR) or Southern-blot analysis.[44] While the efficiency of iPSC generation by plasmid transfection was greatly decreased, and the oncogenic *c-Myc* viral vector transduction was needed, this report provided proof of concept for the generation of iPSCs without transgene integration of the viral vector.

Several recent studies have reported the use of a transient viral vector approach to iPSC generation that would increase efficiency and safety.[45–47] This method begins with the incorporation of all four Yamanaka genes into a single piggyBac (PB) vector, with viral 2A oligopeptides between adjacent genes allowing synthesis of the four factors as a single transcript followed by posttranslational cleavage of the proteins at the appropriate locations. The most important and unique feature of this approach is that the reprogramming genes are removed from the genome by transient transfection of PB transposase.[47] Thus permanent alteration of the genome is avoided and many safety concerns are circumvented. This technique maintains a relatively robust reprogramming efficiency. Using a similar idea, the Cre–loxP recombinase system was used to remove the vector-integrated transgene once reprogramming has been achieved.[46] However, this system is less precise and may leave some residual elements outside the loxP sites including the transposon repeat, which may be mutagenic. These studies confirmed that transient expression of the reprogramming factors is sufficient for reprogramming of somatic cells to acquire pluripotency. They also confirmed that it is possible to remove integrated vector material when it is no longer needed and thus minimize the risk of late cancer development.

D. Recombinant Proteins of Transcription Factors

As an alternate approach, the protein products of the reprogramming genes have been delivered directly to cells, without the viral DNA at all.[48,49] Two independent groups made proteins in which reprogramming factors were fused to poly(arginine), a short basic peptide known as a cell-penetrating peptide (CPP), which can overcome the cell membrane barrier and reprogram human[48] or mouse somatic cells.[49] Zhou and colleagues used purified recombinant proteins containing 11 arginine residues (11R) at the C-terminus of each reprogramming factor to reprogram OG2/Oct4-GFP reporter MEF cells with four cycles of protein treatments plus the histone deacetylase (HDAC) inhibitor, valproic acid (VPA).[49] Kim and colleagues demonstrated reprogrammed human fetal fibroblasts by the treatment of cell extracts from HEK293 cells which expressed each reprogramming factor fused to eight arginine residues (8R), albeit at a very low efficiency (~0.001%) and with prolonged time (~8 weeks). A recent study by Cho and colleagues is also encouraging, as they reported that a single transfer of ESC-extracted proteins to mouse fibroblasts with cell permeabilization can successfully reprogram mouse fibroblasts in relatively short time periods (20–25 days after induction), albeit at a still lower efficiency (0.001%).[50] Thus, although in its present form this protein-based reprogramming technique may fall short of generating patient-specific iPSCs, it is the most promising and realistic technology at the moment to generate virus-free and transgene-free human-patient-specific iPSCs.

E. Synthetic Modified mRNA of Transcription Factors

The efficiency of most of these integration-free systems for iPSC generation is still significantly lower than that of integrating viral vectors. However, more recently, Rossi and colleagues used synthetic modified mRNA to reprogram human fibroblasts into iPSCs and differentiated them into myogenic cells.[51] The resulting iPSCs recapitulate the functional and molecular properties of human ESCs, and were claimed to be generated at much higher efficiencies than standard virus-based techniques or nonintegrating or protein-based methodologies. Although it is technically complicated, the new method can represent an optimal method by avoiding the risks of genomic integration or insertional mutagenesis associated with other virus- or DNA-based methods and overcoming the low efficiency associated with non-DNA methods. In this approach, eukaryotic mRNA was extensively modified by using *in vitro* transcription (IVT) reactions templated by PCR amplicons. To promote efficient translation and boost RNA half-life in the cytoplasm, a 5′ guanine cap was incorporated by inclusion of a synthetic cap analog in the IVT reactions. To reduce the immunogenic profile of synthetic RNA, phosphatase was treated into the capped mRNA. In addition, modified ribonucleoside bases were incorporated into the synthesized mRNAs by substituting 5-methylcytidine (5mC) for cytidine and/or pseudouridine for uridine.[52–56] These modifications remarkably reduced interferon signaling as displayed by Quantitative real time-polymerase chain reaction (qRT-PCR) for a panel of interferon reactive genes. As a preliminary experiment, these modified RNAs encoding the myogenic transcription factor *MYOD* were delivered into C3H10T1/2 cells and were proven to change fibroblast cell fate into myogenic cells.[57] Next, modified RNAs encoding the four Yamanaka factors were synthesized and delivered into fibroblasts every day for 17 days to generate iPSCs. A broad range of transformation in fibroblast morphology to a compact, epithelioid morphology was detected within the first week, followed by the appearance of standard hESC-like colonies toward the end of the second week of transfection. After RNA delivery ended at day 17, colonies were expanded under standard ESC culture conditions and became fully mature iPSCs, named as "RNA-derived iPSCs" (RiPSCs). This protocol showed high reprogramming efficiencies and rapid kinetics with which RiPSCs were converted. This efficiency was two orders of magnitude higher than those typically reported for virus-based generations. Compared to virus-mediated iPSC generation, in which iPSC colonies typically appear around 4 weeks, by day 17 of RNA transfection the plates had become overgrown with ESC-like colonies. This technology meets most of the criteria that are required for patient-specific iPSC generation. Yet, further studies are needed to address the remaining concerns regarding the genomic perturbation and immune response due to the repeated delivery of modified RNAs and the reproducibility by other research groups.

F. microRNAs

microRNAs (miRNAs) are well-known regulators of development and differentiation.[58] Recent studies have reported that specific miRNAs are highly expressed in ESCs and play a crucial role in the regulation of pluripotency-related genes[59,60]. Recently, miRNAs were reported to successfully reprogram of mouse and human somatic cells into iPSCs.[61–63]

III. Nongenetic Approaches for Reprogramming

A. Small Molecules

Huangfu and colleagues tested the effects of small-molecule chemicals involved in chromatin modification on reprogramming using Oct4-GFP reporter mouse cells.[64,65] Treatment of the four factor-infected MEFs with the

DNA methyltransferase inhibitor 5′-azacytidine increased the reprogramming efficiency, and three known HDAC inhibitors—suberoylanilide hydroxamic acid, trichostatin A (TSA), and VPA—also greatly increased the efficiency of reprogramming, with VPA being the most effective compound, showing a >100-fold increase.[64] This demonstration of enhanced reprogramming efficiency by HDAC inhibitors suggests that chromatin modification is a key step in defining a cells' pluripotent state (Fig. 2).

Chemical screens have identified two compounds, BIX01294 and BayK8644, which, in combination with two factors (*Oct4* and *Klf4*), enhanced the reprogramming efficiency of mouse neural progenitors[66] and MEFs[67] (Fig. 2). BIX01294 is an inhibitor of the G9a histone methyltransferases that methylate histone H3 at the position of lysine 9 (H3K9).[68] G9a histone methyltransferase is reported to silence *Oct4* expression during early embryogenesis by subsequent *de novo* DNA methylation at the promoter region, thereby preventing reprogramming.[69–72]

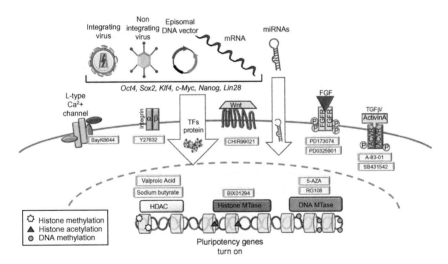

FIG. 2. Potential pathways and epigenetic targets in somatic cell reprogramming. Various signaling pathways and epigenetic targets involved in somatic cell reprogramming that can be modified by small molecules are illustrated. Small-molecule chemicals can enhance reprogramming efficiency through various signaling routes. The DNA methyltransferase inhibitor, histone deacetylase (HDAC) inhibitors, and the histone methyltransferase inhibitor were shown to increase the reprogramming efficiency by epigenetic remodeling of pluripotency-related genes. L-channel calcium agonist, rho-associate kinase (ROCK) inhibitor, Wnt signaling inhibitor, and activators of MEK, FGF, and TGFβ receptor are also known to increase reprogramming efficiency. (See Color Insert.)

Several other small molecules have been found that indirectly influence the epigenetic state of a cell. For example, BayK8644 is an L-channel calcium agonist[66] that exerts its effect through upstream signaling pathways rather than direct epigenetic remodeling. Other small molecules involved in signaling pathways also increase the efficiency of reprogramming, such as the rho-associate kinase (ROCK) inhibitor, Y-27632, which augments human iPSC induction by enhancing cell survival.[30,73] The Wnt signaling inhibitor,[74–76] MEK,[75,76] FGF,[46] and the TGFβ receptor[76,77] also had effects on the generation and maintenance of ground-level pluripotency of iPSCs (Fig. 2). In the near future, small molecules with primary function as epigenetic modulators could facilitate the generation of IPSCs without utilizing genetic material.

B. Altering Cell-Cycle Signal Pathways

Cell-cycle regulatory proteins can also change the efficiency of iPSC generation in ways that are just beginning to be understood. Inhibition of the tumor suppressor protein *p53* related signal pathway facilitates the generation of iPSCs, suggesting that it represses dedifferentiation.[78–80] iPSC technology can be used to study how *p53* modulates the stability of the differentiated state. Inhibition of *p53* directly by *Mdm2* and indirectly by downregulation of *Arf*[80,81] and enhances the progression of normal fibroblasts into iPSCs through its direct target, *p21*, which promotes cellular senescence.[78,79] Deficiency of *p53* improves the efficiency and kinetics of iPSC generation with only two factors, *Oct4* and *Sox2*.[79] Other tumor suppressors, such as *Rb* or *Pten*, are also candidate repressors of dedifferentiation that can be investigated using iPSC technology.[82]

IV. Generation of Human iPSCs from Different Somatic Cell Types

The most significant concern that needs to be solved before human iPSCs can be utilized for clinical purposes is the generation of safe and functional cell types for therapy.[83] Fibroblasts from the embryo, the tail-tip in the mouse, and the dermal fibroblasts in the human have been the most broadly used cell sources for reprogramming, mainly due to their accessibility and relative ease of culture. A widespread study using various mouse iPSCs has confirmed that the origin of the iPSCs has a serious influence on the tumor-forming tendencies in a cell transplantation therapy model.[84] iPSCs derived from the mouse tail-tip fibroblast (mesoderm origin) have shown the highest tumorigenic tendency, whereas gastric epithelial- and hepatocyte-derived iPSCs (both with endodermal origin) have demonstrated significantly lower tumorigenic tendencies.[84]

The molecular mechanism regulating this phenomenon is not yet fully uncovered; however, many reports suggest that the epigenetic memory of the somatic cell of origin is preserved in the iPSCs, and that the memory may affect their directed differentiation potential into blood cells.[85,86] Although it has been reported that human iPSCs keep certain gene expression patterns of the parental cells,[87] it remains unclear whether the cell origin could affect the safety and functionality of human iPSCs. It is consequently important to establish human iPSC lines from different developmental origins and comprehensively investigate the source that might affect both the safety aspects and their differentiation potentials. Human iPSCs have been generated generally from the mesodermal (i.e., fibroblasts and blood cells) or the ectodermal origin cells (i.e., keratinocytes and neural stem cells) (Table I). Recent reports have claimed that the human primary hepatocytes (i.e., endodermal origin) or primary urine cells (i.e., epithelial cells) can be reprogrammed into iPSCs (Table I).[94,96] The technology to develop human endoderm-tissue-derived iPSC lines, together with other established human iPSC lines, provides a foundation to elucidate the mechanisms of cellular reprogramming and study the safety and efficacy of differentially originated human iPSCs for cell therapy. Recently, two groups have reported the generation of human iPSCs from malignant cell lines, indicating that these iPSCs lose certain cancer cell characteristics after reprogramming.[107,108] However, it remains to be investigated whether these cancer-cell-derived iPSCs still have genetic/epigenetic memory of the parental cancer tissue and whether these iPSCs can be used as disease model, in order to study cancer pathogenesis and new drug screening.

V. Characterization of iPSCs

iPSCs express ESC-specific genes that maintain the developmental potential to differentiate into all three primary germ layers, namely, ectoderm, endoderm, and mesoderm. Several functional tests, including *in vitro* differentiation, DNA methylation analysis, *in vivo* teratoma formation, chimera formation, germline transmission, and tetraploid complementation, have been used to define pluripotency of iPSCs. Generally, ESCs and iPSCs share almost identical properties in these assays.[109,110]

A. Genomic Integrity

Maintaining genomic integrity is of crucial importance during the creation of iPSCs, as alterations can cause neoplastic disease and limit therapeutic application. Several groups have investigated the karyotype of mouse[111] and human iPSC lines[5,8] to determine how much genetic alteration is present. One study showed that continuous passaging of human iPSCs resulted in

TABLE I

THE METHODS AND CELL SOURCE FOR iPSC GENERATION

Species	Types	Methods	Cell	TFs	Additional Factors	References
Human	Integrating virus	Retrovirus	Fibroblast	OSKM		5
			Fibroblast	OSNL		8
			Fibroblast	OSKML		88
			Fibroblast	OSKM	hTert, LargeT, ROCKi	30
			Fibroblast	OSKM	p53 shRNA, p53DD	89
			Fibroblast	OSK		10
			Fibroblast	OSK	p53 shRNA	90
			Fibroblast	OS	VPA	
			Keratinocyte	OK	CHIR99021, Parnate, PD325901, SB431542	17
			Cord blood CD34$^+$	OSKM		91,92
			Cord blood CD34$^+$	OSKM	p53 shRNA	89
			Cord blood CD133$^+$	OS		93
			Hepatocyte	OSKM		94
			Mesenchymal stromal cell	OSK		95
			Urine cell	OSKM		96
			Amniotic/yolk-sac cell	OSKM		20
			NSCs	OK		13
			NSCs	O		15
		Lentivirus	Fibroblast	OSKM/OSNL	Large T	97
			Cord blood cells	OSNL		98
			Melanocyte	OSKM/OKM/OSK		18
			Blood cell/T-cell/PB-MNC	OSKM		27
	Nonintegrating	Adenovirus	Keratinocyte	OSKM	Single polycistronic	38
		Sendaivirus	Fibroblast	OSKM		99
			Fibroblast	OSKM		100
			Blood cell, T cell	OSKM		26
		Episomal vector	Fibroblast	OSKMNL		101
		PiggyBag	Fibroblast	OSKM		45
	Other	Protein	Fibroblast	OSKM		48
		mRNA	Keratinocyte/fibroblast	OSKM		51
		miRNA	Fibroblast		miR 200c, 302s, 369	63

Mouse						
	Integrating virus	Retrovirus	Fibroblast	OSKM	5'-azaC, VPA	6
			Fibroblast	OSKM/OSK	Ink4a/ArfshRNA	64
			Fibroblast/keratinocyte	OSKM/OSK		81
			Fibroblast/T lymphocyte	OSKM/OSK		78
			Fibroblast	OSK^{L}M		11
			Fibroblast	OSK		10
			Fibroblast	OSK	mIR 291-3p, 294, 295	60
			Fibroblast, p53−/−, Terc−/−	OSK		79
			Fibroblast, p53−/−	OSK	p53, p21, Ink4a, ArfshRNA	77
			Fibroblast	OKM/OK	RepSox	67
			Fibroblast	OK	BIX01294, BayK8644	17
			Fibroblast	OK	CHIR99021	33
			Hepatocyte	OSKM		102
			BM-MNC	OSKM		20
			Amniotic, yolk-sac cell	OSKM		16
			Adipose-derived cell	OSKM		66
			NPCs	SKM/OK	BIX01293	75
			NSCs	OK	PD0325901, CHIR99021	23
			Pancreatic b-cell	OSKM		31
			Fibroblast	OSKM		103
		Lentivirus	Fibroblast	OSKM/OSK	Single polycistronic	104
		Lentivirus (tet inducible)	Fibroblast	OSKM	Single polycistronic	9
			Fibroblast	OSKM/OKM/OSK	Alk5 inhibitor	74
			Fibroblast	OSK	Wnt3a	105
			B-lymphocyte	OSM	Kenpaullone	24
			Melanocyte	OSKM		18
	Nonintegrating	Adenovirus	Fibroblast, Hepatocyte	OSKM/OKM/OSK		43
		Episomal vector	Fibroblast	OSKM		106
		PiggyBag	Fibroblast	OSKM		45,46
	Other	Protein	Fibroblast		VPA	49
		miRNA	Adipose stromal cells		miR 200c, 302s, 369	63

TFs, transcriptional factors; O, Oct4; S, Sox2; K, Klf4; M, c-Myc; N, Nanog; L, Lin28; ^{L}M, L-Myc.

chromosomal abnormalities starting as early as passage 13.[112] This finding warns that there should be more studies regarding the exact frequency of culture-induced genetic abnormalities in human iPSCs over a long run.

The use of retroviral and lentiviral vectors to express the reprogramming transcription factors has the inherent risk of insertional mutagenesis. However, Aoi and colleagues found no common insertion sites in hepatocyte- and stomach-cell-derived iPSCs.[33] In addition, recent adenoviral and plasmid-based methods have a much lower risk of insertional mutagenesis, because theoretically the genomic DNA is not perturbed by the virus.[43,44] However, even without viral integration, genetic changes might occur as part of the reprogramming process. Another component reflecting DNA integrity, namely, the telomere length, is not altered by the reactivation of mouse[43] and human telomerase reverse transcriptase (hTERT)[5,30] in iPSCs. Studies of iPSCs have suggested that DNA integrity is maintained throughout the generation process; however, only long-term studies will show whether these cells are truly free of malignant potential *in vivo*. In any case, this risk will have to be weighed against the therapeutic potential.

B. Gene Expression Profiles

Comparative global gene expression analyses of the ESC and iPSC transcriptomes using microarrays have been performed for human and mouse lines.[6,111,113] Mikkelsen and colleagues reported that whole-genome expression profiles of iPSCs and ESCs of the same species are no more different than those of individual ESC lines.[114] Nonetheless, other groups noted that iPSCs are not identical to ESCs. Takahashi and colleagues compared the global gene expression profile of human iPSC and human ESCs for 32,266 transcripts.[5] Notably, 1,267 ($\sim$4%) of the genes were detected with more than fivefold difference in up- or downregulation between iPSCs and human ESCs. Soldner and colleagues compared the transcriptional profiles of human iPSC lines, in which the Cre-recombinase excisable exogenous viral sequences had been removed (factor-free human iPSCs), with those of human iPSCs before transgene excision.[115] The transcriptomes of factor-free human iPSCs more closely resembled those of human ESCs than the parental human iPSCs with integrated viral sequences. This could be due to the loss of any downstream gene activation by residual expression of the exogenous transcription factors or by the loss of epigenetic memory of the somatic state after the initial reprogramming event. However, it remains difficult to compare these differences because most groups used genetically unrelated cell lines. More experiments are needed to clarify these discrepancies according to the methods used for reprogramming, sources of iPSCs, disease state of the parental cells, age, and sex.

A well-characterized gene expression pattern occurs after ectopic expression of the four factors in MEFs, including an initial downregulation of cell-type-specific transcription factors[23,114] and upregulation of genes involved in proliferation, DNA replication, and cell-cycle progression.[114] During the reprogramming process, many self-renewal-related genes are reactivated, including fibroblast growth factor 4 (*Fgf4*) as well as polycomb genes.[6,114] However, a large fraction of pluripotency-related genes are only upregulated during the late stages of reprogramming.[23,31,114] In a different study, the expression of key pluripotency-related genes, such as *Oct4*, *Sox2*, and *Rex1*, was approximately twofold lower in the iPSCs compared to two human ESCs lines HSF1 and H9.[88] Pluripotent cells are highly sensitive to the levels of these transcription factors (TFs),[116] and there is a notable amount of normal transcriptional heterogeneity in human ESC cultures.[117] Therefore, the observed variation could reflect the differences in culture conditions rather than incomplete reprogramming. More work on human ESCs is thus required to better understand the extent of normal transcriptional variation within human and also mouse ESCs and to fully understand how iPSCs differ.

C. Epigenetic Status

As the substrate of transcription, chromatin is subjected to various forms of epigenetic regulation including chromatin remodeling, histone modifications, histone variants, and DNA methylation. For example, trimethylation of lysine 9 and lysine 27 of histone 3 (H3K9 and H3K27) correlates with inactive regions of chromatin, whereas H3K4 trimethylation and acetylation of H3 and H4 are associated with active transcription[118] and DNA methylation generally represses gene expression.[119]

By regulating the chromatin structure, epigenetic modifications play an essential role in controlling access to genes and regulatory elements in the genome.[31] The differences in epigenetic status between a somatic cell and a pluripotent stem cell are huge, and dedifferentiation requires global epigenetic reprogramming. For instance, PSCs contain bivalent domains which are characteristic chromatin signatures.[120,121] These are regions enriched for repressive histone H3 lysine 27 trimethylation (H3K27me3) and simultaneously for histone H3 lysine 4 trimethylation (H3K4me3) as an activating signal.[28] It was assumed initially that bivalent domains might be ES-cell-specific because they were first identified using chromatin-immunoprecipitation (ChIP) followed by hybridization to microarrays (ChIP-Chip) that featured key developmental regulators. All of these resolved either to a univalent (H3K4me3 only or H3K27me3 only) state or lost both marks in differentiated cells.[120] Using ChIP followed by high-throughput sequencing (ChIP-seq) technology, Mikkelsen and colleagues showed that bivalent domains are more generally indicative of genes that remain in a poised state. Consequently, pluripotent cells were found to contain large numbers of bivalent domains ($\sim$2,500) compared with multipotent neural progenitor cells (NPCs) ($\sim$200) that still retain multilineage potential but are more restricted than ESCs.[122]

Several studies of the murine iPSC have identified a small number of representative loci that have consistent chromatin and DNA methylation patterns.[6,111,113] Maherali and colleagues used ChIP-Chip to investigate the presence of H3K4me3 and H3K27me3 in the promoter regions of 16,500 genes, and their results showed that iPSCs were highly similar to ESCs in epigenetic state.[113] The H3K4me3 pattern was similar across all samples, indicating that reprogramming was largely associated with changes in H3K27me3 rather than H3K4me3.[113] Mikkelsen and colleagues have used a more comprehensive ChIP-Seq technique to determine genome-wide chromatin maps in several iPS lines that are derived by different methods: drug selection using an *Oct4*–neomycin-resistance gene[111] and a *Nanog*–neomycin-resistance gene,[111] and by morphological appearance.[123] Overall global levels of repressive H3K27me3 and the characteristic bivalent chromatin structure are retained in the various iPSC lines. The restoration of repressive chromatin marks appears crucial to stably silence lineage-specific genes that are active in somatic cells and inactive in undifferentiated pluripotent cells. Failure to establish the repressive marks results in incomplete reprogrammed cells. Activating H3K4me3 patterns are also crucial for complete reprogramming and have been observed to be restored genome-wide, particularly around the promoters of pluripotency-associated genes, such as *Oct4* and *Nanog*, in the fully reprogrammed iPSC lines.[114]

A second component of the epigenetic machinery is DNA methylation, which is a stable and heritable mark that is involved in gene silencing including genomic imprinting and X-chromosome inactivation. DNA methylation patterns are dynamic during early embryonic development and are essential for normal post-implantation development.[124] Overall DNA methylation levels remain stable during ESC differentiation, although they are not static for any given individual gene.[122] The 5′-promoter regions of many transcriptional units contain clusters of the dinucleotide CpG, which are methylated at transcriptionally silent genes and demethylated upon activation. In differentiated cells, the *Oct4, Nanog*, and *Sox2* promoter regions are highly methylated and in an inactivated state, whereas in ESCs these promoters are unmethylated to be activated. During reprogramming, almost complete demethylation of these promoters has been observed.[29,111,113,114] Therefore, the loss of DNA methylation at the promoters of pluripotency-related genes appears essential for achieving complete reprogramming. Interestingly, loss of DNA methylation at this class of genes seems to be a rather late event in the reprogramming process because cells that have already acquired self-renewing properties still showed high levels of DNA methylation.[114]

D. Developmental Potential: Pluripotency

Research on the transcriptional and epigenetic state of iPSCs is highly informative and it might ultimately be possible to characterize newly derived iPSC lines based on their genomic profiling alone. Before selecting the most informative markers, it is important to use *in vivo* assays to analyze the interplay between transcriptome, epigenome, and developmental potential. Recently, Jaenisch and Young provided a detailed comparison of the different strategies for assessing the developmental potential and their stringency.[125] *In vitro* differentiation is the least stringent assay, whereas tetraploid-embryo complementation is the most stringent assay for testing developmental potential.[125,126] These strategies could be used to determine the pluripotency of mouse iPSCs, but only *in vitro* differentiation and teratoma formation could be applied to test human iPSCs. Mouse iPSCs appear to have developmental potential similar to that of ESCs as confirmed by teratoma formation capability and by the high contribution to chimera formation with germline transmission.[6,111,113] To show the final step of developmental potential of iPSCs as equivalent to that of ESCs, three separate groups injected mouse iPSCs (2N) into tetraploid blastocysts (4N), which are capable of producing placental and other extraembryonic tissues but not the embryo itself, and have created live mice.[109,110,127] The procedure, called "tetraploid complementation," is the most stringent test for pluripotency. If the stem cells that are injected into the tetraploid blastocyst differentiate into embryonic tissues that produce a mouse, then the stem cells are considered truly pluripotent.

E. Differences Between iPSCs and ESCs

Comparative studies on mouse iPSCs and ESCs have demonstrated that the two cell types are largely indistinguishable with regard to the expression of cell-surface markers, gene expression profiles, and capacity to form teratomas when injected into immunodeficient hosts.[112,128,129] On the other hand, emerging studies have indicated that iPSCs are not identical to ESCs. Lowry and colleagues compared the expression of pluripotency-related genes in human iPSCs against two human ESC lines, HSF1 and H9, and reported that expression levels were approximately twofold lower in the iPSCs.[88] Wernig and colleagues found quantitative differences in promoter methylation of pluripotency-related genes including *Oct-4*, *Sox-2*, and *Nanog*.[111] ChIP detected some differences in "bivalent" loci between the two cell types.[5] Recently, a careful transcriptional profiling analysis revealed that human iPSCs have a distinct gene expression pattern compared to human ESCs due to the "footprint" existence of their original tissues.[130] Human iPSCs tend to be closer to their corresponding donor cell types than to other donor cell types. Interestingly, human iPSCs derived from different donor cells have varying distances from human ESCs. Furthermore, the efficiency and stability of sublineage differentiation are different between human iPSCs and human ESCs.[131] These results were consistent across iPSC lines and independent of the core set of reprogramming transgenes used to derive iPSCs as well as the presence or absence of reprogramming transgenes in iPSCs. Whether the different gene expression profile of iPSCs contributes to the differentiation variability needs to be further investigated. Comparison of different iPSC clones with ESCs could allow the identification of clones that are most similar to ESC, and comparative studies of the two cell types will enhance our understanding of the molecular mechanisms underlying differentiation potential. Different lines of both iPSCs and ESCs have distinct characteristics, but no molecular criteria are yet available to determine whether a given line is suitable for clinical application. Indeed, it is possible that different iPSC lines might lend themselves to different therapeutic applications. The identification of molecular markers that can indicate the suitability of a line for a particular application will be an important area of future research. The establishment of methods of iPSC generation for clinical applications and biomedical researches is an ongoing process. More comprehensive knowledge of the reprogramming process is therefore crucial for clinical applications and reprogramming biology of iPSCs.

VI. Conclusion

Since the first success of iPSC generation, iPSCs has been established by a variety of methods. Induction of pluripotency with viral vectors is sufficient for *in vitro* use of iPSCs when the remaining reprogramming factors do not significantly interrupt the designed assays. However, the transgene integration and alteration of the endogenous genomic organization could cause a negative safety issue for clinical applications. For these reasons, much effort has been made to identify safer strategies. Later, a transgene-removable strategy for generating "transgene-free" iPSCs was established, which demonstrated a similar efficiency with the transgene strategy.[115,132] Factor-free iPSCs maintain a pluripotent state and show a global gene expression profile more closely related to ESCs than that of iPSCs carrying transgenes. However, the current excising approach may still leave a short sequence of exogenous DNA, primarily viral LTR, in the genome of iPSCs, and hence not completely eliminate the theoretical risk of insertional mutagenesis. This minimal risk may be further reduced in the future by targeting the removable vector into a safe genomic locus. Recently, emphasis has been placed on the development of nonintegrating vectors including episomal plasmids, peptides, and recombinant proteins for delivery of pluripotency factors.[14,48,49] However, compared to virus-mediated methods, these alternatives showed low efficiency and slow process, and the number of potential iPSC clones generated was markedly reduced.[133] Moreover, human iPSCs are still not established with methods using purified peptides or recombinant proteins. Most recently, synthetic RNA-based reprogramming has been developed in which RNA is translated into protein in the cytoplasm and host DNA remains unaltered. Synthetic RNA-based reprogramming is clean, safe, and fast, and iPSCs are genetically identical to their source cells. The reprogramming process takes a little over 2 weeks with 4% of the cells being reprogrammed. This is about 100-fold more efficient than reprogramming by the gene transfer technique.[51] However, mRNA still has a number of serious problems. It is difficult to synthesize long mRNA chemically, and researchers have not been able to make large mRNAs. The expression window of mRNA is only 2–3 days; so repeated transfection is needed to achieve reprogramming and innate immune response must be suppressed alongside mRNA transfection. Other progress has been made in which miRNAs alone are sufficient to induce pluripotency in mouse and human cells.[61,63] Compared to coding-gene-based reprogramming approaches, miRNA-based reprogramming offers several clear advantages. It completely avoids using oncogenic transcription factors such as *c-Myc* and *Oct4* and does not need to introduce genetic changes into cell genomes. However, these new reprogramming approaches need to be repeated by other laboratories, and the molecular mechanism of activating the entire pluripotency network is yet to be revealed.

iPSCs are similar to ESCs in both the capacity for self-renewal *in vitro* and pluripotency and can avoid ethical issues associated with ESCs. For these reasons, iPSCs have great potential in cell-replacement therapies and tissue-engineering applications.[134] One of the most exciting aspects of the development of iPSCs was their potential use for patient-specific autologous transplants without using a human blastocyst and without immunosuppressive therapy, although this enthusiasm was tempered a bit by recent studies in which teratomas derived from iPSCs were rejected with massive CD4[+] T-cell infiltration.[135] The immunogenicity was apparently caused by overexpression of a few specific genes in iPSC-derived teratomas, suggesting that subtle epigenetic changes could have important therapeutic consequences. Growing evidence suggests that there are difference between human iPSCs and human ESCs and among human iPSCs lines according to their reprogramming methods and donor cells of origin. Therefore, further investigations are needed to address their differentiation potential and therapeutic applicability. Nevertheless, iPSCs provide an unprecedented tool for studying the pathophysiology of disease, pharmacological and toxicological testing, identification of new therapeutic targets, and investigating developmental and reprogramming biology.[136] iPSCs will continue to make a significant impact upon many areas of biomedical research.

REFERENCES

1. Briggs R, King TJ. Transplantation of living nuclei from blastula cells into enucleated frogs' eggs. *Proc Natl Acad Sci USA* 1952;**38**:455–63.
2. Campbell KH, McWhir J, Ritchie WA, Wilmut I. Sheep cloned by nuclear transfer from a cultured cell line. *Nature* 1996;**380**:64–6.

3. Eminli S, Jaenisch R, Hochedlinger K. Strategies to induce nuclear reprogramming. *Ernst Schering Found Symp Proc* 2006;**5**:83–98.

4. Hochedlinger K, Jaenisch R. Nuclear reprogramming and pluripotency. *Nature* 2006;**441**: 1061–7.

5. Takahashi K, Tanabe K, Ohnuki M, Narita M, Ichisaka T, Tomoda K, et al. Induction of pluripotent stem cells from adult human fibroblasts by defined factors. *Cell* 2007;**131**:861–72.

6. Takahashi K, Yamanaka S. Induction of pluripotent stem cells from mouse embryonic and adult fibroblast cultures by defined factors. *Cell* 2006;**126**:663–76.

7. Tokuzawa Y, Kaiho E, Maruyama M, Takahashi K, Mitsui K, Maeda M, et al. Fbx15 is a novel target of Oct3/4 but is dispensable for embryonic stem cell self-renewal and mouse development. *Mol Cell Biol* 2003;**23**:2699–708.

8. Yu J, Vodyanik MA, Smuga-Otto K, Antosiewicz-Bourget J, Frane JL, Tian S, et al. Induced pluripotent stem cell lines derived from human somatic cells. *Science* 2007;**318**:1917–20.

9. Maherali N, Hochedlinger K. Tgfbeta signal inhibition cooperates in the induction of iPSCs and replaces Sox2 and cMyc. *Curr Biol* 2009;**19**:1718–23.

10. Nakagawa M, Koyanagi M, Tanabe K, Takahashi K, Ichisaka T, Aoi T, et al. Generation of induced pluripotent stem cells without Myc from mouse and human fibroblasts. *Nat Biotechnol* 2008;**26**:101–6.

11. Nakagawa M, Takizawa N, Narita M, Ichisaka T, Yamanaka S. Promotion of direct reprogramming by transformation-deficient Myc. *Proc Natl Acad Sci USA* 2010;**107**:14152–7.

12. Wernig M, Meissner A, Cassady JP, Jaenisch R. c-Myc is dispensable for direct reprogramming of mouse fibroblasts. *Cell Stem Cell* 2008;**2**:10–2.

13. Hester ME, Song S, Miranda CJ, Eagle A, Schwartz PH, Kaspar BK. Two factor reprogramming of human neural stem cells into pluripotency. *PLoS One* 2009;**4**:e7044.

14. Kim JB, Greber B, Arauzo-Bravo MJ, Meyer J, Park KI, Zaehres H, et al. Direct reprogramming of human neural stem cells by OCT4. *Nature* 2009;**461**:649–53.

15. Kim JB, Sebastiano V, Wu G, Arauzo-Bravo MJ, Sasse P, Gentile L, et al. Oct4-induced pluripotency in adult neural stem cells. *Cell* 2009;**136**:411–9.

16. Tat PA, Sumer H, Jones KL, Upton K, Verma PJ. The efficient generation of induced pluripotent stem (iPS) cells from adult mouse adipose tissue-derived and neural stem cells. *Cell Transplant* 2010;**19**:525–36.

17. Li W, Zhou H, Abujarour R, Zhu S, Young Joo J, Lin T, et al. Generation of human-induced pluripotent stem cells in the absence of exogenous Sox2. *Stem Cells* 2009;**27**:2992–3000.

18. Utikal J, Maherali N, Kulalert W, Hochedlinger K. Sox2 is dispensable for the reprogramming of melanocytes and melanoma cells into induced pluripotent stem cells. *J Cell Sci* 2009;**122**: 3502–10.

19. Sugii S, Kida Y, Kawamura T, Suzuki J, Vassena R, Yin YQ, et al. Human and mouse adipose-derived cells support feeder-independent induction of pluripotent stem cells. *Proc Natl Acad Sci USA* 2010;**107**:3558–63.

20. Nagata S, Toyoda M, Yamaguchi S, Hirano K, Makino H, Nishino K, et al. Efficient reprogramming of human and mouse primary extra-embryonic cells to pluripotent stem cells. *Genes Cells* 2009;**12**:1395–404.

21. Li C, Zhou J, Shi G, Ma Y, Yang Y, Gu J, et al. Pluripotency can be rapidly and efficiently induced in human amniotic fluid-derived cells. *Hum Mol Genet* 2009;**18**:4340–9.

22. Zhao HX, Li Y, Jin HF, Xie L, Liu C, Jiang F, et al. Rapid and efficient reprogramming of human amnion-derived cells into pluripotency by three factors OCT4/SOX2/NANOG. *Differentiation* 2010;**80**:123–9.

23. Stadtfeld M, Brennand K, Hochedlinger K. Reprogramming of pancreatic beta cells into induced pluripotent stem cells. *Curr Biol* 2008;**18**:890–4.

24. Hanna J, Markoulaki S, Schorderet P, Carey BW, Beard C, Wernig M, et al. Direct reprogramming of terminally differentiated mature B lymphocytes to pluripotency. *Cell* 2008;**133**: 250–64.

25. Staerk J, Dawlaty MM, Gao Q, Maetzel D, Hanna J, Sommer CA, et al. Reprogramming of human peripheral blood cells to induced pluripotent stem cells. *Cell Stem Cell* 2010;**7**:20–4.

26. Seki T, Yuasa S, Oda M, Egashira T, Yae K, Kusumoto D, et al. Generation of induced pluripotent stem cells from human terminally differentiated circulating T cells. *Cell Stem Cell* 2010;**7**:11–4.

27. Loh YH, Hartung O, Li H, Guo C, Sahalie JM, Manos PD, et al. Reprogramming of T cells from human peripheral blood. *Cell Stem Cell* 2010;**7**:15–9.

28. Mikkelsen TS, Ku M, Jaffe DB, Issac B, Lieberman E, Giannoukos G, et al. Genome-wide maps of chromatin state in pluripotent and lineage-committed cells. *Nature* 2007;**448**:553–60.

29. Okita K, Ichisaka T, Yamanaka S. Generation of germline-competent induced pluripotent stem cells. *Nature* 2007;**448**:313–7.

30. Park IH, Zhao R, West JA, Yabuuchi A, Huo H, Ince TA, et al. Reprogramming of human somatic cells to pluripotency with defined factors. *Nature* 2008;**451**:141–6.

31. Brambrink T, Foreman R, Welstead GG, Lengner CJ, Wernig M, Suh H, et al. Sequential expression of pluripotency markers during direct reprogramming of mouse somatic cells. *Cell Stem Cell* 2008;**2**:151–9.

32. Wernig M, Lengner CJ, Hanna J, Lodato MA, Steine E, Foreman R, et al. A drug-inducible transgenic system for direct reprogramming of multiple somatic cell types. *Nat Biotechnol* 2008;**26**:916–24.

33. Aoi T, Yae K, Nakagawa M, Ichisaka T, Okita K, Takahashi K, et al. Generation of pluripotent stem cells from adult mouse liver and stomach cells. *Science* 2008;**321**:699–702.

34. Akagi K, Suzuki T, Stephens RM, Jenkins NA, Copeland NG. RTCGD: retroviral tagged cancer gene database. *Nucleic Acids Res* 2004;**32**:D523–7.

35. Hawley RG. Does retroviral insertional mutagenesis play a role in the generation of induced pluripotent stem cells? *Mol Ther* 2008;**16**:1354–5.

36. Varas F, Stadtfeld M, de Andres-Aguayo L, Maherali N, di Tullio A, Pantano L, et al. Fibroblast-derived induced pluripotent stem cells show no common retroviral vector insertions. *Stem Cells* 2009;**27**:300–6.

37. Wu X, Li Y, Crise B, Burgess SM. Transcription start regions in the human genome are favored targets for MLV integration. *Science* 2003;**300**:1749–51.

38. Carey BW, Markoulaki S, Hanna J, Saha K, Gao Q, Mitalipova M, et al. Reprogramming of murine and human somatic cells using a single polycistronic vector. *Proc Natl Acad Sci USA* 2009;**106**:157–62.

39. Sommer CA, Stadtfeld M, Murphy GJ, Hochedlinger K, Kotton DN, Mostoslavsky G. Induced pluripotent stem cell generation using a single lentiviral stem cell cassette. *Stem Cells* 2009;**27**:543–9.

40. Li Z, Dullmann J, Schiedlmeier B, Schmidt M, von Kalle C, Meyer J, et al. Murine leukemia induced by retroviral gene marking. *Science* 2002;**296**:497.

41. Hacein-Bey-Abina S, Von Kalle C, Schmidt M, McCormack MP, Wulffraat N, Leboulch P, et al. LMO2-associated clonal T cell proliferation in two patients after gene therapy for SCID-X1. *Science* 2003;**302**:415–9.

42. Howe SJ, Mansour MR, Schwarzwaelder K, Bartholomae C, Hubank M, Kempski H, et al. Insertional mutagenesis combined with acquired somatic mutations causes leukemogenesis following gene therapy of SCID-X1 patients. *J Clin Invest* 2008;**118**:3143–50.

43. Stadtfeld M, Nagaya M, Utikal J, Weir G, Hochedlinger K. Induced pluripotent stem cells generated without viral integration. *Science* 2008;**322**:945–9.

44. Okita K, Nakagawa M, Hyenjong H, Ichisaka T, Yamanaka S. Generation of mouse induced pluripotent stem cells without viral vectors. *Science* 2008;**322**:949–53.
45. Woltjen K, Michael IP, Mohseni P, Desai R, Mileikovsky M, Hamalainen R, et al. piggyBac transposition reprograms fibroblasts to induced pluripotent stem cells. *Nature* 2009;**458**:766–70.
46. Kaji K, Norrby K, Paca A, Mileikovsky M, Mohseni P, Woltjen K. Virus-free induction of pluripotency and subsequent excision of reprogramming factors. *Nature* 2009;**458**:771–5.
47. Yusa K, Rad R, Takeda J, Bradley A. Generation of transgene-free induced pluripotent mouse stem cells by the piggyBac transposon. *Nat Methods* 2009;**6**:363–9.
48. Kim D, Kim CH, Moon JI, Chung YG, Chang MY, Han BS, et al. Generation of human induced pluripotent stem cells by direct delivery of reprogramming proteins. *Cell Stem Cell* 2009;**4**:472–6.
49. Zhou H, Wu S, Joo JY, Zhu S, Han DW, Lin T, et al. Generation of induced pluripotent stem cells using recombinant proteins. *Cell Stem Cell* 2009;**4**:381–4.
50. Cho HJ, Lee CS, Kwon YW, Paek JS, Lee SH, Hur J, et al. Induction of pluripotent stem cells from adult somatic cells by protein-based reprogramming without genetic manipulation. *Blood* 2010;**116**:386–95.
51. Warren L, Manos PD, Ahfeldt T, Loh YH, Li H, Lau F, et al. Highly efficient reprogramming to pluripotency and directed differentiation of human cells with synthetic modified mRNA. *Cell Stem Cell* 2010;**7**:618–30.
52. Kariko K, Buckstein M, Ni H, Weissman D. Suppression of RNA recognition by Toll-like receptors: the impact of nucleoside modification and the evolutionary origin of RNA. *Immunity* 2005;**23**:165–75.
53. Kariko K, Muramatsu H, Welsh FA, Ludwig J, Kato H, Akira S, et al. Incorporation of pseudouridine into mRNA yields superior nonimmunogenic vector with increased translational capacity and biological stability. *Mol Ther* 2008;**16**:1833–40.
54. Kariko K, Weissman D. Naturally occurring nucleoside modifications suppress the immunostimulatory activity of RNA: implication for therapeutic RNA development. *Curr Opin Drug Discov Devel* 2007;**10**:523–32.
55. Nallagatla SR, Bevilacqua PC. Nucleoside modifications modulate activation of the protein kinase PKR in an RNA structure-specific manner. *RNA* 2008;**14**:1201–13.
56. Uzri D, Gehrke L. Nucleotide sequences and modifications that determine RIG-I/RNA binding and signaling activities. *J Virol* 2009;**83**:4174–84.
57. Davis RL, Weintraub H, Lassar AB. Expression of a single transfected cDNA converts fibroblasts to myoblasts. *Cell* 1987;**51**:987–1000.
58. Ruvkun G. Molecular biology. Glimpses of a tiny RNA world. *Science* 2001;**294**:797–9.
59. Houbaviy HB, Murray MF, Sharp PA. Embryonic stem cell-specific MicroRNAs. *Dev Cell* 2003;**5**:351–8.
60. Judson RL, Babiarz JE, Venere M, Blelloch R. Embryonic stem cell-specific microRNAs promote induced pluripotency. *Nat Biotechnol* 2009;**27**:459–61.
61. Anokye-Danso F, Trivedi CM, Juhr D, Gupta M, Cui Z, Tian Y, et al. Highly efficient miRNA-mediated reprogramming of mouse and human somatic cells to pluripotency. *Cell Stem Cell* 2011;**8**:376–88.
62. Lin SL, Chang DC, Lin CH, Ying SY, Leu D, Wu DT. Regulation of somatic cell reprogramming through inducible mir-302 expression. *Nucleic Acids Res* 2011;**39**:1054–65.
63. Miyoshi N, Ishii H, Nagano H, Haraguchi N, Dewi DL, Kano Y, et al. Reprogramming of mouse and human cells to pluripotency using mature microRNAs. *Cell Stem Cell* 2011;**8**:633–8.

64. Huangfu D, Maehr R, Guo W, Eijkelenboom A, Snitow M, Chen AE, et al. Induction of pluripotent stem cells by defined factors is greatly improved by small-molecule compounds. *Nat Biotechnol* 2008;**26**:795–7.
65. Szabo PE, Hubner K, Scholer H, Mann JR. Allele-specific expression of imprinted genes in mouse migratory primordial germ cells. *Mech Dev* 2002;**115**:157–60.
66. Shi Y, Do JT, Desponts C, Hahm HS, Scholer HR, Ding S. A combined chemical and genetic approach for the generation of induced pluripotent stem cells. *Cell Stem Cell* 2008;**2**:525–8.
67. Shi Y, Desponts C, Do JT, Hahm HS, Scholer HR, Ding S. Induction of pluripotent stem cells from mouse embryonic fibroblasts by Oct4 and Klf4 with small-molecule compounds. *Cell Stem Cell* 2008;**3**:568–74.
68. Chang Y, Zhang X, Horton JR, Upadhyay AK, Spannhoff A, Liu J, et al. Structural basis for G9a-like protein lysine methyltransferase inhibition by BIX-01294. *Nat Struct Mol Biol* 2009;**16**:312–7.
69. Feldman N, Gerson A, Fang J, Li E, Zhang Y, Shinkai Y, et al. G9a-mediated irreversible epigenetic inactivation of Oct-3/4 during early embryogenesis. *Nat Cell Biol* 2006;**8**:188–94.
70. Epsztejn-Litman S, Feldman N, Abu-Remaileh M, Shufaro Y, Gerson A, Ueda J, et al. De novo DNA methylation promoted by G9a prevents reprogramming of embryonically silenced genes. *Nat Struct Mol Biol* 2008;**15**:1176–83.
71. Dong KB, Maksakova IA, Mohn F, Leung D, Appanah R, Lee S, et al. DNA methylation in ES cells requires the lysine methyltransferase G9a but not its catalytic activity. *EMBO J* 2008;**27**:2691–701.
72. Tachibana M, Matsumura Y, Fukuda M, Kimura H, Shinkai Y. G9a/GLP complexes independently mediate H3K9 and DNA methylation to silence transcription. *EMBO J* 2008;**27**: 2681–90.
73. Krawetz RJ, Li X, Rancourt DE. Human embryonic stem cells: caught between a ROCK inhibitor and a hard place. *Bioessays* 2009;**31**:336–43.
74. Marson A, Foreman R, Chevalier B, Bilodeau S, Kahn M, Young RA, et al. Wnt signaling promotes reprogramming of somatic cells to pluripotency. *Cell Stem Cell* 2008;**3**:132–5.
75. Silva J, Barrandon O, Nichols J, Kawaguchi J, Theunissen TW, Smith A. Promotion of reprogramming to ground state pluripotency by signal inhibition. *PLoS Biol* 2008;**6**:e253.
76. Li W, Wei W, Zhu S, Zhu J, Shi Y, Lin T, et al. Generation of rat and human induced pluripotent stem cells by combining genetic reprogramming and chemical inhibitors. *Cell Stem Cell* 2009;**4**:16–9.
77. Ichida JK, Blanchard J, Lam K, Son EY, Chung JE, Egli D, et al. A small-molecule inhibitor of Tgf-beta signaling replaces Sox2 in reprogramming by inducing nanog. *Cell Stem Cell* 2009;**5**:491–503.
78. Hong H, Takahashi K, Ichisaka T, Aoi T, Kanagawa O, Nakagawa M, et al. Suppression of induced pluripotent stem cell generation by the p53-p21 pathway. *Nature* 2009;**460**:1132–5.
79. Kawamura T, Suzuki J, Wang YV, Menendez S, Morera LB, Raya A, et al. Linking the p53 tumour suppressor pathway to somatic cell reprogramming. *Nature* 2009;**460**:1140–4.
80. Utikal J, Polo JM, Stadtfeld M, Maherali N, Kulalert W, Walsh RM, et al. Immortalization eliminates a roadblock during cellular reprogramming into iPS cells. *Nature* 2009;**460**:1145–8.
81. Li H, Collado M, Villasante A, Strati K, Ortega S, Canamero M, et al. The Ink4/Arf locus is a barrier for iPS cell reprogramming. *Nature* 2009;**460**:1136–9.
82. Zheng H, Ying H, Yan H, Kimmelman AC, Hiller DJ, Chen AJ, et al. p53 and Pten control neural and glioma stem/progenitor cell renewal and differentiation. *Nature* 2008;**455**:1129–33.

83. Chun YS, Chaudhari P, Jang YY. Applications of patient-specific induced pluripotent stem cells; focused on disease modeling, drug screening and therapeutic potentials for liver disease. *Int J Biol Sci* 2010;**6**:796–805.

84. Miura K, Okada Y, Aoi T, Okada A, Takahashi K, Okita K, et al. Variation in the safety of induced pluripotent stem cell lines. *Nat Biotechnol* 2009;**27**:743–5.

85. Kim K, Doi A, Wen B, Ng K, Zhao R, Cahan P, et al. Epigenetic memory in induced pluripotent stem cells. *Nature* 2010;**467**:285–90.

86. Polo JM, Liu S, Figueroa ME, Kulalert W, Eminli S, Tan KY, et al. Cell type of origin influences the molecular and functional properties of mouse induced pluripotent stem cells. *Nat Biotechnol* 2010;**28**:848–55.

87. Marchetto MC, Yeo GW, Kainohana O, Marsala M, Gage FH, Muotri AR. Transcriptional signature and memory retention of human-induced pluripotent stem cells. *PLoS One* 2009;**4**: e7076.

88. Lowry WE, Richter L, Yachechko R, Pyle AD, Tchieu J, Sridharan R, et al. Generation of human induced pluripotent stem cells from dermal fibroblasts. *Proc Natl Acad Sci USA* 2008;**105**:2883–8.

89. Takenaka C, Nishishita N, Takada N, Jakt LM, Kawamata S. Effective generation of iPS cells from CD34 + cord blood cells by inhibition of p53. *Exp Hematol* 2010;**38**:154–62.

90. Marion RM, Strati K, Li H, Murga M, Blanco R, Ortega S, et al. A p53-mediated DNA damage response limits reprogramming to ensure iPS cell genomic integrity. *Nature* 2009;**460**:1149–53.

91. Loh YH, Agarwal S, Park IH, Urbach A, Huo H, Heffner GC, et al. Generation of induced pluripotent stem cells from human blood. *Blood* 2009;**113**:5476–9.

92. Ye Z, Zhan H, Mali P, Dowey S, Williams DM, Jang YY, et al. Human-induced pluripotent stem cells from blood cells of healthy donors and patients with acquired blood disorders. *Blood* 2009;**114**:5473–80.

93. Giorgetti A, Montserrat N, Aasen T, Gonzalez F, Rodriguez-Piza I, Vassena R, et al. Generation of induced pluripotent stem cells from human cord blood using OCT4 and SOX2. *Cell Stem Cell* 2009;**5**:353–7.

94. Liu H, Ye Z, Kim Y, Sharkis S, Jang YY. Generation of endoderm-derived human induced pluripotent stem cells from primary hepatocytes. *Hepatology* 2010;**51**:1810–9.

95. Oda Y, Yoshimura Y, Ohnishi H, Tadokoro M, Katsube Y, Sasao M, et al. Induction of pluripotent stem cells from human third molar mesenchymal stromal cells. *J Biol Chem* 2010;**285**:29270–8.

96. Zhou T, Benda C, Duzinger S, Huang Y, Li X, Li Y, et al. Generation of induced pluripotent stem cells from urine. *J Am Soc Nephrol* 2011;**22**:1221–8.

97. Mali P, Ye Z, Hommond HH, Yu X, Lin J, Chen G, et al. Improved efficiency and pace of generating induced pluripotent stem cells from human adult and fetal fibroblasts. *Stem Cells* 2008;**26**:1998–2005.

98. Haase A, Olmer R, Schwanke K, Wunderlich S, Merkert S, Hess C, et al. Generation of induced pluripotent stem cells from human cord blood. *Cell Stem Cell* 2009;**5**:434–41.

99. Zhou W, Freed CR. Adenoviral gene delivery can reprogram human fibroblasts to induced pluripotent stem cells. *Stem Cells* 2009;**27**:2667–74.

100. Fusaki N, Ban H, Nishiyama A, Saeki K, Hasegawa M. Efficient induction of transgene-free human pluripotent stem cells using a vector based on Sendai virus, an RNA virus that does not integrate into the host genome. *Proc Jpn Acad Ser B Phys Biol Sci* 2009;**85**: 348–62.

101. Yu J, Hu K, Smuga-Otto K, Tian S, Stewart R, Slukvin II, et al. Human induced pluripotent stem cells free of vector and transgene sequences. *Science* 2009;**324**:797–801.

102. Kunisato A, Wakatsuki M, Kodama Y, Shinba H, Ishida I, Nagao K. Generation of induced pluripotent stem cells by efficient reprogramming of adult bone marrow cells. *Stem Cells Dev* 2010;**19**:229–38.
103. Shao L, Feng W, Sun Y, Bai H, Liu J, Currie C, et al. Generation of iPS cells using defined factors linked via the self-cleaving 2A sequences in a single open reading frame. *Cell Res* 2009;**19**:296–306.
104. Chang CW, Lai YS, Pawlik KM, Liu K, Sun CW, Li C, et al. Polycistronic lentiviral vector for "hit and run" reprogramming of adult skin fibroblasts to induced pluripotent stem cells. *Stem Cells* 2009;**27**:1042–9.
105. Lyssiotis CA, Foreman RK, Staerk J, Garcia M, Mathur D, Markoulaki S, et al. Reprogramming of murine fibroblasts to induced pluripotent stem cells with chemical complementation of Klf4. *Proc Natl Acad Sci USA* 2009;**106**:8912–7.
106. Gonzalez F, Barragan Monasterio M, Tiscornia G, Montserrat Pulido N, Vassena R, Batlle Morera L, et al. Generation of mouse-induced pluripotent stem cells by transient expression of a single nonviral polycistronic vector. *Proc Natl Acad Sci U S A* 2009;**106**:8918–22.
107. Carette JE, Pruszak J, Varadarajan M, Blomen VA, Gokhale S, Camargo FD, et al. Generation of iPSCs from cultured human malignant cells. *Blood* 2010;**115**:4039–42.
108. Miyoshi N, Ishii H, Nagai K, Hoshino H, Mimori K, Tanaka F, et al. Defined factors induce reprogramming of gastrointestinal cancer cells. *Proc Natl Acad Sci USA* 2010;**107**:40–5.
109. Kang L, Wang J, Zhang Y, Kou Z, Gao S. iPS cells can support full-term development of tetraploid blastocyst-complemented embryos. *Cell Stem Cell* 2009;**5**:135–8.
110. Zhao XY, Li W, Lv Z, Liu L, Tong M, Hai T, et al. iPS cells produce viable mice through tetraploid complementation. *Nature* 2009;**461**:86–90.
111. Wernig M, Meissner A, Foreman R, Brambrink T, Ku M, Hochedlinger K, et al. In vitro reprogramming of fibroblasts into a pluripotent ES-cell-like state. *Nature* 2007;**448**:318–24.
112. Aasen T, Raya A, Barrero MJ, Garreta E, Consiglio A, Gonzalez F, et al. Efficient and rapid generation of induced pluripotent stem cells from human keratinocytes. *Nat Biotechnol* 2008;**26**:1276–84.
113. Maherali N, Sridharan R, Xie W, Utikal J, Eminli S, Arnold K, et al. Directly reprogrammed fibroblasts show global epigenetic remodeling and widespread tissue contribution. *Cell Stem Cell* 2007;**1**:55–70.
114. Mikkelsen TS, Hanna J, Zhang X, Ku M, Wernig M, Schorderet P, et al. Dissecting direct reprogramming through integrative genomic analysis. *Nature* 2008;**454**:49–55.
115. Soldner F, Hockemeyer D, Beard C, Gao Q, Bell GW, Cook EG, et al. Parkinson's disease patient-derived induced pluripotent stem cells free of viral reprogramming factors. *Cell* 2009;**136**:964–77.
116. Niwa H, Miyazaki J, Smith AG. Quantitative expression of Oct-3/4 defines differentiation, dedifferentiation or self-renewal of ES cells. *Nat Genet* 2000;**24**:372–6.
117. Osafune K, Caron L, Borowiak M, Martinez RJ, Fitz-Gerald CS, Sato Y, et al. Marked differences in differentiation propensity among human embryonic stem cell lines. *Nat Biotechnol* 2008;**26**:313–5.
118. Jenuwein T, Allis CD. Translating the histone code. *Science* 2001;**293**:1074–80.
119. Santos F, Dean W. Epigenetic reprogramming during early development in mammals. *Reproduction* 2004;**127**:643–51.
120. Bernstein BE, Meissner A, Lander ES. The mammalian epigenome. *Cell* 2007;**128**:669–81.
121. Bernstein BE, Mikkelsen TS, Xie X, Kamal M, Huebert DJ, Cuff J, et al. A bivalent chromatin structure marks key developmental genes in embryonic stem cells. *Cell* 2006;**125**:315–26.
122. Meissner A, Mikkelsen TS, Gu H, Wernig M, Hanna J, Sivachenko A, et al. Genome-scale DNA methylation maps of pluripotent and differentiated cells. *Nature* 2008;**454**:766–70.

123. Meissner A, Wernig M, Jaenisch R. Direct reprogramming of genetically unmodified fibroblasts into pluripotent stem cells. *Nat Biotechnol* 2007;**25**:1177–81.
124. Rougier N, Bourc'his D, Gomes DM, Niveleau A, Plachot M, Paldi A, et al. Chromosome methylation patterns during mammalian preimplantation development. *Genes Dev* 1998;**12**: 2108–13.
125. Jaenisch R, Young R. Stem cells, the molecular circuitry of pluripotency and nuclear reprogramming. *Cell* 2008;**132**:567–82.
126. Eggan K, Akutsu H, Loring J, Jackson-Grusby L, Klemm M, Rideout 3rd WM, et al. Hybrid vigor, fetal overgrowth, and viability of mice derived by nuclear cloning and tetraploid embryo complementation. *Proc Natl Acad Sci USA* 2001;**98**:6209–14.
127. Boland MJ, Hazen JL, Nazor KL, Rodriguez AR, Gifford W, Martin G, et al. Adult mice generated from induced pluripotent stem cells. *Nature* 2009;**461**:91–4.
128. Blelloch R, Venere M, Yen J, Ramalho-Santos M. Generation of induced pluripotent stem cells in the absence of drug selection. *Cell Stem Cell* 2007;**1**:245–7.
129. Nishikawa S, Goldstein RA, Nierras CR. The promise of human induced pluripotent stem cells for research and therapy. *Nat Rev Mol Cell Biol* 2008;**9**:725–9.
130. Ghosh Z, Wilson KD, Wu Y, Hu S, Quertermous T, Wu JC. Persistent donor cell gene expression among human induced pluripotent stem cells contributes to differences with human embryonic stem cells. *PLoS One* 2010;**5**:e8975.
131. Hu BY, Weick JP, Yu J, Ma LX, Zhang XQ, Thomson JA, et al. Neural differentiation of human induced pluripotent stem cells follows developmental principles but with variable potency. *Proc Natl Acad Sci USA* 2010;**107**:4335–40.
132. Sommer CA, Sommer AG, Longmire TA, Christodoulou C, Thomas DD, Gostissa M, et al. Excision of reprogramming transgenes improves the differentiation potential of iPS cells generated with a single excisable vector. *Stem Cells* 2010;**28**:64–74.
133. Ramalho-Santos M. iPS cells: insights into basic biology. *Cell* 2009;**138**:616–8.
134. Rolletschek A, Wobus AM. Induced human pluripotent stem cells: promises and open questions. *Biol Chem* 2009;**390**:845–9.
135. Zhao T, Zhang ZN, Rong Z, Xu Y. Immunogenicity of induced pluripotent stem cells. *Nature* 2011;**474**:212–5.
136. Belmonte JC, Ellis J, Hochedlinger K, Yamanaka S. Induced pluripotent stem cells and reprogramming: seeing the science through the hype. *Nat Rev Genet* 2009;**10**:878–83.

Induced Pluripotent Cells in Cardiovascular Biology: Epigenetics, Promises, and Challenges

ENEDA HOXHA AND RAJ KISHORE

Feinberg Cardiovascular Research Institute, Feinberg School of Medicine, Northwestern University, Chicago, Illinois, USA

Cardiovascular diseases are still the leading cause of death worldwide. Despite the improvement shown in the prognosis of patients with acute MI, there remains still a significant mortality risk. Since the main underlying problem after an MI is the loss of cardiomyocytes and microvasculature, treatment strategies aimed at preserving or regenerating myocardial tissue have been examined as potential therapeutic modalities. Toward this goal, many cell types are being investigated as potent sources of cardiomyocytes for cell transplantation. The progress made toward the generation of induced Pluripotent Stem (iPS) cells hold great potential for future use in myocardial repair. We review critical aspects of these cell's potential, such as their generation, their differentiating ability, the known epigenetic mechanisms that allow for their reprogramming, maintenance of pluripotency, their cardiovascular differentiation and therapeutic potential, and the possibility of an epigenetic memory. Understanding the molecular circuitry of these cells will provide a better understanding of their potential as well as limitations in future clinical use.

"Diseases desperate grown,
By desperate appliance are relieved,
Or not at all"
William Shakespeare in Hamlet

Progress in Molecular Biology
and Translational Science, Vol. 111
http://dx.doi.org/10.1016/B978-0-12-398459-3.00002-2

I. Introduction

Cardiovascular diseases are still the leading cause of death worldwide. Of the almost 17 million people who die each year from cardiovascular causes, over 11 million die as a result of cardiac disease and stroke.[1] Myocardial infarction (MI) carries a short-term mortality rate of about 7% (with aggressive therapy), and congestive heart failure accounts for an even more distressing 20% one-year mortality. Despite the improvement shown in the prognosis of patients with acute MI with the use of available therapies including thrombolysis and urgent coronary revascularizations, there remains still a significant mortality risk, and a significant proportion of MI survivors are at risk for developing heart failure. Post-MI left ventricular (LV) remodeling—a process characterized by a mechanical expansion of the infarcted wall followed by progressive LV dilation and dysfunction[2]—plays an important role in the progressive nature of postinfarct heart failure. Since the main underlying trigger of this process is the loss of cardiomyocytes and microvasculature in the infarcted wall, treatment strategies aimed at preserving or regenerating myocardial tissue have been examined as potential therapeutic modalities. Currently, both basic scientists and clinician researchers are trying to develop a substitute for CM that can take charge of cardiomyocyte in functional heart. Toward this goal, many cell types are being investigated as potent sources of cardiomyocytes for cell transplantation therapy to repair and regenerate the damaged myocardium. The goal of regenerative medicine for clinical applications is twofold: first, the identification of a stem cell population that is derived from an autologous source, thereby overriding the concerns of immunological rejection; and, second, ensuring that such identified stem cell population is plastic enough to differentiate into any desired cell type.

II. Non-iPS-Cell-Based Therapies and Their Limitations

A. Autologous Adult Bone-Marrow-Derived Stem Cells for Myocardial Repair

Half a century ago, McCulloch and Till introduced the concept of the adult stem cell by identifying the bone marrow (BM) as a repository of cells having the capability of reconstituting the entire hematopoietic system following lethal irradiation.[3] More than 30 years later, Asahara and Isner extended this concept, revealing that BM-derived cells were also capable of vasculogenesis, a process previously considered to be restricted to embryonic life.[4] Evidence has also continued to accumulate, indicating the remarkable ability of adult stem cells to produce differentiated cells from embryologically unrelated tissues. There

are also convincing data regarding the existence of cardiac stem cells, supporting the notion that the heart has its own intrinsic regenerative capacity.[5] In
addition, both cardiac stem cells and extracardiac BM/BM-derived progenitors
have been shown to regenerate myocytes and vasculature.[6–9] In particular,
when BM-derived cells, both of murine and of human origin, are transplanted
or mobilized in animal models of MI, they have been shown to home to infarct
and periinfarct myocardium, inducing both myogenesis and angiogenesis,
thereby improving cardiac function and survival.[10] Thus, experimental evidence supports the notion that transplanted BM-derived cells participate
directly and indirectly in the regeneration of cardiac myocytes and the microvasculature post MI. Three available lines of evidence converge in favor of this
interpretation: (1) the recent observations that cardiac endogenous stem cells
are present in the normal myocardium and are involved in the maintenance of
cardiac cellular homeostasis, with the ability to expand and regenerate myocytes and microvasculature in the infarcted myocardium,[5] (2) the evidence that
in humans cardiomyocyte repopulation by BM-derived progenitors of hematopoietic origin can take place,[11,12] and (3) the demonstration that it is possible
to increase the efficiency of the intrinsic cardiac regenerative capacity of
animals with AMI by the local delivery or mobilization of BM-derived cells,
resulting in a reduction of infarct size and an improvement in LV performance
and survival.[6,13–15] It is therefore reasonable to propose that BM-derived cells
are part of an endogenous repair/regeneration process. These observations also
provide evidence that BM-derived cells act through a combination of paracrine
effects that stimulate the expansion, homing, and differentiation of endogenous
cardiac stem cells on the one hand and the transdifferentiation of the BM-
derived cells toward cardiac and vascular cells on the other.[16,17] The capability
of a subpopulation of BM cells to differentiate into cardiomyocytes is a crucial
but controversial biological issue. Experimental reports have documented that
new myocytes and vascular cells are formed from labeled cells injected into
damaged myocardium.[5,14,15] Clinical studies, particularly of heart and BM sex-
mismatched organ transplants in humans, have provided clear evidence to
support this paradigm.[7] In contrast to these findings, other investigators have
been unable to document myocyte differentiation in the mouse model and,
therefore, question the potential of BM cells to regenerate myocardial tissue.[18,19] Several other potential limitations of using autologous BM-derived
stem/progenitor cells have been identified and therefore must be acknowl-
edged. Risk factors for coronary artery disease are reported to be associated
with a reduced number and functional activity of EPCs in the peripheral blood
of patients.[20–22] Likewise, patients with diabetes showed lower EPC numbers.[23] Whether these noted defects in EPCs from patients with preexisting
risk factors or systemic diseases such as diabetes also translate to other BM-
derived stem cells has not been studied and would need to be investigated,

although the notion of modifying the EPC phenotype in order to overcome differences in potency has already been successfully tested[24,25]. MI is primarily a disease of older persons, and the senescent myocardium may differ biologically from the myocardium of young persons and from that of the small adult animals typically used in initial stem cell experiments. In particular, the cells (myocytes) themselves and the intercellular messaging milieu in the interstitial space may be profoundly different under each clinical condition. Senescence in the aging myocardium has been characterized by the predominance of large myofibers expressing p19INK4—a marker of cellular aging and apoptosis.[25] It is likely that the molecular signals produced by such cells and their extracellular environment are not as favorable for stem cell differentiation, migration, and integration as are the signals present in younger hearts. These conditions need to be further characterized, as this knowledge may allow physicians to modify the environment, making it more conducive to successful stem cell treatment.

Thus, the availability of adult stem cells that can regenerate multiple cell types, including the heart muscle, remains a significant challenge, and availability of a more plastic, pluripotent and yet autologous stem cell source that can regenerate microvasculature and large vessels as well as lost cardiomyocytes is continuously being pursued.

III. Therapeutic Cloning and Embryonic Stem Cells

The inherent plasticity of embryonic stem cells (ESCs) is therefore argued to be an advantage for their potential application in regenerative medicine. The successful animal cloning and generation of ESC lines by somatic cell nuclear transfer has generated immense enthusiasm for therapeutic cloning. The goal of therapeutic cloning is to produce pluripotent stem cells with the nuclear genome of the patient and induce the cells to differentiate into replacement cells, for example, cardiomyocytes for repairing damaged heart tissue. Reports on the generation of pluripotent stem cells[26,27] or histocompatible tissues[28] by nuclear transplantation and on the correction of a genetic defect in cloned embryonic stem (ES) cells[29] suggest that therapeutic cloning could in theory provide a source of cells for regenerative therapy. Robl and colleagues were the first to report the production of transgenic ES or ES-like cells cloned from fibroblasts.[26,30] These cells were shown to differentiate into the three germ layers and to contribute to chimeric bovine offspring. Mouse ES-like cells have also been derived by somatic cell nuclear transplantation, which could be differentiated into multiple cell types.[27] Interestingly, nonhuman-primate pluripotent stem cells derived by parthenogenesis could be differentiated into various cell types.[31] This raised intriguing possibilities for producing human "stem-like cells" from unfertilized eggs. Recent well-publicized events on the validity of human therapeutic cloning reports, however, underscore the difficulties associated with the generation of human ES lines for therapeutic purposes. However, a number of limitations may hinder the strategy of therapeutic cloning for future clinical applications. Extremely low efficiency of somatic nuclear transfer is a major concern. A survey of mouse ES lines derived by SNT indicates an overall efficiency of 82%.[27,29,32] These data suggest that, for the generation of a single nuclear transfer embryonic stem (ntES) cell line, 12 cloned blastocysts would be required and the number of oocytes required to generate one such ES line, a number that is critical for human applications, varies from as high as 1000 to as low as 15.[33] A cautious objective for clinical applications is to produce two to three ES cell lines per individual, realizing that some of these may be aneuploid or not sufficiently pluripotent. This would require, in the best case and in the best hands, 60–90 successfully reconstructed oocytes and the availability of many more for micromanipulation. Analysis of the literature on mouse SNT-derived ES lines raises concerns about the feasibility and relevance of therapeutic cloning, in its current embodiment, for human clinical practice. Nuclear transfer is unlikely to be much more efficient in humans than in mice. Optimistically, about 100 human oocytes would be required to generate customized ES cell lines for a single individual. A crucial difference is that, although 100 mouse oocytes can be obtained from a few super-ovulated females at a cost of $40, human oocytes must be harvested from super-ovulated volunteers, who are reimbursed for their participation. Add to this the complexity of the clinical procedure, and the cost of a human oocyte could be $1000–2000 in the United States. Thus, to generate a set of customized ES cell lines for an individual, the budget for the human oocyte material alone would be ~$100,000–200,000. This is a prohibitively high sum that will impede the widespread application of this technology in its present form. This limitation might be alleviated with oocytes from other species,[34] but mitochondrial genome differences between species are likely to pose a problem. Another current challenge of therapeutic cloning is to overcome abnormalities encountered in cloned animals,[31] as these may reflect defects in cloned ES cells. In spite of the production of cloned animals and ES-like cells by nuclear transplantation, reports of unstable or abnormal gene expression patterns in cloned embryos and fetuses[35,36] suggest incomplete reprogramming. Finally, ethical debate related to human oocyte manipulations add to the limitations. It is therefore imperative to develop alternative strategies to oocyte-independent therapeutic cloning. Although ESCs have been used in animal studies of cardiac repair,[37,38] ethical, technical, and regulatory issues, as well as unavailability of autologous human ESCs for cell therapy applications, limit the potential therapeutic utility of ESCs in humans.

IV. Induced Pluripotent Stem cells from Differentiated Somatic Cells

A. Generation of iPS Cells

The remarkable discovery by Yamanaka and colleagues that adult cells can be reprogrammed to become ESC-like iPS cells by retroviral transduction of defined transcription factors, namely, Oct4, Sox2, Klf4, and c-Myc (referred to as Yamanaka factors), has revolutionized stem cell research and led to the generation of mouse and human iPS cells from different laboratories using the same or different sets of transgene.[39–43] Within 5 years of this original discovery, the field of iPS cells has moved forward with unprecedented speed, and we have come a long way in both the methods available to generate iPS cells and our understanding of their molecular biology.[44,45] Many of the shortcomings of the original approach, such as the use of a retrovirus that may integrate into the reprogrammed cells' genome, use of an oncogene as one of the reprogramming factors, etc., have been overcome: reprogramming can now be achieved with plasmid transfection of defined factors, without the use of c-Myc[33,46] and with as few as two factors.[45] The generation of iPS cells by adenovirus or recombinant protein or PiggyBac system,[45,47] as well as the use of cell and transgene-free ESC protein extracts,[48] has further alleviated certain original concerns regarding the potential use of iPS cells in regenerative medicine. It is now well established that iPS cells can differentiate into all three germ-layer-derived cells. The seminal achievement of induced pluripotency holds great promise for regenerative medicine. Patient-specific iPS cells could provide useful platforms for drug discovery and offer unprecedented insights into disease mechanisms and, in the long term, may be used for cell and tissue replacement therapies. The discovery of the proliferative capacity and plasticity of various stem cell populations, especially iPS cells, has sparked much interest and debate regarding their use as a potential therapy. One of the most valuable aspects of these cells is the ability to derive them from autologous sources while still having the ability to differentiate them in cells of any of the three germ lineages. However, while the derivation of iPS cells has been studied and various methods exist,[49] the efficiency of deriving such cells is low and the process of doing so for therapeutic purposes long. Additionally methods to increase the efficiency and reduce the derivation time can affect the downstream use of the generated cells. Thus, the tissue of origin, the cell type, and methodology involved need to be carefully considered and optimized for each individual application of iPS cells.[49] While all of these parameters are important in designing the most efficient and reproducible protocol of generating iPS cells to be used clinically, the somatic cell of origin used requires special consideration. The cell type used not only determines things such as reprogramming efficiency and tendency to form terratomas[49] but also, due to epigenetic memory, determines the differentiation preference of these cells.[50]

iPS cells are syngeneic, indicating that they can become an ideal cell source for regenerative medicine. However, detailed differentiation properties and the directional differentiation system of iPS cells have not been fully characterized. Similarly, the mechanistic insights into epigenetic reprogramming of somatic cells during induced pluripotency have just begun to emerge. Understanding of these epigenetic mechanisms and their retention after reprogramming is crucial to an efficient clinical application of these cells.

B. Epigenetic Mechanisms of Induced Pluripotency

The mechanism of nuclear reprogramming via modifications in a cells epigenetic makeup is a complex process the exact mechanism of which remains largely elusive. Epigenetic and chromatin modifications are critical in iPS cell biology, as these marks play an important role in what defines a stem cell. Modifications in this code directly influence gene expression or repression.[51–60] The epigenetics of pluripotent cells is extremely complex, as ES and iPS cells need to fulfill three very important requirements in order to be pluripotent. First, due to continuous proliferation, pluripotent cells need to continuously and actively maintain their epigenetic code.[51] In addition, being pluripotent means that genes important in maintaining an undifferentiated state need to stay active while genes important in development need to be repressed.[51–58] Third, the chromatin should be open enough for differentiation into a particular cell lineage to occur quickly once the differentiating machinery in the cell has started.[51,56] Epigenetic modifying factors and enzymes are critical players in all these processes.[54,58]

While iPS cells have been shown to possess a pluripotency potential very similar to that of ES cells, very little is known about epigenetic remodeling that has to occur for each cell type to exist. ES cells, which are derived from the inner cell mass of the early embryo, rely on epigenetic modifications that occur both during gamete formation and postfertilization. In the case of ES cells, this process starts with the determination of the germline and continues at each stage during gamete development and fertilization.[61] For iPS cells, things are a little different. The journey toward becoming an iPS cell can start from a fully differentiated state. The epigenetic code of a fully differentiated state has been characterized during development where genes not important for the specific cell fate are silenced through DNA methylation and histone posttranslational modifications, whereas other genes important for the specific cell fate are activated through permissive histone modification in the presence of specific transcription factors. Reprogramming to an iPS cell requires full epigenetic reprogramming of these cells in a short window of time. We understand very little of how this reprogramming occurs and whether or not erasing of the previous epigenetic state occurs in full or there is residual information that persists in the pluripotent iPS cell. It is unclear how each of the transcription factors used to generate iPS cells induces epigenetic changes that lead to the induction of pluripotency-associated transcription factors and silencing of endogenous genes in a somatic cell (Fig. 1). Epigenetic changes are

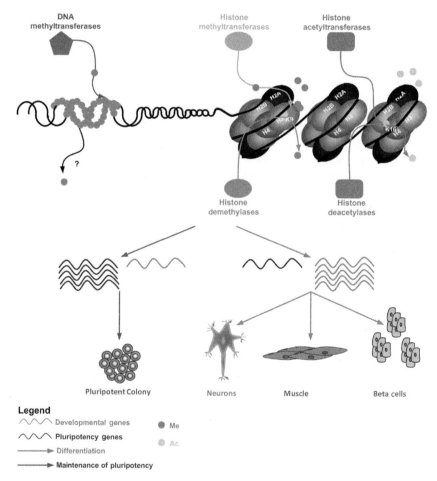

FIG. 1. Epigenetic modifications play a key role in defining iPS cell identity. Epigenetic modifications such as DNA methylation/demethylation, histone methylation/demethylation, and histone acetylation/deacetylation affect expression levels of pluripotent and developmental genes, thus playing a critical role in defining whether a cell remains pluripotent or differentiates into a particular lineage. (See Color Insert.)

modifications of DNA or chromatin that do not involve DNA sequence alteration or deletion.[62–64] Adult mammals contain hundreds of cell types distributed among their organs, each with identical DNA content. Yet, each of these cell types has a unique pattern of gene expression that occurs independently of DNA sequence *per se*. Though each cell has identical DNA content, the way in which it is packaged with chromosomal proteins differs greatly from cell to cell. Indeed, recent findings from chromatin research and animal cloning (nuclear transfer) studies suggest that much of the molecular basis of tissue-specific gene expression is rooted in the details of chromatin structure.[62,63] Much of the information obtained from studies on nuclear cloning and somatic-ES cell fusion on the epigenetic changes in somatic nuclei points toward somatic cell chromatin remodeling mediated via the process of chromatin condensation, DNA methylation/demethylation, and histone modifications (for detailed review, see Refs. 47,65). Histone modifications and DNA methylation are functionally linked.[62–64] DNA methylation has for long been recognized as an essential epigenetic mediator of gene turn on/off during development. The majority of DNA methylation occurs in 5′-CpG-3′ dinucleotides; a process during which DNA methyl transferase enzymes (dnmt1, dnmt3a, dnmt3b) attach a methyl group on the fifth carbon position of cytosine residues only, where they appear as CG dinucleotides and this modification is associated with a repressed chromatin state and inhibition of gene expression. Many studies have detailed the repressive effects that promoter-associated DNA methylation has on gene expression.[64,65] Notably, a large fraction of normal genomic DNA methylation is found in genes, within both intronic and exonic regions. Experiments using plasmid-based reporter genes have indicated that methylation over the body of a gene can result in reduced gene expression.[66] Considering that DNA methylation is involved in various biological phenomena, such as tissue-specific gene expression, cell differentiation, X-chromosome inactivation, genomic imprinting, changes in chromatin structure, and tumorigenesis,[62–64] it is conceivable that the changes in specific DNA methylation patterns is one of the principal epigenetic events underlying induced pluripotency in somatic cells. In the case of reprogramming, however, how DNA methylation patterns are changed is not fully understood. While the importance and implications of DNA methyl transferases in the maintenance of iPS cells and differentiation has been studied, mechanisms through which DNA is demethylated and how methylation is lost during reprogramming are not fully understood. A molecule, AID, which is critical for V(D)J[67] recombination in the immune system, has been proposed to play a key role in demethylating pluripotency-associated genes during reprogramming of somatic cells to iPS cells. Despite this, we still do not understand what molecules initially trigger the events through which methylation at the pluripotency-associated gene promoters is lost and what factors target these molecules to these sites.

While loss of methylation and the molecular mechanism involved in it during reprogramming need to be further investigated, the similarity at the DNA methylation level between iPS cells and ES cells is still a matter of debate. Some studies indicate that some iPS cells show very different genome-wide methylation patterns when compared to ES cells.[68] After comparing methylation levels of iPS cells derived from umbilical cord blood cells and human neonatal keratinocytes, a recent study found differences in genome-wide methylation levels of both cell types when compared to ES cells.[68] Both types of cells showed both incomplete *de novo* methylation of somatic genes as well as incomplete demethylation of pluripotency-associated genes.[68] This supports further the notion of lingering methylation marks after reprogramming that constitute epigenetic memory are important factors that could lead to preferences in differentiation of these cells. If this were indeed true, it would provide an important tool rather than an obstacle in the use of these cells

In addition to DNA methylation, posttranslational modifications in the histone proteins have emerged to play certain key roles in regulating gene activity, most likely through modulation of the chromatin structure. Several histone modifications are known to affect gene transcription, such as methylation, acetylation, phosphorylation, ubiquination, and sumolyation. The majority of these modifications occur on lysine and arginine residues on the histone tails. To date, at least eight lysine residues that can be acetylated in the N-termini of histone H3 (K9, K14, K18, and K23) and H4 (K5, K8, K12, and K16) are known, and six methylatable lysine positions exist in those of histone H3 (K4, K9, K27, K36, and K79) and H4 (K20). In general, the acetylation of histone H3 and H4 correlates with gene activation, while deacetylation correlates with gene silencing.[69] Methylation of H3-K4 also marks active chromatin, which contrasts with the modulation of inactive chromatin by methylation of H3-K9.[70] Methylation of H3-K27 is an epigenetic mark for recruitment of polycomb group (Pc-G) complexes and is prominent in the inactivated X-chromosome of female mammalian somatic cells.[70] The amino-terminal tail of histone H3 is subject to three distinctive methylation states: mono-, di-, and trimethylation. Pericentric heterochromatin is enriched for trimethylated H3-K9, while the centromeric regions are enriched for the dimethylated state.[70] At H3-K27, both di- and trimethylation are observed across several nucleosomes, and it is the trimethylated state that has been found to induce stable recruitment of Pc-G complexes. At H3-K4, fully activated promoters are enriched for the trimethylated state, while H3-K4 dimethylation correlates with the basal transcription-permissive state.[71] Thus, it appears that dimethylation activity prepares histones for a trimethylating activity, which then propagates stably as activated or silenced chromatin domains.

The similarity or difference at the histone level between ES cells and iPS cells is also a matter of debate. While DNA methylation is usually associated with repression or activation, any one histone modifications alone rarely determines the status of a gene. It is usually a combination of different histone modifications (e.g., methylation and acetylation), modification of different amino acid residues (e.g., histone3 lysine9 and histone3–lysine27), and a cross talk between both DNA methylation and histone modifications. Additionally, the chromatin state in iPS cells can change as the cells are passaged, which might result in slightly different results regarding the chromatin state depending on the passage number of the cells used for analysis.[61,72] We have only scratched the tip of the iceberg, and further research is needed to understand chromatin modifications that lead to reprogramming, changes that occur during differentiation of these cells, and the extent of similarity between these cells and ES cells.

In addition to DNA methylation and chromatin remodeling, micro RNAs (miRNA) are emerging as key players in both reprogramming to iPS cells and differentiation of these cells. miRNAs are small noncoding RNAs transcribed from both intragenic and intergenic regions. They play critical roles in a variety of different processes and we are starting to understand better their role in pluripotent cells. The role of miRNA in pluripotency has been studied using Dicer-null and DGCR8-null ES cells, which lack all mature miRNAs.[73–76] When differentiation of ES cells lacking Dicer is induced via embryoid body formation, cells show only a slight decrease in Oct4 levels and only a slight increase in the expression of early differentiation genes.[73] This indicates a role for miRNA in early differentiation of these cells. Additionally, miRNAs seem to play an important role in the pluripotency network. Two miRNA clusters (290 and 302 clusters) have binding sites on their promoter regions for pluripotency-associated genes. Moreover, members of these clusters have been shown to be regulated by pluripotency-associated genes such as *Oct4*, *Sox2*, and *Nanog*.[77,78] Expression of pluripotency-associated miRNA clusters (miR 520) and inhibition of tissue-specific miRNAs (let-7) during reprogramming can increase the efficiency of the dedifferentiation process.[73]

Further understanding of these epigenetic mechanisms and the way they cross talk with each other and intersect in the regulation of pluripotency-associated genes during reprogramming and maintenance of pluripotency and development during differentiation is crucial to our understanding of iPS cell biology. While differences in these pathways between iPS and ES cells make standardizing their therapeutic use more challenging, the epigenetic memory found in iPS cells and their preference to a specific lineage can be exploited in directing the differentiation of these iPS cells to a particular cell type for use in regenerative medicine.

V. Cardiovascular Lineage Differentiation of iPS Cells

Based on the available information on the iPS cells, including their ability to generate cells representative of all three germ layers and to contribute to chimera formation, we can conclude that iPS cells have a developmental potency comparable to that of ES cells. Experimental evidence from various laboratories suggests that iPS cells can be differentiated into various functional cell types, including neurons, cardiomyocytes, and hematopoietic cells.[79–86,89,100] A summary of studies demonstrating the generation of functional cardiovascular lineage cells from murine and human iPS cells is shown in Table I. We will focus on the derivation and functional aspects of cardiovascular cell types derived from iPS cells. It has been shown that cardiomyocytes and endothelial cells (ECs) can be successfully differentiated from murine iPS cells *in vitro* as well as *in vivo* when these iPS cells are transplanted in a murine MI model.[48] Another recent study has demonstrated the potency of cardiac repair potential of iPS cells in MI model.[63] Taura *et al.*[87] determined the features of the directed differentiation of human iPS cells into vascular ECs and mural cells (MCs), and compared that process with human embryonic stem (hES) cells.[64] The authors succeeded in inducing and isolating human vascular cells from iPS cells and showed that the properties of human iPS cell differentiation into vascular cells are nearly identical to those of hES cells. Zhang *et al.* (2009) successfully characterized the cardiac differentiation potential of human iPS cells compared to hES cells.[61] Electrophysiology studies indicated that iPS cells have a capacity similar to ES cells for differentiation into nodal-, atrial-,

TABLE I

CARDIOVASCULAR CELLS DIFFERENTIATED FROM IPS CELLS

S no.	Method	Sources	Types of cardiovascular cells	In vitro/ in vivo	References
1	Retroviral-mediated	MEF, hEF	Cardiac and hematopoietic lineages	Both	60,62,63,66,76,77
2	Lentivirus	MEF, hEF	Cardiac myocytes, vascular endothelial, cells, and mural cells	*In vitro*	61,64,65,78
3	Chemical (DMSO, LIF)	P19 ES cell, MES cells	Cardiac myocytes	*In vitro*	67,79
4	ES cell extracts	NIH3T3 cells	Cardiac myocytes and endothelial cells	Both	38

and ventricular-like phenotypes based on their action potential characteristics. Both iPS and ES cell-derived cardiomyocytes exhibited responsiveness to beta-adrenergic stimulation, manifested by an increase in spontaneous rate and a decrease in action potential duration. Study by Martinez-Fernandez[88] demonstrated reproducible cardiogenic differentiation of mouse iPS cells, bioengineered independently of the c-MYC oncogene, throughout embryonic development and into adulthood, and yielded proficient cardiogenic progeny. Lineage specification *in vitro* allowed isolation of cardiomyocytes that inherited properties of functional excitation–contraction coupling. These were validated *in vivo* by sustained chimerism into adulthood, which supported the high demands of normal heart performance. Thus, this three-factor iPS cell demonstrated lineage-specific production of bona fide cardiac tissues using a nuclear reprogramming approach that minimized oncogenic exposure and fulfilled high-stringency criteria for derivation of functional cardiomyocytes.[88,101] Another study showed that iPS-cell-derived cardiomyocytes display spontaneous rhythmic intracellular Ca^{2+} fluctuations with amplitudes of Ca^{2+} transients comparable to those of ES-cell-derived cardiomyocytes, indicating functional coupling of the newly generated cardiomyocytes. Electrophysiological studies with multielectrode arrays demonstrated the functionality and presence of the beta-adrenergic and muscarinic signaling cascade in these cells.[66] Similarly, Narazaki et al.[90] showed that purified Flk1$^+$ cells isolated from iPS cells give rise to MCs, arterial, venous, and lymphatic ECs, as well as self-beating cardiomyocytes, and that the time course and efficiency of the differentiation were comparable to those of mouse ES cells.[67] Xie et al.[91] successfully differentiated mouse iPS cells into SMCs *in vitro*.[68] Also, the iPS-derived SMCs acquired SMC functional characteristics including contraction and calcium influx in response to stimuli. Together, these studies have demonstrated that cells of cardiovascular lineages can be derived from the iPS cells and these cells closely resemble those derived from ES cells in their phenotypic and functional properties (Fig. 2). Given that, unlike ES cells, iPS cells retain the autologous source requirement, cells derived from iPS cells may have a better and safer cell therapy potential.

However, a comparative analysis of the molecular, epigenetic, and biological properties of cells differentiated from iPS cells and those of the same lineage/phenotype of somatic cells from which the iPS cells originated from is essential to understand the translational potential of these cells. Lee et al.[92] generated iPS cells from human aortic vascular smooth muscle cells (HASMCs) and redifferentiated these iPS cells back into SMCs.[69] The authors characterized and compared the gene expression and cellular functionality of iPS-cell-derived SMCs with the parental HASMCs from which the iPS cells were derived. Established iPS cells were shown to possess properties equivalent to those of hES cells, in terms of the cell surface markers, global mRNA

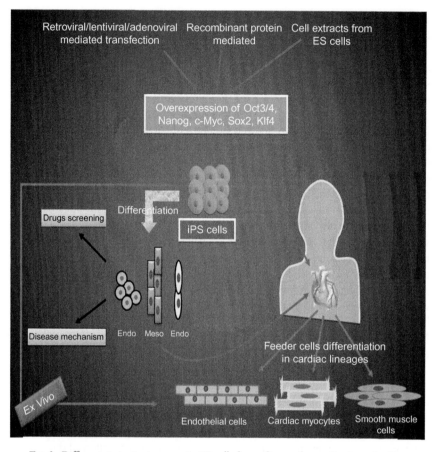

Fig. 2. Different strategies to generate iPS cells for cardiovascular applications. Cardiovascular lineage cells differentiated from iPS cells could then be used for drug screening, to study disease mechanisms and potentially for regenerative purposes. (See Color Insert.)

and miRNA expression patterns, epigenetic status of Oct4, Rex1, and Nanog promoters, as well as *in vitro/in vivo* pluripotency. The cells were differentiated into SMCs to enable a direct, comparative analysis with parental HASMCs. The authors observed that iPS-cell-derived SMCs were very similar to parental HASMCs in gene expression patterns, epigenetic modifications of pluripotency-related genes, and *in vitro* functional properties. These studies provided proof of concept that iPS cells generated from a particular somatic cell phenotype can be successfully redifferentiated into their original identity and, based on *in vitro* molecular and epigenetic characteristics, resemble the parental cell. When translated to the need for replenishment of one particular type of cell, these data may have important implications as to the source of autologous cells to be used for reprogramming to iPS. The stimulation of microvascular repair and collateral artery growth is a promising alternative approach for noninvasive treatment of arterial obstructive disease, such as coronary, peripheral, or cerebral artery disease. With the growing knowledge of the mechanisms involved and the factors that influence these processes, an increasing number of clinical trials are being performed to stimulate neovascularization. The expression of growth factors and the cooperation of surrounding and infiltrating cells seem to be essential in orchestrating the complex processes involved. In light of the above developments, attempts to stimulate the growth of collateral arteries using the iPS cells described by Lee *et al.*[92] might ultimately lead to new treatment options for patients with vascular occlusive diseases.[69]

However, caution should be used while determining whether differentiated cells derived from iPS cells, such as smooth muscle cells, have the same functional properties as their physiological *in vivo* counterparts. Also, it is important to determine the factors involved in maintaining such physiological functions *in vivo*. Recent studies describe the contribution of various growth factors and the corresponding inhibitors to heart development during embryogenesis. Bone morphogenetic proteins, Wnt protein, and Notch signals play critical roles in heart development in a context- and time-dependent manner. Although several signaling protein families are involved in the development of the heart, limited evidence is available about the exact signals that mediate the differentiation of ES cells into cardiomyocytes. There is no single growth factor that acts constantly throughout the entire process of organ induction during the development of multiple organ systems; it is the time- and context-dependent expression of multiple factors that is critical for proper development. Consistent with ES cells, the exposure of iPS cells to such growth factors is hypothesized to augment differentiation into cardiomyocytes. Use of such appropriate developmental signal information for functional blood vessel development using iPS cells has the potential for providing the foundations for future cardiovascular regenerative medicine.

A. Applications of iPS Cells

Based on the literature available, including the ability of iPS cells to generate all lineages of the embryo and to contribute to chimera formation, we can conclude that iPS cells have a developmental potency comparable to that of ES cells. Although far from being ready for widespread clinical applications, the utility and potential of iPS cells in moving the stem cell and regenerative medicine fields forward is already evident. One of the obvious benefits over ES-cell-based therapies would be that iPS cells tailored to specific individuals should provide the opportunity for cell replacement therapy without the need for immunosuppressants, as autologous transplantation of genetically identical cells, and potentially tissues and organs, overcomes the issue of immune rejection. Importantly, in cases of a known genetic defect, utilizing gene therapy approaches could, in principle, restore cellular function. Shortly after the first report describing iPS cell methodology, a seminal proof-of-principle study demonstrated the potential of iPS-cell-based cell replacement therapy in a humanized mouse model of sickle cell anemia.[79] More recently, a similar approach was taken with human individuals with Fanconi anemia, a disease characterized by severe genetic instability.[93] In this case, the mutant gene was replaced using lentiviral vectors prior to epigenetic reprogramming of the patient fibroblasts and keratinocytes, as the genetic instability of the mutant fibroblasts made them nonpermissive for iPS cell generation. Studies have also illustrated therapeutic transplantation applications for healthy iPS-cell-derived somatic cells in mouse models of disease. In particular, iPS-cell-derived dopamine neurons functionally integrated into adult brain in a rat model of PD, leading to an improvement of the phenotype,[80] while hemophilia A mice injected with iPS-cell-derived ECs into their livers were protected in a death-inducing bleeding assay.[94]

Although further work needs to be done toward generating and extensively characterizing clinical-grade iPS cells before human cell replacement therapies can be attempted, disease modeling and drug screening are two immediate applications for reprogramming technology. iPS cells provide useful tool for disease modeling *in vitro*, facilitating studies of mechanisms underlying disease development, drug screening, and development of new therapeutic strategy. Some human iPS cell lines have been established from patients with neural diseases, such as amyotrophic lateral sclerosis and spinal muscular atrophy.[81,95] These ALS and SMA patient-specific iPS cells can be directed to differentiate into motor neurons and such neurons originating from ALS and SMA patient-specific iPS cells provide *in vitro* disease models to study these two neural diseases. Similarly, Park *et al.*[96] reported the generation of iPS cells from patients with a variety of genetic diseases with either Mendelian or complex inheritance; these diseases include adenosine deaminase deficiency-related severe combined immunodeficiency (ADA-SCID), Shwachman–Bodian–Diamond syndrome (SBDS), Gaucher disease (GD) type III, Duchenne (DMD) and Becker muscular dystrophy (BMD), PD, Huntington's disease (HD), juvenile-onset, type 1 diabetes mellitus (JDM), Down syndrome (DS)/trisomy 21, and the carrier state of Lesch–Nyhan syndrome. Such disease-specific stem cells offer an unprecedented opportunity to recapitulate both normal and pathologic human tissue formation *in vitro*, thereby enabling disease investigation and drug development. The unique properties of ES and iPS cells also provide for practical approaches in pharmaceutical toxicology and pharmacogenomics. In particular, hepatotoxicity and cardiotoxicity are two principal causes of drug failure during preclinical testing, while the variability in individual responses to potential therapeutic agents is also a major problem in effective drug development.[97–99] The advantage of iPS cell technology is that it allows for the first time the generation of a library of cell lines that may to a substantial extent represent the genetic and potentially epigenetic variation of a broad spectrum of the population. The use of this tool in high-throughput screening assays could allow better prediction of the toxicology caused by and therapeutic responses induced by newly developed dugs and offer insight into the underlying mechanisms.

B. Major Limitations and Challenges for Clinical Application of iPS Cells

Given the many potential risks of applying autologous iPS cell treatment to human subjects, iPS cell therapies in regenerative medicine may encounter strict regulatory restrictions. For patient-specific iPS cells, there is still a lack of consensus in deriving, culturing, and differentiating iPS cells among different laboratories. The inconsistency in their derivation, maintenance, and staging also has led to conflicting data regarding the molecular circuitry that governs iPS cell reprogramming and differentiation. Furthermore, individual iPS cell lines may vary in their ability to differentiate into different cell lineages, making testing and approval by the FDA in a timely fashion even more difficult. Another issue that may hinder the clinical translation of iPS cell therapies is the economic feasibility of producing individualized iPS cell therapeutic products. The viability of a business model for patient-specific iPS treatment is still unknown. It may well be the case that few if any pharmaceutical companies will be able to produce cost-effective individualized iPS cell products tailored for a single patient at a time. To be commercially feasible, these cells will need to be made in standardized, large-scale production, and the individual needs or profiles of patients will need to be easily assessed to allow matching and wide distributions.

Several issues will need a more comprehensive resolution before iPS cells can find clinical applications. Many of these issues are, in fact, not unique to iPS cells but have already been noted for human ESCs as well. Tumorigenicity remains a major concern in the use of iPS cells in regenerative medicine. Making a somatic cell pluripotent predisposes that cell to cause a tumor. There are compelling reasons for worrying about iPS cell tumorigenicity based on actual published data. Of greatest concern is that nearly all iPS cells described in published studies have been demonstrated to cause teratomas, proving pluripotency but also tumorigenicity, and that mice genetically derived to contain some tissues from iPS cells exhibit an incidence of malignant tumors. Genetic changes inherent in the iPS cell generation process may pose the risk of enhancing tumorigenesis both through the introduced genes themselves and, in theory, via potential changes at specific integration sites. Moving away from methods of induction that rely on genetic changes (e.g., using protein factors rather than exogenous retrovirus-based nucleic acid transduction) may reduce the tumorigenic potential, but this remains to be confirmed. Another question that needs a more comprehensive analysis is whether all iPS lines derived from the same parent somatic cells are exact epigenetic and molecular signatures or whether they vary from one another to a certain extent, and whether these variations influence their tumorigenicity and developmental potential. Additionally, the long-term stability of induced pluripotency and epigenetic status of a given iPS cell line needs to be validated.

Mostly under the radar in the field of stem cell safety are potential undesirable side effects of epigenetic changes in the developed iPS cell lines. Because epigenetic changes are postulated to play a key role in the reprogramming and are at the heart of iPS cell formation, it is critically important to more fully study the global epigenetic changes associated with pluripotency, especially induced pluripotency. An ever-changing or unstable epigenome can pose an unanticipated risk of enhancing tumorigenicity. Characterizing the relationships among the epigenome, pluripotency, and tumorigenicity should prove of great benefit for developing safe regenerative medicine. The road from where we are today to a future with iPS and human ESC-based regenerative medicine therapies being safe and more common treatment modalities is not a clear, linear one. Beyond tumorigenicity issues, it still needs to be determined how comparable the *in vitro*-derived cell types are to their *in vivo* counterparts and how to isolate them at sufficient purity. Regardless of the type of stem cells used, high-quality transplantation methods using tissue engineering must be developed. The existing and developing evidence from fields as diverse as developmental biology, stem cell biology, and tissue engineering must be integrated to achieve the full potential of iPS cells in cardiovascular regenerative medicine. Moreover, a general and efficient way of characterizing the large numbers of human iPS cells and iPS-cell-derived cells also needs to be developed to provide a routine, high-throughput method for quality control and for their immediate application in drug screens and basic research.

References

1. Joggerst SJ, Hatzopoulos AK. Stem cell therapy for cardiac repair: benefits and barriers. *Expert Rev Mol Med* 2009;**11**:e20.
2. Pfeffer MA. Left ventricular remodeling after acute myocardial infarction. *Annu Rev Med* 1995;**46**:455–66.
3. McCulloch EA, Till JE. The radiation sensitivity of normal mouse bone marrow cells, determined by quantitative marrow transplantation into irradiated mice. *Radiat Res* 1960;**13**:115–25.

4. Asahara T, Murohara T, Sullivan A, Silver M, van der Zee R, Li T, et al. Isolation of putative progenitor endothelial cells for angiogenesis. *Science* 1997;**275**:964–7.
5. Beltrami AP, Barlucchi L, Torella D, Baker M, Limana F, Chimenti S, et al. Adult cardiac stem cells are multipotent and support myocardial regeneration. *Cell* 2003;**114**:763–76.
6. Kawamoto A, Gwon HC, Iwaguro H, Yamaguchi JI, Uchida S, Masuda H, et al. Therapeutic potential of ex vivo expanded endothelial progenitor cells for myocardial ischemia. *Circulation* 2001;**103**:634–7.
7. Quaini F, Urbanek K, Beltrami AP, Finato N, Beltrami CA, Nadal-Ginard B, et al. Chimerism of the transplanted heart. *N Engl J Med* 2002;**346**:5–15.
8. Kawamoto A, Tkebuchava T, Yamaguchi J, Nishimura H, Yoon YS, Milliken C, et al. Intramyocardial transplantation of autologous endothelial progenitor cells for therapeutic neovascularization of myocardial ischemia. *Circulation* 2003;**107**:461–8.
9. Yoon YS, Wecker A, Heyd L, Park JS, Tkebuchava T, Kusano K, et al. Clonally expanded novel multipotent stem cells from human bone marrow regenerate myocardium after myocardial infarction. *J Clin Invest* 2005;**115**:326–38.
10. Losordo DW, Dimmeler S. Therapeutic angiogenesis and vasculogenesis for ischemic disease: part II: cell-based therapies. *Circulation* 2004;**109**:2692–7.
11. Muller P, Pfeiffer P, Koglin J, Schafers HJ, Seeland U, Janzen I, et al. Cardiomyocytes of noncardiac origin in myocardial biopsies of human transplanted hearts. *Circulation* 2002;**106**:31–5.
12. Laflamme MA, Myerson D, Saffitz JE, Murry CE. Evidence for cardiomyocyte repopulation by extracardiac progenitors in transplanted human hearts. *Circ Res* 2002;**90**:634–40.
13. Kocher AA, Schuster MD, Szabolcs MJ, Takuma S, Burkhoff D, Wang J, et al. Neovascularization of ischemic myocardium by human bone-marrow-derived angioblasts prevents cardiomyocyte apoptosis, reduces remodeling and improves cardiac function. *Nat Med* 2001;**7**:430–6.
14. Orlic D, Kajstura J, Chimenti S, Jakoniuk I, Anderson SM, Li B, et al. Bone marrow cells regenerate infarcted myocardium. *Nature* 2001;**410**:701–5.
15. Orlic D, Kajstura J, Chimenti S, Bodine DM, Leri A, Anversa P. Transplanted adult bone marrow cells repair myocardial infarcts in mice. *Ann N Y Acad Sci* 2001;**938**:221–9, discussion 229-230.
16. Iwasaki H, Kawamoto A, Ishikawa M, Oyamada A, Nakamori S, Nishimura H, et al. Dose-dependent contribution of CD34-positive cell transplantation to concurrent vasculogenesis and cardiomyogenesis for functional regenerative recovery after myocardial infarction. *Circulation* 2006;**113**:1311–25.
17. Badorff C, Brandes RP, Popp R, Rupp S, Urbich C, Aicher A, et al. Transdifferentiation of blood-derived human adult endothelial progenitor cells into functionally active cardiomyocytes. *Circulation* 2003;**107**:1024–32.
18. Balsam LB, Wagers AJ, Christensen JL, Kofidis T, Weissman IL, Robbins RC. Haematopoietic stem cells adopt mature haematopoietic fates in ischaemic myocardium. *Nature* 2004;**428**:668–73.
19. Murry CE, Soonpaa MH, Reinecke H, Nakajima H, Nakajima HO, Rubart M, et al. Haematopoietic stem cells do not transdifferentiate into cardiac myocytes in myocardial infarcts. *Nature* 2004;**428**:664–8.
20. Walter DH, Haendeler J, Reinhold J, Rochwalsky U, Seeger F, Honold J, et al. Impaired CXCR4 signaling contributes to the reduced neovascularization capacity of endothelial progenitor cells from patients with coronary artery disease. *Circ Res* 2005;**97**:1142–51.
21. Schmidt-Lucke C, Rossig L, Fichtlscherer S, Vasa M, Britten M, Kamper U, et al. Reduced number of circulating endothelial progenitor cells predicts future cardiovascular events: proof of concept for the clinical importance of endogenous vascular repair. *Circulation* 2005;**111**:2981–7.

22. Urbich C, Dimmeler S. Risk factors for coronary artery disease, circulating endothelial progenitor cells, and the role of HMG-CoA reductase inhibitors. *Kidney Int* 2005;**67**:1672–6.

23. Tepper OM, Galiano RD, Capla JM, Kalka C, Gagne PJ, Jacobowitz GR, et al. Human endothelial progenitor cells from type II diabetics exhibit impaired proliferation, adhesion, and incorporation into vascular structures. *Circulation* 2002;**106**:2781–6.

24. Murasawa S, Llevadot J, Silver M, Isner JM, Losordo DW, Asahara T. Constitutive human telomerase reverse transcriptase expression enhances regenerative properties of endothelial progenitor cells. *Circulation* 2002;**106**:1133–9.

25. Nadal-Ginard B, Kajstura J, Leri A, Anversa P. Myocyte death, growth, and regeneration in cardiac hypertrophy and failure. *Circ Res* 2003;**92**:139–50.

26. Cibelli JB, Stice SL, Golueke PJ, Kane JJ, Jerry J, Blackwell C, et al. Transgenic bovine chimeric offspring produced from somatic cell-derived stem-like cells. *Nat Biotechnol* 1998;**16**:642–6.

27. Munsie MJ, Michalska AE, O'Brien CM, Trounson AO, Pera MF, Mountford PS. Isolation of pluripotent embryonic stem cells from reprogrammed adult mouse somatic cell nuclei. *Curr Biol* 2000;**10**:989–92.

28. Lanza RP, Chung HY, Yoo JJ, Wettstein PJ, Blackwell C, Borson N, et al. Generation of histocompatible tissues using nuclear transplantation. *Nat Biotechnol* 2002;**20**:689–96.

29. Rideout 3rd WM, Hochedlinger K, Kyba M, Daley GQ, Jaenisch R. Correction of a genetic defect by nuclear transplantation and combined cell and gene therapy. *Cell* 2002;**109**:17–27.

30. Robl JM, Gilligan B, Critser ES, First NL. Nuclear transplantation in mouse embryos: assessment of recipient cell stage. *Biol Reprod* 1986;**34**:733–9.

31. Cibelli JB, Grant KA, Chapman KB, Cunniff K, Worst T, Green HL, et al. Parthenogenetic stem cells in nonhuman primates. *Science* 2002;**295**:819.

32. Kawase E, Yamazaki Y, Yagi T, Yanagimachi R. Pedersen RA Mouse embryonic stem (ES) cell lines established from neuronal cell-derived cloned blastocysts. *Genesis* 2000;**28**:156–63.

33. Mombaerts P. Therapeutic cloning in the mouse. *Proc Natl Acad Sci USA* 2003;**100**(Suppl 1):11924–5.

34. Lanza RP, Cibelli JB, West MD. Prospects for the use of nuclear transfer in human transplantation. *Nat Biotechnol* 1999;**17**:1171–4.

35. Humpherys D, Eggan K, Akutsu H, Friedman A, Hochedlinger K, Yanagimachi R, et al. Abnormal gene expression in cloned mice derived from embryonic stem cell and cumulus cell nuclei. *Proc Natl Acad Sci USA* 2002;**99**:12889–94.

36. Bortvin A, Eggan K, Skaletsky H, Akutsu H, Berry DL, Yanagimachi R, et al. Jaenisch R Incomplete reactivation of Oct4-related genes in mouse embryos cloned from somatic nuclei. *Development* 2003;**130**:1673–80.

37. Kofidis T, de Bruin JL, Yamane T, Tanaka M, Lebl DR, Swijnenburg RJ, et al. Stimulation of paracrine pathways with growth factors enhances embryonic stem cell engraftment and host-specific differentiation in the heart after ischemic myocardial injury. *Circulation* 2005;**111**:2486–93.

38. Rajasingh J, Bord E, Hamada H, Lambers E, Qin G, Losordo DW. Kishore R STAT3-dependent mouse embryonic stem cell differentiation into cardiomyocytes: analysis of molecular signaling and therapeutic efficacy of cardiomyocyte precommitted mES transplantation in a mouse model of myocardial infarction. *Circ Res* 2007;**101**:910–8.

39. Park IH, Zhao R, West JA, Yabuuchi A, Huo H, Ince TA, et al. Reprogramming of human somatic cells to pluripotency with defined factors. *Nature* 2008;**451**:141–6.

40. Takahashi K, Okita K, Nakagawa M, Yamanaka S. Induction of pluripotent stem cells from fibroblast cultures. *Nat Protoc* 2007;**2**:3081–9.

41. Takahashi K, Tanabe K, Ohnuki M, Narita M, Ichisaka T, Tomoda K, et al. Induction of pluripotent stem cells from adult human fibroblasts by defined factors. *Cell* 2007;**131**:861–72.

42. Takahashi K, Yamanaka S. Induction of pluripotent stem cells from mouse embryonic and adult fibroblast cultures by defined factors. *Cell* 2006;**126**:663–76.
43. Yu J, Vodyanik MA, Smuga-Otto K, Antosiewicz-Bourget J, Frane JL, Tian S, et al. Induced pluripotent stem cell lines derived from human somatic cells. *Science* 2007;**318**:1917–20.
44. Wernig M, Meissner A, Foreman R, Brambrink T, Ku M, Hochedlinger K, et al. In vitro reprogramming of fibroblasts into a pluripotent ES-cell-like state. *Nature* 2007;**448**:318–24.
45. Kiskinis E, Eggan K. Progress toward the clinical application of patient-specific pluripotent stem cells. *J Clin Invest* 2010;**120**:51–9.
46. Nakagawa M, Koyanagi M, Tanabe K, Takahashi K, Ichisaka T, Aoi T, et al. Generation of induced pluripotent stem cells without Myc from mouse and human fibroblasts. *Nat Biotechnol* 2008;**26**:101–6.
47. Amabile G, Meissner A. Induced pluripotent stem cells: current progress and potential for regenerative medicine. *Trends Mol Med* 2009;**15**:59–68.
48. Rajasingh J, Lambers E, Hamada H, Bord E, Thorne T, Goukassian I, et al. Cell-free embryonic stem cell extract-mediated derivation of multipotent stem cells from NIH3T3 fibroblasts for functional and anatomical ischemic tissue repair. *Circ Res* 2008;**102**:e107–17.
49. Stadtfeld M, Hochedlinger K. Induced pluripotency: history, mechanisms, and applications. *Genes Dev* 2010;**24**(20):2239–63.
50. Kim K, Doi A, Wen B, Ng K, Zhao R, Cahan P, et al. Epigenetic memory in induced pluripotent stem cells. *Nature* 2010;**467**(7313):285–90.
51. Bernstein BE, Mikkelsen TS, Xie X, Kamal M, Huebert DJ, Cuff J, et al. A bivalent chromatin structure marks key developmental genes in embryonic stem cells. *Cell* 2006;**125**(2):315–26.
52. Holliday R. Epigenetics: a historical overview. *Epigenetics* 2006;**1**(2):76–80.
53. Lee TI, Jenner RG, Boyer LA, Guenther MG, Levine SS, Kumar RM, et al. Control of developmental regulators by polycomb in human embryonic stem cells. *Cell* 2006;**125**(2):301–13.
54. Ma P, Schultz RM. Histone deacetylase 1 (Hdac1) regulates histone acetylation, development, and gene expression in preimplantation mouse embryos. *Dev Biol* 2008;**319**(1):110–20.
55. Marmorstein R. Protein modules that manipulate histone tails for chromatin regulation. *Nat Rev Mol Cell Biol* 2001;**2**(6):422–32.
56. Mikkelsen TS, Ku M, Jaffe DB, Issac B, Lieberman E, Giannoukos G, et al. Genome-wide maps of chromatin state in pluripotent and lineage-committed cells. *Nature* 2007;**448**(7153):553–60.
57. Ren X. Comments on control of developmental regulators by polycomb in human embryonic stem cells. *Med Hypotheses* 2006;**67**(6):1469–70.
58. Simon JA, Kingston RE. Mechanisms of polycomb gene silencing: knowns and unknowns. *Nat Rev Mol Cell Biol* 2009;**10**(10):697–708.
59. Vakoc CR, Sachdeva MM, Wang H, Blobel GA. Profile of histone lysine methylation across transcribed mammalian chromatin. *Mol Cell Biol* 2006;**26**(24):9185–95.
60. Xi S, Geiman TM, Briones V, Guang Tao Y, Xu H, Muegge K. Lsh participates in DNA methylation and silencing of stem cell genes. *Stem Cells* 2009;**27**(11):2691–702.
61. Hajkova P. Epigenetic reprogramming in the germline: towards the ground state of the epigenome. *Philos Trans R Soc Lond B Biol Sci* 2011;**366**(1575):2266–73.
62. Trojer P, Reinberg D. Histone lysine demethylases and their impact on epigenetics. *Cell* 2006;**125**:213–7.
63. Eden S, Hashimshony T, Keshet I, Cedar H, Thorne AW. DNA methylation models histone acetylation. *Nature* 1998;**394**:842.
64. Klose RJ, Bird AP. Genomic DNA methylation: the mark and its mediators. *Trends Biochem Sci* 2006;**31**:89–97.

65. Hemberger M, Dean W, Reik W. Epigenetic dynamics of stem cells and cell lineage commitment: digging Waddington's canal. *Nat Rev Mol Cell Biol* 2009;**10**:526–37.
66. Hsieh CL. Stability of patch methylation and its impact in regions of transcriptional initiation and elongation. *Mol Cell Biol* 1997;**17**:5897–904.
67. Dudley DD, Chaudhuri J, Bassing CH, Alt FW. Mechanism and control of V(D)J recombination versus class switch recombination: similarities and differences. *Adv Immunol* 2005;**86**:43–112.
68. Kim K, Zhao R, Doi A, Ng K, Unternaehrer J, Cahan P, et al. Donor cell type can influence the epigenome and differentiation potential of human induced pluripotent stem cells. *Nat Biotechnol* 2011;**29**:1117–9.
69. Fry CJ, Peterson CL. Chromatin remodeling enzymes: who's on first? *Curr Biol* 2001;**11**: R185–97.
70. Lachner M, O'Sullivan RJ, Jenuwein T. An epigenetic road map for histone lysine methylation. *J Cell Sci* 2003;**116**:2117–24.
71. Santos-Rosa H, Schneider R, Bannister AJ, Sherriff J, Bernstein BE, Emre NC, et al. Kouzarides T Active genes are tri-methylated at K4 of histone H3. *Nature* 2002;**419**:407–11.
72. Meissner A. Epigenetic modifications in pluripotent and differentiated cells. *Nat Biotechnol* 2010;**28**(10):1079–88.
73. Mallanna SK, Rizzino A. Emerging roles of microRNAs in the control of embryonic stem cells and the generation of induced pluripotent stem cells. *Dev Biol* 2010;**344**(1):16–25.
74. Lee Y, Jeon K, Lee JT, Kim S, Kim VN. MicroRNA maturation: stepwise processing and subcellular localization. *EMBO J* 2002;**17**:4663–70.
75. Lewis BP, Burge CB, Bartel DP. Conserved seed pairing, often flanked by adenosines, indicates that thousands of human genes are microRNA targets. *Cell* 2005;**1**:15–20.
76. Zeng Y, Cullen BR. Sequence requirements for micro RNA processing and function in human cells. *RNA* 2003;**1**:112–23.
77. Barroso-delJesus A, Romero-Lopez C, Lucena-Aguilar G, Melen GJ, Sanchez L, Ligero G, et al. Embryonic stem cell-specific miR302-367 cluster: human gene structure and functional characterization of its core promoter. *Mol Cell Biol* 2008;**21**:6609–19.
78. Card DA, Hebbar PB, Li L, Trotter KW, Komatsu Y, Mishina Y, et al. Oct4/Sox2-regulated miR-302 targets cyclin D1 in human embryonic stem cells. *Mol Cell Biol* 2008;**20**:6426–38.
79. Hanna J, Wernig M, Markoulaki S, Sun CW, Meissner A, Cassady JP, et al. Treatment of sickle cell anemia mouse model with iPS cells generated from autologous skin. *Science* 2007;**318**:1920–3.
80. Wernig M, Zhao JP, Pruszak J, Hedlund E, Fu D, Soldner F, et al. Neurons derived from reprogrammed fibroblasts functionally integrate into the fetal brain and improve symptoms of rats with Parkinson's disease. *Proc Natl Acad Sci USA* 2008;**105**:5856–61.
81. Dimos JT, Rodolfa KT, Niakan KK, Weisenthal LM, Mitsumoto H, Chung W, et al. Induced pluripotent stem cells generated from patients with ALS can be differentiated into motor neurons. *Science* 2008;**321**:1218–21.
82. Kuzmenkin A, Liang H, Xu G, Pfannkuche K, Eichhorn H, Fatima A, et al. Functional characterization of cardiomyocytes derived from murine induced pluripotent stem cells in vitro. *FASEB J* 2009;**23**:4168–80.
83. Zhang J, Wilson GF, Soerens AG, Koonce CH, Yu J, Palecek SP, et al. Functional cardiomyocytes derived from human induced pluripotent stem cells. *Circ Res* 2009;**104**:e30–41.
84. Pfannkuche K, Liang H, Hannes T, Xi J, Fatima A, Nguemo F, et al. Cardiac myocytes derived from murine reprogrammed fibroblasts: intact hormonal regulation, cardiac ion channel expression and development of contractility. *Cell Physiol Biochem* 2009;**24**:73–86.
85. Yoshida Y, Yamanaka S. iPS cells: a source of cardiac regeneration. *J Mol Cell Cardiol* 2011;**50** (2):327–32 32.85.

86. Nelson TJ, Martinez-Fernandez A, Yamada S, Perez-Terzic C, Ikeda Y, Terzic A. Repair of acute myocardial infarction by human stemness factors induced pluripotent stem cells. *Circulation* 2009;**120**:408–16.
87. Taura D, Sone M, Homma K, Oyamada N, Takahashi K, Tamura N, et al. Induction and isolation of vascular cells from human induced pluripotent stem cells–brief report. *Arterioscler Thromb Vasc Biol* 2009;**29**:1100–3.
88. Martinez-Fernandez A, Nelson TJ, Yamada S, Reyes S, Alekseev AE, Perez-Terzic C, et al. iPS programmed without c-MYC yield proficient cardiogenesis for functional heart chimerism. *Circ Res* 2009;**105**:648–56.
89. Mauritz C, Schwanke K, Reppel M, Neef S, Katsirntaki K, Maier LS, et al. Generation of functional murine cardiac myocytes from induced pluripotent stem cells. *Circulation* 2008;**118**:507–17.
90. Narazaki G, Uosaki H, Teranishi M, Okita K, Kim B, Matsuoka S, et al. Directed and systematic differentiation of cardiovascular cells from mouse induced pluripotent stem cells. *Circulation* 2008;**118**:498–506.
91. Xie CQ, Huang H, Wei S, Song LS, Zhang J, Ritchie RP, et al. A comparison of murine smooth muscle cells generated from embryonic versus induced pluripotent stem cells. *Stem Cells Dev* 2009;**18**:741–8.
92. Lee TH, Song SH, Kim KL, Yi JY, Shin GH, Kim JY, et al. Functional recapitulation of smooth muscle cells via induced pluripotent stem cells from human aortic smooth muscle cells. *Circ Res* 2010;**106**:120–8.
93. Eminli S, Foudi A, Stadtfeld M, Maherali N, Ahfeldt T, Mostoslavsky G, et al. Differentiation stage determines potential of hematopoietic cells for reprogramming into induced pluripotent stem cells. *Nat Genet* 2009;**41**:968–76.
94. Xu D, Alipio Z, Fink LM, Adcock DM, Yang J, Ward DC, et al. Phenotypic correction of murine hemophilia A using an iPS cell-based therapy. *Proc Natl Acad Sci USA* 2009;**106**:808–13.
95. Ebert AD, Yu J, Rose Jr. FF, Mattis VB, Lorson CL, Thomson JA, et al. Induced pluripotent stem cells from a spinal muscular atrophy patient. *Nature* 2009;**457**:277–80.
96. Park IH, Arora N, Huo H, Maherali N, Ahfeldt T, Shimamura A, et al. Disease-specific induced pluripotent stem cells. *Cell* 2008;**134**:877–86.
97. Rubin LL. Stem cells and drug discovery: the beginning of a new era?. *Cell* 2008;**132**:549–52.
98. Davila JC, Cezar GG, Thiede M, Strom S, Miki T, Trosko J. Use and application of stem cells in toxicology. *Toxicol Sci* 2004;**79**:214–23.
99. Stefanovic S, Abboud N, Desilets S, Nury D, Cowan C, Puceat M. Interplay of Oct4 with Sox2 and Sox17: a molecular switch from stem cell pluripotency to specifying a cardiac fate. *J Cell Biol* 2009;**186**:665–73.
100. Schenke-Layland K, Rhodes KE, Angelis E, Butylkova Y, Heydarkhan-Hagvall S, Gekas C, et al. Reprogrammed mouse fibroblasts differentiate into cells of the cardiovascular and hematopoietic lineages. *Stem Cells* 2008;**26**:1537–46.
101. Martinez-Fernandez A, Nelson TJ, Ikeda Y, Terzic A. c-MYC independent nuclear reprogramming favors cardiogenic potential of induced pluripotent stem cells. *J Cardiovasc Transl Res* 2010;**3**:13–23.

Reprogramming of Somatic Cells

JOHNSON RAJASINGH

Cardiovascular Research Institute, Division of Cardiovascular Diseases, Department of Internal Medicine, University of Kansas Medical Center, Kansas City, Kansas, USA

Reprogramming of adult somatic cells into pluripotent stem cells may provide an attractive source of stem cells for regenerative medicine. It has emerged as an invaluable method for generating patient-specific stem cells of any cell lineage without the use of embryonic stem cells. A revolutionary study in 2006 showed that it is possible to convert adult somatic cells directly into pluripotent stem cells by using a limited number of pluripotent transcription factors and is called as iPS cells. Currently, both genomic integrating viral and nonintegrating nonviral methods are used to generate iPS cells. However, the viral-based technology poses increased risk of safety, and more studies are now focused on nonviral-based technology to obtain autologous stem cells for clinical therapy. In this review, the pros and cons of the present iPS cell technology and the future direction for the successful translation of this technology into the clinic are discussed.

I. Introduction

Reprogramming of adult somatic cells into pluripotent stem cells may provide an attractive source of stem cells for regenerative medicine, including postinfarct cardiac and other ischemic tissue repair. Recent experimental evidence has revealed that nuclear reprogramming of terminally differentiated adult mammalian cells leading to their dedifferentiation is possible. The best example of somatic cell reprogramming to the totipotent stage comes from reproductive and therapeutic cloning experiments using somatic nuclear transfer (SNT), wherein transplantation of somatic nuclei into enucleated oocyte

Progress in Molecular Biology
and Translational Science, Vol. 111
http://dx.doi.org/10.1016/B978-0-12-398459-3.00003-4

51

cytoplasm can extensively reprogram somatic cell nuclei with new patterns of gene expression, new pathways of cell differentiation, successful generation of embryonic stem cells (ESCs), and birth of cloned animals. Therapeutic cloning by SNT for clinical application, although conceptually attractive, is, to date, not practical, given the technical difficulties: oocyte dependence, ethical and legal concerns, and prohibitive cost associated with the process. It is, therefore, imperative to develop alternative strategies for somatic cell reprogramming. In 2006, a revolutionary study showed that it is possible to convert adult somatic cells directly into pluripotent stem cells by using a limited number of ESC transcription factors and culture under specific ESC culture conditions. The following year, two other studies reported the generation of ES-like cells from terminally differentiated human fibroblasts by the retroviral transduction of defined ESC-specific transcription factors, called induced pluripotent stem (iPS) cells. Currently, both genomic integrating viral and nonintegrating nonviral methods are used to generate iPS cells. However, the viral-based technology posed increased risk of safety and studies are now focused on nonviral-based technology to obtain autologous stem cells for clinical therapy.

II. Reprogramming of Somatic Cells into Pluripotent Stem Cells

The currently available technology of generating pluripotent stem cells can be broadly classified into two categories, namely, oocyte-dependent and oocyte-independent nuclear reprogramming.

A. Oocyte-Dependent Nuclear Reprogramming

Initially, nuclear reprogramming was achieved by SNT. The idea of ooplasmic molecules that reprogram mature somatic cells has triggered the development, understanding, and identification of numerous key reprogramming molecules. A better understanding of the mechanisms underlying SNT will contribute to the improvement of the newly developed iPS technology. This SNT method requires the oocyte to generate pluripotent stem cells, which remains an ethical issue in addition to being technologically challenging.

1. Somatic Cell Nuclear Transfer

The successful animal cloning and the birth of the sheep Dolly in 1997[1] as well as the generation of pluripotent ESC lines by SNT generated immense enthusiasm a decade ago. The goal of SNT is to generate pluripotent stem cells from the nucleus of donor somatic cells with the help of oocytes. Basically, SNT is the transplantation of a diploid nucleus from the donor somatic cell to the enucleated oocyte. This recipient diploid oocyte mimics as zygote and begins to divide, specialize, and form a hollow sphere of cells, called a blastocyst. The blastocyst has an outer layer of cells called the trophectoderm, a hallow space inside called the blastocoel, and cluster of cells called the inner cell mass (ICM). The transplantation of the blastocyst into the uterus of an animal can lead to the successful birth of a cloned animal, and is called reproductive cloning. The ICM forms virtual pluripotent ESCs during *in vitro* culture and is capable of producing every type of cell found in an organism—but it cannot form a whole organism—and is called therapeutic cloning (Fig. 1). Studies have shown that ESCs generated through SNT technology are functionally identical to the donor nucleus of a fully grown organism except the mitochondrial genome, which is derived from the oocyte.[2–4] The SNT-derived ESC lines are transcriptionally and genetically indistinguishable from normally fertilized ESCs, presumably because of the selection of faithfully reprogrammed cells in culture.[5,6] This close genetic identity holds the promise of circumventing

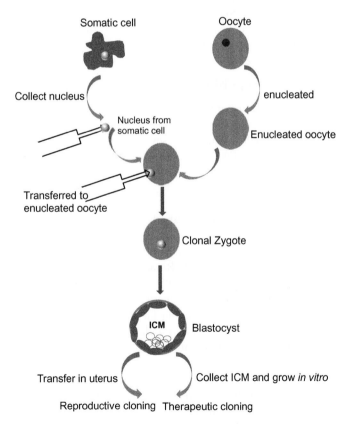

FIG. 1. Process of somatic cell nuclear transfer. (For color version of this figure, the reader is referred to the online version of this chapter.)

immune rejection of cells during SNT-derived cell differentiation.[7] SNT technology can be applied to produce autologous differentiated cells for therapeutic purposes, called therapeutic cloning, as well as for generating an animal, called reproductive cloning. To date, the production of cloned offspring by nuclear transfer had been achieved in rodents, rabbits, nonhuman primates, and several species of domesticated animals using totipotent embryonic cells as the nuclear donors.[8–12] SNT can lead to efficient reprogramming of pluripotent gene expression. The SNT reprogramming process includes the exchange of somatic proteins for oocyte proteins, the posttranslational modification of histones, and the demethylation of DNA. These events occur in an ordered manner and on a defined timescale, indicating that reprogramming by SNT relies on deterministic processes.[13] At present, SNT is an efficient method; around 40–50% embryos develop to normal blastocyst and about 30–40% of these can give rise to normal ESCs with an overall efficiency of 12–20% within a limited time.[5,14] Several limitations may hinder the strategy of therapeutic cloning for its clinical applications: nuclear transfer is unlikely to be more efficient in humans than in mice; reproductive cloning is expensive and highly inefficient; and more than 90% of cloning attempts fail to produce viable offspring.

B. Oocyte-Independent Nuclear Reprogramming

The discovery of oocyte-independent methods of reprogramming has provided for the utilization of stem cell technology without the ethical controversy surrounding traditional ESCs. There are several methods available to generate oocyte-independent pluripotent stem cells, but each has its own drawbacks in producing clinical-grade cells.

1. CELL FUSION

One of the oocyte-independent methods of somatic cell reprogramming is cell fusion. A number of studies have demonstrated that somatic cell nuclear reprogramming can be achieved by *in vitro* cell fusion (hybridization) with both mouse[15,16] and human ES cells.[17] Remarkably, similar experiments using cell fusion with embryonic carcinoma cells (ECCs) and embryonic germ cells (EGCs) demonstrated that the fused somatic cells could revert to pluripotent stem cells and suggested that transacting factors, with similar activity as the oocyte cytoplasm, are present in other cell types.[17–19] Epigenetic nuclear reprogramming of mammalian somatic nuclei was also reported when the somatic nuclei were fused to xenopus eggs.[20,21] These studies provide evidence that a differentiated nucleus can be reprogrammed to a pluripotent state by the fusion of a somatic cell with a stem or germ cell. Tada and colleagues have shown that, when mouse thymic lymphocytes are fused with embryonic germ cells, demethylation of several imprinted and nonimprinted genes in the thymocyte nucleus takes place.[18] The epigenetic changes correlate with the expression of a silent maternal allele of the peg1/mest imprinted gene in the thymocyte nucleus. Furthermore, the EGC–thymocyte heterokaryons display pluripotency because they contribute to various tissues in chimeric embryos.[18] Reversion of lymphocyte nuclei also occurs after fusion, evidenced by the reactivation of the inactive X-chromosome from female thymocytes, induction of a normally repressed Oct4-GFP transgene in the thymocyte nuclei, and contribution of the ESC–thymocyte hybrids to the three primary germ layers in chimeras.[15] ESCs fused with mouse neuronal stem cells also reactivate silent ESC-specific genes in the neuronal genome and confer pluripotency of the hybrids.[16] Similarly, fusion of neuronal progenitor cells[22] or bone-marrow-derived cells[23] with ESCs results in hybrids that express markers of pluripotency and contribute to chimeras and teratoma formation.[23] Fusion of ECCs with T-lymphoma cells also promotes the formation of colonies expressing pluripotent cell transcript from the lymphoma genome.[24]

Recently, another study has shown that human ES cells, when fused with human fibroblast, result in hybrid cells that maintain a suitable tetraploid DNA content and have morphology, growth rate, and antigen expression patterns characteristic of human ESCs.[17] The differentiation of these tetraploid hybrid cells yielded all three germ layers cell types both *in vitro* and *in vivo*. These studies clearly demonstrate that epigenetic modifications can be induced in a differentiated somatic cell genome without the requirement of an oocyte. It has been shown that fusion-induced reprogramming is an efficient method for reprogramming differentiated somatic cells to a pluripotential state. *Oct4* gene reactivation occurs within 1–2 days postfusion of somatic cells with pluripotent stem cells. Reactivation of *Oct4* can be monitored by the detection of the green fluorescent protein (GFP) signal from the *Oct4-GFP* transgene of somatic cells. In this report, the authors fused double transgenic (OG2/ROSA26) somatic cells with ESCs, and demonstrated the presence of the somatic cell genome in all GFP-positive ESC-like colony-forming cells, confirming their identity as the cell fusion hybrids.[25] Thus, components of pluripotent EGCs, ESCs, or ECCs have the potential of eliciting reprogramming events in a somatic genome. However, this strategy, although excellent for mechanistic studies, retains certain drawbacks that are associated with oocyte-dependent reprogramming of somatic cells. First, the two cells used to generate hybrid cells are not derived from an autologous source; second, the efficiency of fusion remains low; and third, the genetic stability and long-term survival of hybrids remains to be established.

2. Somatic Cell Reprogramming by Transcription Factors (iPS cells)

Cloning and cell fusion studies have demonstrated that reprogramming of mature somatic cells into pluripotent stem cells could be mediated by uncharacterized transacting factors present in the cytoplasm oocyte or ESCs. In 2006,

Takahashi and Yamanaka described the generation of ES-like cells by an elegant screen for 24 selected genes as candidate factors to induce pluripotency, based on their known role in the maintenance of ES cell identity, and introduced them into mouse fibroblasts that could activate pluripotent genes. These activated cells formed colonies with the characteristic ESC morphology, which they called "induced pluripotent stem" cells.[26] Next, they narrowed the set of factors necessary and sufficient to obtain iPS cells down to four: *OCT3* (also known as *POU5F1* or *OCT4*), Sry box-containing factor 2 (*SOX2*), Krüppel-like factor 4 (*KLF4*), and *c-Myc*. Cells expressing these four factors were able to induce the formation of teratomas with all three germ layers *in vivo* and embryoid bodies *in vitro*, showing the successful reprogramming of differentiated cells into pluripotent cells. The following year, two other studies from the Yamanaka (*Oct4, Sox2, c-Myc*, and *Klf4*) and Thompson (*Oct4, Nanog, Sox2*, and *Lin28*) groups showed that transcription-factor-mediated reprogramming to a pluripotent state using just four factors can also be achieved in human cells.[27,28] Although these studies did provide the evidence that overexpression of these ESC-specific genes leads to the derivation of ESC-like cells from primary human fibroblasts, they raised several critical questions, which were answered during the subsequent studies by various authors.

Recently, significant improvements have been made in iPS cell reprogramming technology. Following the successful reprogramming of fibroblasts, iPS cells have been derived from other somatic cell populations such as keratinocytes,[29,30] stomach cells,[31] liver cells,[31,32] neural cells,[33,34] lymphocytes,[35] B cells,[36] human blood cells,[37–39] human cord blood cells,[40,41] pancreatic β cells,[42] melanocytes,[43] human amniotic cells,[44] human platelets,[45] human astrocytes,[46] and human adipose tissues.[47,48] All the above studies point toward the universality of induced pluripotency and further studies are needed to explain the molecular mechanism.

a. Methods of iPS Cell Generation.   Currently, owing to advances in technology, the generation of iPS cells is less problematic and technically more feasible than SNT. A number of different methods have been devised to generate the iPS cells (Fig. 2). Although different reprogramming protocols have been reported, each has some drawbacks to translate this technology into clinical applications. Therefore, it is crucial to identify suitable techniques for factor delivery and obtain colonies that are faithfully reprogrammed during the generation of iPS cells. The pros and cons of the different methods of iPS cell generation in relation to SNT and cell fusion are summarized in Table I.

i. Viral-Integration-Based Gene Delivery System.   Several experimental strategies have been reported to activate the endogenous silent pluripotent genes in the somatic cells. The initial approaches in the generation of iPS cells

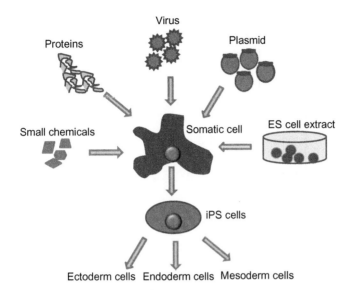

FIG. 2. Various methods of induced pluripotent stem cell generation. (For color version of this figure, the reader is referred to the online version of this chapter.)

employed a retrovirus to transport the four important pluripotent transcription factors (*Oct4, Sox2, Klf4,* and *c-Myc*) into mouse and human fibroblasts.[26,27] Retroviruses are highly efficient gene-transfer vehicles because they provide prolonged transgene expression and less immunogenicity. Subsequently, other groups used the lentivirus to shuttle another set of four transcription factors (*Oct4, Sox2, Nanog,* and *Lin28*) in human fibroblast cells to generate iPS cells.[28] Lentiviruses belong to the subclass of retroviruses capable of transducing a wide range of cells and are more efficient in iPS cell generation. Transcription factor delivery into somatic cells and various other studies have been carried out using either retrovirus or lentivirus to establish the iPS cells.[29,31,34,49–54] It has been shown that the ability of a gene delivery vector to transfer reprogramming factors into a given cell type is dependent on the vector concentration.[54,55] The process of reprogramming requires controlled stoichiometric expression of transgenes for a transient period to activate the endogenous pluripotent genes.[49,56–61] Even though retroviruses and lentiviruses are efficient in delivering transgenes, there are some disadvantages that hinder the practical use of iPS cells in clinical settings.[6] The potential of stable random integration in the host genome making leaky expression and leading to oncogenesis has been identified as a major limitation.[50] The random integration also makes heterogeneous iPS cells that need to be screened for proper and complete reprogrammed clones.

TABLE I
SUMMARY OF DIFFERENT TYPES OF REPROGRAMMING METHODS AND THEIR EFFICIENCY

Methods	Somatic nuclear transfer	Cell fusion	Induced pluripotent stem cells					ES extract
			Viral	Nonviral	RNA	Protein	Chemical	
Techniques	Technically challenging	Simple	Lentivirus, piggyBac, Cre/loxP	Episomal vector, minicircle plasmid	mRNA, miRNA	Proteins	Small molecules	ES cell whole extract
Reprogramming factors delivery	Whole oocyte reprogramming factors available	Whole factors available	Vector integration with limited factors	Vector nonintegration with limited factors	No integration with limited factors	No integration with limited factors	No integration and no exogenous factor	Whole factors available
Efficiency (%)	12	0.005	0.1–1.0	0.001	1	0.001	0.01	0.01
Complete reprogramming	Yes	Yes	Yes	Yes	Yes	Yes	Partial	Yes
Advantages	Reliable, normal, most efficient,	No integration, oocyte independent	Moderate efficiency	No integration	No integration, moderate efficiency	No integration	No integration	No integration, oocyte independent
Disadvantages	Oocyte dependent and ethical issues; immunogenic	Polyploidy and abnormal cells	Genomic integration, Incomplete silencing, possibility of tumor	Laborious screening for integration free cells, slow and low efficiency, incomplete silencing	Multiple rounds of transfection is needed	Low efficiency	Low efficiency, multiple time treatment is needed	Low efficiency

The major difference between retroviral and lentiviral vectors is that lentiviruses are more efficient in the generation of iPS cells but less so in the degree of silencing the transgenes when compared to retroviruses in pluripotent cells, and can cause differentiation block.[32,54,62] To overcome this problem, doxycycline-inducible vector systems have been employed to generate iPS cell reprogramming systems, which provide a more attractive approach as they permit temporal control over factor expression.[54,63] These systems involve differentiating primary iPS cell clones generated with a doxycyline-inducible system into somatic cells using either *in vitro* differentiation (for human cells)[63,64] or blastocyst injection (for mice).[65,66] These somatic cells are then cultured in doxycycline-containing media, thus triggering the formation of "secondary" iPS cells at higher efficiency than the primary iPS cells. Secondary systems therefore allow the reprogramming of large quantities of genetically homogeneous cells for biochemical studies as well as cells that are difficult to culture or transduce, and facilitate the comparison of genetically matched iPS cells derived from different somatic cell types.[6] Even with doxycycline-inducible systems, low levels of transgene expression have been found to affect the transcriptome of iPS cells, which severely impairs differentiation both *in vitro* and *in vivo*.[67,68] This approach is not applicable for therapeutic purposes because of genomic integration of the vector in the reprogrammed cells.

ii. Viral-Nonintegration-Based Gene Delivery System. Studies are aimed toward nonintegrating viral approaches to make iPS cell technology more therapeutically applicable. Techniques to generate integration-free iPS cells can be subdivided into two categories: (1) use of vectors that do not integrate into the host cell genome and (2) use of integrating vectors that can be subsequently removed from the genome.[6] The adenovirus delivery system has shown that the iPS cells could be generated without transgene integration into the host genome.[69] The success of nonintegrating transient transgene expression in generating iPS cells has provided an opportunity to potentially develop a safe, cost effective, and easier to generate and manipulate method by a nonintegrating viral delivery system. The adenoviral integration-free iPS cells were generated from adult mouse hepatocytes[32] and mouse embryonic fibroblasts (MEFs) transfected with two plasmids, one expressing *c-Myc* and the second construct expressing *Oct4, Klf4,* and *Sox2*[70], providing proof of principle that proviral insertions are not required for iPS cell generation. Later, it was simplified by the delivery of all-in-one single plasmid containing all the four reprogramming factors using nucleofection transfection technology.[71] Nucleofection technology is provided by Amaxa (Cologne, Germany) and is based on the electroporation technique which delivers DNA directly to the nucleus with the help of the electric field generated by the nucleofector. In contrast to the regular plasmid vector, episomal plasmid vectors are able to replicate themselves as extrachromosomal elements and exhibit prolonged expression of the reprogramming genes in target cells. The first human iPS cells were successfully generated from neonatal foreskin fibroblasts that did not have genomic integration, with the help of three combinations of the Epstein–Barr virus oriP/EBNA1, all of which include *Oct4, Sox2, c-Myc,* and *Klf4,* together with *Nanog, Lin28,* and *SV40LT.*[72] In this technique, three individual plasmids carrying a total of seven factors, including the oncogene SV40, are required to generate iPS cells. It has not been shown to successfully reprogram cells from adult donors, who constitute a more clinically relevant target population. Further, expression of the EBNA1 protein, as was required for this technique, may increase immune cell recognition of transfected cells, thus potentially limiting clinical application if the transgene is not completely removed.[55]

Human fibroblasts have also been reprogrammed into iPS cells with adenoviral vectors[73] and the Sendai virus (SeV).[74] Adenovirus is generally poor in genetransfer efficiencies to generate iPS cells (0.005%), which may be due to the limited availability of receptors or coreceptors required for binding and internalizing in the cell types used.[75] The use of the SeV, which is an RNA virus and carries no risk of altering host genome, is an efficient solution for generating safe iPS cells. Sendai viral vectors achieved reprogramming of human somatic cells and were able to derive transgene-free hiPS cells by antibody-mediated negative selection utilizing the cell-surface marker haemagglutinin–neuraminidase (HN) that is expressed on SeV-infected cells.[74] It is also very difficult to control the level of gene expression in infected cells, and repeated infections are required, thereby delaying the reprogramming kinetics.[32] It is difficult to rule out the possibility of vector particle contamination in the reprogrammed cell genome by polymerase chain reaction (PCR), and may require the expensive technology of whole-genome sequencing.[76] Owing to these drawbacks, it remains questionable whether these approaches could be translated to the clinic.

Several laboratories have developed integration-dependent gene delivery vectors with incorporated loxP sites that can be subsequently excised from the host genome by transient expression of Cre recombinase to avoid issues during nonintegration.[67,77] Cre–loxP recombination and piggyBac transposition are the two strategies to excise the exogenous reprogramming factors from genomic integration sites. In the Cre–loxP strategy, a loxP site was positioned in the 3′ long terminal repeat (LTR) of lentivirus vectors that contain a dox-inducible minimal cytomegalovirus (CMV) promoter to drive the expression of the reprogramming factors.[67] During replication, the loxP was duplicated into the 5′ LTR, resulting in genomic integration of transgenes flanked by two loxP sites. Subsequent transient expression of Cre enabled the excision of the floxed reprogramming factors. Interestingly, the iPS cells generated using this strategy displayed a gene expression profile closer to that of human ESCs (hESCs). Unfortunately, Cre-mediated excisions of transgene can lead to genomic instability and genome rearrangements due to multiprotein vectors and multiple integrations. To over-come these potential drawbacks, a single polycistronic vector containing the four reprogramming factors connected with 2A peptide linkers was developed.[77,78] The Cre expression resulted in efficient excision of the integrated transgenes, with the possibility of retaining the vector DNA in the genome, interrupting pro-moters, coding sequences, and regulatory elements.[55] On the other hand, the piggyBac (PB) transposon is capable of excising itself without leaving any rem-nants of exogenous DNA in the cell genome.[79,80] Recently, transgene-free iPS cells could also be generated with piggyBac transposons, which are mobile genetic elements that can be introduced into and removed from the host genome by transient expression of the transposase.[66,81] The low error rate of this process allows for a seamless excision, but requires characterization of integration sites in iPS cells before and after the transposon removal. It also remains unclear whether transposase expression can induce unpredictable genomic alterations in the iPS cell phenotype.

iii. Nonviral Vector Method of iPS Cells. There have been several methods of nonviral vectors designed for the delivery of pluripotent factors in somatic cell reprogramming technology. Most human iPS cell derivation tech-niques use integrating viruses, which may leave residual transgene sequences as part of the host genome, thereby unpredictably altering the cell phenotype in downstream applications. Recently, a study that did not use a nucleic acid-based vector showed the use of a single minicircle vector to generate transgene-free iPS cells from human adipose stem cells.[82] Minicircle is a circular nonviral DNA, a novel compact vector that is free of bacterial DNA and capable of persistent high-level expression in cells. In this study, the authors constructed a plasmid (P2PhiC31-LGNSO) that contained a single cassette of four repro-gramming factors (*Oct4, Sox2, Lin28,* and *Nanog*) plus a GFP reporter gene, each separated by the self-cleavage peptide 2A sequences.[83] Then the plasmid backbone was removed with the advantage of PhiC31-based intramolecular recombination technology to obtain the minicircle. Minicircle DNA benefits from higher transfection efficiencies and longer ectopic expression due to its lower activation of exogenous silencing mechanisms[84,85] and thus may repre-sent an ideal mechanism for generating iPS cells. The minicircle, when used to transfect the human adipose stem cells, needs repeated transfection and gradually becomes ES-type morphology without changing the global gene expression. There are no further studies on minicircle-based iPS cells gener-ated from other somatic cell types.

One possible way to induce pluripotency in somatic cells to avoid the risk of genomic modifications is through direct delivery of reprogramming proteins. Studies have shown that iPS cells can also be derived from both mouse and human fibroblast by delivering reprogramming factors as purified recombination proteins[86] or genetically engineered HEK293 cells.[87] Importantly, methods that rely on repeat administration of transient vectors, whether DNA based or protein based, have so far shown very low iPS cell derivation efficiencies,[32,70,72,73,82,87] presumably because of weak or inconstant expression of reprogramming factors. A simple nonintegrating strategy for reprogramming cell fate study has been demonstrated, which is based on the administration of synthetic mRNA approach and can reprogram multiple human cell types to pluripotency with efficiencies that greatly surpass those of established protocols. They further show that the same technology can be used to efficiently direct the differentiation of RNA-induced pluripotent stem cells (RiPSCs) into terminally differentiated myogenic cells.[88] This also has its own drawbacks, such as the difficulty to synthesize long mRNA chemically, proneness to degradation, the need for rigorous quality control, and, finally, the recognition of RNA by the host cell as foreign agents and degradation by the RISC machinery that is available inside the cells. Recently, studies have been reported on eluding this problem by modifying the nucleotides in the RNA, thereby avoiding RNA degradation and their recognition as foreign agents.[89,90] However, the efficiency of protein-induced reprogramming of somatic cells is as low as 0.001% of the starting cell number, and the procedure takes about 8 weeks. More recently, a greatly improved study has shown episomal reprogramming efficiency by using a cocktail containing the MEK inhibitor PD0325901, the GSK3b inhibitor CHIR99021, the TGF-b/Activin/Nodal receptor inhibitor A-83-01, the ROCK inhibitor HA-100, and the human leukemia inhibitory factor. Moreover, the authors have successfully established a feeder-free reprogramming condition using a chemically defined medium with bFGF and N2B27 supplements and a chemically defined hESC medium mTeSR1 for the derivation of footprint-free human iPS cells. These improvements enabled the routine derivation of footprint-free human iPS cells from skin fibroblasts, adipose-tissue-derived cells, and cord blood cells.[91] This technology will likely be valuable for the production of clinical-grade human iPS cells, but the possibility of tumor formation after transplantation needs to be ruled out.

iv. ESC Extract iPS Cells. ESC-extract-induced reprogramming of somatic cells is another method of integration-free generation of iPS cells. It has been shown that nuclear and cytoplasmic extracts from several cell types elicit changes in cell fate.[92,93] The procedure involves reversible permeabilization of a somatic cell, exposure of the permeabilized cells to the reprogramming extract, and resealing of the cells. The first report of using ES cell extract showed that transient exposure of reversibly permeabilized fibroblasts led to the activation of pluripotent stem cells.[94–96] However, the expression and the precise timing of other genes involved in the cell-free-extract-induced reprogramming have not been addressed. We have reported that exposure to mouse ESC extracts induces marked epigenetic reprogramming in NIH3T3 fibroblasts, including the reactivation of ESC-specific gene expression. These reprogrammed cells possess pluripotent stem cell-like characteristics including the multilineage differentiation potential.[97] A recent study has demonstrated that a single transfer of ES-cell-derived protein is able to induce pluripotency in adult mouse fibroblasts.[98]

   v. *Chemically Induced Pluripotent Stem Cells.*   Even though several approaches have been shown to overcome the problem of transgene integration into the cell genome, the possibility of reprogramming cells by a chemical approach has been predicted recently by several scientists.[76,99] It has been reported that chemical reprogramming has several advantages over other reprogramming technologies: (1) a drug can reach its target very quickly and selectively; (2) the concentration of drugs can be easily varied to reach the desired effect in the most efficient way; (3) chemical treatments can be simple as compared to other reprogramming techniques; and (4) the new reprogramming drugs once identified can be easily synthesized on a large scale at low cost.[100] The chemical modulation of cells is not a new approach and, in fact, many new synthetic and natural products have been extensively used over the past decades, even long before the high-throughput screening technology became available. It was shown a decade ago that retinoic acid and its derivatives can be used to induce mouse ES cells differentiated into multiple phenotypes.[101] Another molecule, 5-aza-2-deoxycytidine Aza, which is a potent DNA methylation inhibitor, was first synthesized 40 years ago[102] and it was shown to be an effective drug for acute myelogenous leukemia.[103] A recent study has shown that transient induction of the transcription factor is more than enough to generate iPS cells.[70] More recent studies have shown that iPS cells can be generated by a combination of both transcription factors and small molecules.[104–106] Several studies have reported that the use of small molecules is a crucial tool for understanding the intricate mechanisms that lead to cellular reprogramming. It has also been shown that small molecules have the potential to effectively replace viral transduction of key transcription factors such as *Sox2*, *Klf4*, and *c-Myc*.[107–109] Recently, studies have shown that 5-aza-2-deoxycytidine alone or in combination with other histone deacetylase (HDAC) inhibitors can alter cell fate.[110–112] More recently, we have shown that combined Aza and TSA (trichostatin A, a potent HDAC inhibitor) treatment of bone marrow progenitor cells (BMPs)-induced epigenetic changes characterized by greater AceH3K9 expression and reduced histone deacetylase 1 (HDAC1) and lysine-specific demethylase 1 (LSD1) expression compared to untreated BPCs. These epigenetic changes induced expression of *Oct4*, *Nanog*, and *Sox2* transcription factors in BPCs.[113] In this study, we took a different temporary nonintegrating reprogramming approach and found a transient activation of pluripotent genes by small molecules in BPCs just long enough to be transformed into cardiomyocytes.[113]

vi. miRNA-Induced iPS Cells.   microRNAs (miRNAs) are single-stranded RNAs (around 18–24 nucleotides) associated with a protein complex called the "RNA-induced silencing complex." Small RNAs are usually generated from the noncoding regions of gene transcripts and function to suppress gene expression by translational repression.[114–117] In recent years, the crucial role of miRNAs in the control of pluripotent stem cells has been clearly demonstrated by the discovery that ES cells lacking mature miRNAs exhibit defects in proliferation and differentiation.[118,119] Recently, it has been shown that miRNAs miR-290 and miR-302 play an important part in the gene networks controlled by the pluripotency factor *Sox2, Oct4,* and *Nanog.*[120] miRNAs have been found to function in many important processes, such as expression of self-renewal genes in hESCs,[121] cell-cycle control of ESCs,[122] alternative splicing,[123] and heart development.[124] Furthermore, it was recently reported that ESC-specific miRNAs (miR-291-3p, miR-294, or miR-294) enhanced mouse iPS cell derivation and replaced the function of c-Myc during reprogramming[125] and that hESC-specific miR-302 could alleviate the senescence response due to the four-factor expression in human fibroblast.[126] More recently, Li *et al.* showed that miRNAs function directly in iPS cell induction and that interference with the miRNA biogenesis machinery significantly decreases reprogramming efficiency. They also identified three clusters of miRNAs, namely, miR-17–92, miR-106b–25, and miR-106a–363, which are highly induced during early stages of reprogramming. Functional analysis demonstrated that miRNA-derived clones are fully pluripotent. Overall, they proposed that miR-93 and miR-106b are key regulators of the reprogramming activity.[127]

3. Cell Type and Reprogramming Efficiency

Derivation of iPS cells from a number of different species such as humans,[28,128,129] rats,[104] rhesus monkeys,[130] and bovines[131] has demonstrated that the fundamental features of the transcriptional network governing pluripotency have remained conserved during evolution. For any somatic cell reprogramming, information about the plasticity of the starting cell is helpful, as cellular reprogramming is an interplay between plasticity and environmental factors such as epigenetic modifications. The choice of cell source is an important criterion to consider and should be thoughtfully explored prior to initiating reprogramming experiments. The efficiency, time period, and extent of reprogramming vary among different starting cell types, for example, both terminally differentiated cells such as keratinocytes,[29,64] neural cells,[34,132] stomach and liver cells,[31] melanocytes,[43] pancreatic cells,[42] lymphocytes,[36,133] cord blood mononuclear cells[134], and adult stem cells such as adipose-derived stem cells[47], further underpinning the universality of induced pluripotency.

The starting cell also has an effect on the requirement of reprogramming factors. The neuronal stem cells require only the ectopic expression of *Oct4* along with *Klf* or *c-Myc* due to the abundance of endogenous *Sox2* already present in the cells.[34,132,135] More recently, a study has shown that iPS cells generated without *c-Myc* are capable of producing full-term mice.[136] A comparative study has shown that iPS cells obtained from mouse fibroblasts, hematopoietic cells, and myogenic cells exhibit distinct transcriptional and epigenetic patterns. Moreover, it has been demonstrated that the cellular origin influences the *in vitro* differentiation potentials of iPS cells into embryoid bodies and different hematopoietic cell types. Notably, continuous passaging of iPS cells largely attenuates these differences. This study also suggested that early passage iPS cells retain a transient epigenetic memory of their somatic cells of origin, which manifests as differential gene expression and altered differentiation capacity.[137] Recently, a study has shown that different somatic cell types do not have the same potential to be reprogrammed following fusion with ESCs. Furthermore, the *in vitro* age of somatic cells can also affect the reprogramming efficiency. These findings constitute an important set of considerations for cell selection during reprogramming studies.[138] Owing to technology development, more research is needed to compare the iPS cells derived from different cell types to get clinical-grade iPS cells.

4. EPIGENETIC REPROGRAMMING MECHANISM OF iPS CELLS

To reprogram the somatic cells to iPS cells, we need to understand the cellular and molecular mechanisms of the reprogramming process as well as epigenetic changes of the genome that follows the reprogramming steps. Epigenetic changes are defined as modifications of DNA or chromatin that do not involve alterations of the DNA sequence or genetic deletions.[139] ESCs carry several epigenetic characteristics that distinguish them from differentiated somatic cells and contribute to stem cell identity.[140] Recent studies into the mechanisms by which nuclear cloning and somatic–ESC fusion induce epigenetic changes in somatic cell nuclei suggest that the somatic cell chromatin is remodeled through chromatin condensation, DNA methylation/demethylation, and histone modifications (acetylation/deacetylation, phosphorylation, and methylation/demethylation).[141,142] These histone and DNA modifications are functionally linked.[143,144] DNA demethylation/methylation is an essential epigenetic process required for gene activation or inactivation during development.[145] A change in the DNA methylation pattern is also a principal epigenetic event underlying reprogramming of somatic cells.[146–148] DNA methylation most often occurs on 5'-CpG-3' dinucleotides.[149] DNA methyl transferase enzymes (dnmt 1, dnmt3a, dnmt 3b) attach a methyl group to the fifth carbon position of cytosine residues in the CG dinucleotide,[148] and the protein condenses into a protein complex consisting of methyl-binding proteins, HDACs, and repressor proteins at the methylated CpG sites,[147] which leads to the inhibition of gene expression.[150] In addition to DNA methylation, posttranslational modifications of histone proteins regulate gene activity by modulating the chromatin structure.[151]

Gene silencing is mediated by DNA methylation and histone deacetylation at the promoter region of genes[152], whereas gene activation is mediated by DNA demethylation and histone acetylation.[153] DNA demethylation and methylation are mediated and balanced by DNA demethylase (DeMT) and DNA methyltransferases (Dnmts), respectively. Likewise, histone acetylation and deacetylation are mediated and balanced by the enzymes histone acetyltransferases (HATs) and HDACs, respectively (Fig. 3). Transcriptional activation of pluripotent genes such as *Oct4* requires acetylation of histone H3 at lysine 9 (aceH3K9), whereas histone deacetylation by HDACs 1, 2, 3, and 8 induces closed chromatin confirmations leading to repression of gene activity.[97,113,154,155] Transcriptional activation is also modulated by histone demethylase activity of LSD1, which is a homolog of nuclear amine oxidase; however, depending on the cellular context, LSD1 can either repress transcription as a component of the CoREST transcriptional corepressor complex[156,157] or activate transcription by functioning as a coactivator.[113,156] Studies have shown that both activating H3K4me3 and repressive H3K27me3 histone modifications are found at pluripotent genes.[158,159] It has also been shown that Oct4 regulates transcription of the histone H3K9 demethylases *JMJD1a* and *JMJD2c*.[160] Possibly, exogenous Oct4 is able to induce expression of these demethylases, which in turn remove the repressive H3K9me3 mark from endogenous Oct4 loci.[140]

Recently, a study has shown that epigenetic modifying drugs such as 5-azacytidine and RG108 (inhibitor of DNMT) could increase reprogramming efficiency during iPS generation.[105] On the contrary, a global survey of DNA methylation in iPS cells and their parental origins (mostly fibroblasts) shows

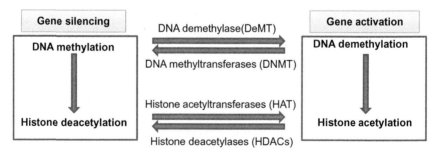

FIG. 3. Pluripotent gene induction mechanism during somatic cell reprogramming. (For color version of this figure, the reader is referred to the online version of this chapter.)

that iPS cells have higher DNA methylation than their origins.[161,162] The upregulation of DNMT3b and DNMT3l occurs during the reprogramming process, and might be responsible for de novo methylation during iPS cell generation. The abnormal DNA methylation pattern, including hypermethylation and hypomethylation, can cause abnormal differentiation properties of iPS cells. Other epigenetic states, such as histone methylation and acetylation, probably participate in iPS cell reprogramming. Addition of an HDAC inhibitor—valproic acid—improves reprogramming efficiency in both mouse and human fibroblasts.[51,107] Other HDAC inhibitors, such as suberoylanilide hydroxamic acid and TSA, also increase iPS cell generation from mouse fibroblasts.[51,163] In case of an SNT aberrant, DNA methylation (epigenetic) pattern has resulted in the inability to reach its pluripotent state during development into blastocyst in cattle.[164] In such cases, histone deacetylase inhibitor TSA treatment increases the efficiency of development to term of SNT embryos.[165] The promoters of *Oct3/4* and *Nanog* are highly methylated and silent in starting fibroblast cells but demethylated and active in iPS cells.[166]

More recently, we have shown that BPCs treated with a combination of Aza and TSA in a dose-dependent manner activated the pluripotent genes Oct4, *Nanog*, and *Sox2* and resulted in expression of these proteins. However, we do not think that these cells become pluripotent as defined earlier,[27] since transplantation of these cells did not result in the formation of teratomas. Our results also show that treatment of BPCs with Aza plus TSA induced modifications resulting in expression of genes associated with a pluripotent phenotype. We observed decreases in HDAC1 and LSD1 protein expression and increase in AceH3K9 protein expression, indicative of epigenetic changes at the level of histones. We also observed increased aceH3K9 protein expression and its interaction with *Oct4* promoter, demonstrating that the *Oct4* promoter region was acetylated in the Aza- and TSA-treated BPCs. Thus, the chromatin-modifying agents Aza and TSA induced epigenetic modification of BPCs at least in part by histone modifications and chromatin remodeling, leading to the activation of dormant pluripotent genes.[113]

5. MOLECULAR MECHANISM OF REPROGRAMMING

The molecular mechanism of how the reprogramming factors induce the epigenetic changes that are associated with reprogramming is not yet known. A live imaging study has revealed the reprogramming process involved in the conversion of the MEF to iPS cells.[167] In this study, the authors observed that the MEF cells, after transduction of reprogramming factors, started dividing symmetrically several times and maintaining fibroblastic shape. In most cases, after 48 h of transgene transduction, the ancestral cells gradually transformed into ES-cell-like morphology. Importantly, time lapse analysis uncovered frequent failure in reprogramming at the late stage of iPS cell induction. During the asymmetrical division of iPS cell reprogramming process, one of the descendants became an iPS cell and the other descendant with same retroviral insertion underwent cell death, which was largely dependent on the c-Myc transgene. During reprogramming, the ancestral cells lost their mesenchymal character after the induction of transgenes and transformed into epithelial-like cells; the phenomenon was called the "mesenchymal-to-epithelial transition" (MET).[168,169] In the meantime, *Oct4* and *Sox2* suppressed a key embryonic zinc finger transcriptional repressor (*SNAIL*), a key factor of the epithelial-to-mesenchymal transition (EMT), which is the opposite of MET. It is known that TGF-β induces EMT through the activation of SNAIL and thereby negatively regulates MET. The *c-Myc* transgene enhances MET through the downregulation of TGF-β signals by suppression of *TGF-βR1* and *TGF-βR2* expression. In addition, *Klf4* upregulates *E-cadherine*, a gene associated with epithelial cells. A temporal gene analysis revealed that the BMP signals promote MET through the miR-200 family. The inhibition of MET by TGF-β or by siRNA greatly reduces reprogramming efficiency, suggesting that MET is an important cellular event during the reprogramming process.[163]

Several reprogramming factors work with protein–protein interaction and bind with promoters of several genes. Epigenetic modification facilitates enzyme/protein access to the promoter. There are two sets of genes with opposing functions: upregulation of genes that are involved in stemness-like *STAT3* and *ZIC3*[170]; and downregulation of genes that are responsible for differentiation such as *PAX6*, *ATBF1*, and *SUZ12*.[171] During reprogramming of MEF, induction of *SSEA-1* and repression of *Thy-1* gene are noticed at the early stage of reprogramming.[54] At a later stage, endogenous pluripotent markers *OCT3/4* and *SOX2*, as well as telomerase, are expressed, which add telomere at the end of chromosomes. Recently, a study highlighted that fully reprogrammed cells are positive for the pluripotent marker *SSEA-4* and *TRA-1-60*, and negative for fibroblast marker CD13, and showed the inactivated retroviral promoter.[172] Work on the signaling pathway shows that modulation of Wnt/β-catenin, MAPK/ERK, TGF-β, and PI3K/AKT signaling pathway increases the likelihood of somatic cell reprogramming.[173] Prigione and Adjaye[174] have demonstrated that hESCs and iPS cells both in the undifferentiated state and all stages of differentiation have similar mitochondrial properties that are distinct from those of fibroblasts. This was done by global transcriptional profiling and suggests that the mitochondrial profile remains similar upon differentiation.[166,174]

C. Incomplete Programming

Human iPS cells are remarkably similar to ES cells, but recent reports indicate that there may be important differences between them.[137,175,176] Recently, Ohi et al. carried out a systematic comparison of human iPS cells generated from hepatocytes (endoderm), skin fibroblasts (mesoderm), and melanocytes (ectoderm). All low-passage iPS cells analyzed retained the transcriptional memory of the original cells. The persistent expression of somatic genes can be partially explained by incomplete promoter DNA methylation. This epigenetic mechanism underlies a robust form of memory that can be found in iPS cells generated by multiple laboratories using different methods, including RNA transfection. Incompletely silenced genes tend to be isolated from other genes that are repressed during reprogramming, indicating that recruitment of the silencing machinery may be inefficient at isolated genes. Knockdown of the incompletely reprogrammed gene *C9orf64* (chromosome 9, open reading frame 64) reduces the efficiency of human iPS cell generation, indicating that somatic memory genes may be functionally relevant during reprogramming.[177] Chan *et al.* found that there are three types of hiPS cells based on cell-surface markers. These cells differ from each other in the methylation status of the promoter region of the *Nanog* and *Oct4* loci. The fully reprogrammed cells are positive for pluripotent markers *SSEA-4* and *TRA-1-60*, and negative for the fibroblast marker CD13. Together with these markers and growth properties, they help to identify the real fully reprogrammed iPS cell lines.[172] The incompletely reprogrammed cells may self-renew as a result of the presence of *c-Myc*. Removal of *c-Myc* from reprogramming factors has shown a significantly decreased number of incompletely reprogrammed colonies in mice and humans.[178] For that reason, Hanna *et al.* used a soluble wnt3a that promotes iPS regeneration in the absence of *c-Myc*.[179] Wnt-3a-conditioned media can reprogram MEF cells.[180] Wnt signaling results in the inhibition of GSK-3 and stabilization of cytoplasmic β-catenin. Li *et al.* have created iPS cells from somatic cells by genetic transduction that remains homogeneous. They developed a method by which both human and rat iPS cells (riPSCs) could be maintained by LIF and a cocktail of the ALK5 inhibitor, the GSK3 inhibitor, and the MEK inhibitor.[166,181]

D. Direct Reprogramming of Specific Lineage

ESCs and iPS cells, although attractive cell sources, have limited lineage-specific differentiation potential and carry the risk of posttransplantation tumor formation.[182] As discussed above, several studies have demonstrated that one cell type can convert to another directly, without the need to first revert to an undifferentiated pluripotent state (Fig. 4). The first evidence supporting transgene-induced direct lineage reprogramming of fibroblast to myoblast is by overexpression of *MyoD*.[183] Recently, a number of studies have used lineage-specific transgenes to transdeterminate/transdifferentiate various cell types such as cardiomyocytes,[113,184–188] neurons,[189–191] macrophages, and dendrites[192,193] (Table II). Direct reprogramming studies are still in their infancy and have a long way to go before they can be put to clinical use. Chambers and Studer have summarized a common roadmap for evaluating current and future directed reprogramming studies.[194] The major roadmap for reprogramming

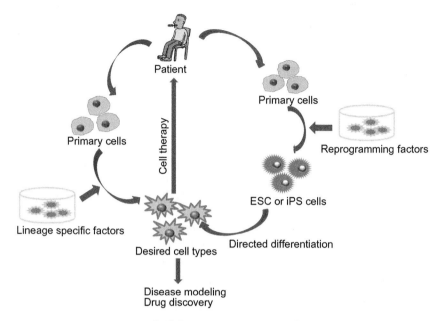

Fig. 4. Current strategies of cell therapy in regenerative medicine. (For color version of this figure, the reader is referred to the online version of this chapter.)

TABLE II

Advances in Direct Lineage-Specific Reprogramming

Authors	Year	TF used	Cell type involved
Davis et al.	1987	Myo D	Fibroblasts to skeletal muscle cells
Laiosa et al.	2006	C/EBP α, PU.1	T lymphocytes to macrophages and dendrites
Zhou et al.	2008	Ngn3, Pdx1, Mafa	Pancreatic cells to β-cells
Takeuchi et al.	2009	Gata4, Tbx5, Baf60c	Embryonic noncardiac mesoderm to cardiomyocytes
Ieda et al.	2010	Tbx5, Mef2c, Gata4	Cardiac fibroblasts to cardiomyocytes
Vierbuchen et al.	2010	Ascl1, Brn2, Mytl1	MEFs to neurons
Efe et al.	2011	iPSC factors	MEFs to cardiomyocytes
Kim et al.	2011	iPSC factors	MEFs to neural progenitors
Outani et al.	2011	C-myc, Kif4, Sox9	Dermal fibroblasts to chondrogenic cells
Pasha et al.	2011	DNMT inhibitor RG108	Skeletal myoblast to cardiac progenitor cells
Rajasingh et al.	2011	Epigenetic modifiers Aza and TSA	Marrow progenitor cells to cardiac progenitor cells

studies consists of (1) identification of the cell origin that is to be reprogrammed and the cell population which might be homogeneously derived from clonal isolation; (2) identification of the minimal set of factors sufficient for reprogramming by "leave-one-out" experiments; (3) measurement of inductive efficiency; (4) comparison of the molecular state of reprogrammed cells to the target cell; (5) measurement of functionalities such as survival, integration, and physiological response both *in vivo* and *in vitro*; and (6) assessment of the stability of directed reprogrammed cell fate. Recently, we have shown that directed differentiation of BMPs into cardiac progenitor cells expressing the markers *Flk-1, Gata4,* and *Nkx2.5,* is possible, using the chromatin-modifying agents Aza and TSA alone.[113] The major advantages to this approach are the following: (i) Patient-specific scalable supply of defined cell types can be generated; (ii) Minimal time is required for lineage reprogramming compared to iPS cell reprogramming and lineage differentiation; (iii) Disease-specific cells can be generated for the patient-specific needs; (iv) Direct reprogramming methods provide significantly higher efficiency of cell number for cell transplantation; (v) They reduce the risk of tumor formation during and after cell transplantation (Table III). Thus direct lineage-specific reprogramming is a powerful tool in modern regenerative medicine.

1. IMMUNOGENICITY OF iPS CELLS

The generation of iPS cell-derived differentiated cells has been expected to provide autologous cells for cell therapy. However, the immunogenicity of these cells is yet to be strictly examined. Recently, it has been shown that the transplantation of iPS cells induced a T-cell-dependent immune response in syngenic mouse and formed teratomas in immunodeficient mice.[195] In this study, the authors investigated the immunoreactions to iPS cells and ESCs in

TABLE III

MAJOR ADVANTAGES OF DIRECTED LINEAGE-SPECIFIC REPROGRAMMING IN COMPARISON WITH iPS CELLS

Induced pluripotent stem cells	Directed reprogrammed cells
iPS cells offer greater scalability, flexibility, and plasticity	Directed towards scalable particular lineage
More time required to generate, expand, characterize, and differentiate	Minimum time required
Complex procedure	Simple procedure
Tumor formation is a major concern	Tumor formation is reduced
Incomplete reprogramming	Incomplete reprogramming
Insertional mutagenesis caused by transgenes	Insertional mutagenesis caused by transgenes
In situ conversion to specific lineage is not possible	Effective in situ conversion to specific lineage

immunocompetent mice using three types of experimental models: an iPS cell autograft, an ESC autograft, and an ESC allograft. The syngenic transplantation of C57BL/6 (B6)-derived ESCs into B6 mice showed a high frequency of teratoma formation (97%), whereas 129/SvJ-derived nonsyngenic ESCs did not develop teratomas except in one mouse, in which they found vigorous T-cell infiltration, a landmark of immune system rejection. On the other hand, syngenic iPS cells established by retroviral vectors (ViPSCs) generated teratomas in 65% of the transplantation sites after 30 days. All the ViPSC-derived teratomas they examined showed T-cell infiltration. The retroviral vectors integrated the reprogramming factors into the genome of the ViPSCs, and the authors found that the teratomas expressed Oct3/4 from their integrated loci. A recent report revealed the existence of T cells that naturally target *Oct3/4* proteins in peripheral blood.[196] Therefore, the authors suggested that the high T-cell infiltration of the ViPSC teratomas was due to the ectopic expression of *Oct3/4*. To avoid this issue, they used three lines of nonintegrated episomally derived iPS cells (EiPSCs) for further experiments. The tumor formation rate of these EiPSCs was 84%. However, more than half of them still showed T-cell infiltration and some tumors regressed during the observation period. The authors subsequently observed that nine genes were commonly expressed in the regressed teratomas derived from two lines of EiPSCs and demonstrated that three of these genes, namely, *Zg6*, *Hormad1*, and *Cyp3a11*, interfered with the teratoma formation of syngenic ESCs when overexpressed. *Hormad1* had previously been identified as a tumor antigen[197] and was also expressed in most teratomas developed from integration-free iPS cells, which were independently induced using different methods involving adenoviral vectors, plasmids, and recombinant proteins. Finally, the authors used $CD4^{-/-}$ or $CD8^{-/-}$ mice to confirm that the interference with tumor formation was due to the T-cell activity. As expected, the robust immune reaction of ViPSCs and the ESCs expressing Hormad1 was abolished. In addition, no regression was observed in the teratomas formed by EiPSCs. Therefore, both $CD4^-$ helper T cells and $CD8^-$ cytotoxic T cells are essential for the observed immune reactions.[195,198]

III. Future Perspectives

Currently, hiPS cells are easy to create in the lab from different types of cells by a variety of methods. But several obstacles must be overcome before the cells generated can be used widely in the clinic. Different cell types have different levels of reprogramming potency and different levels of requirements of reprogramming factors. This may suggest that there are still undiscovered factors that are needed for efficient and quick generation of iPS cells. Future challenges lie in the generation of clinical-grade iPS cells in an effort to advance their use from the lab to the clinic. Many protocols available now have been developed in the mouse model, and they need to be translated to the human system. Development of a robust production protocol and an automated method for screening and selecting iPS cells during reprogramming would facilitate rapid screening of iPS cells to identify those that lack potentially harmful mutations and are readily differentiated into various cell types. Finally, manipulations that render the cells unsuitable for use in humans, such as viral integration and coculture with animal cells, must be eliminated.

REFERENCES

1. Wilmut I, Schnieke AE, McWhir J, Kind AJ, Campbell KH. Viable offspring derived from fetal and adult mammalian cells. *Nature* 1997;**385**:810–3.
2. Cibelli JB, Stice SL, Golueke PJ, Kane JJ, Jerry J, Blackwell C, et al. Transgenic bovine chimeric offspring produced from somatic cell-derived stem-like cells. *Nat Biotechnol* 1998;**16**:642–6.
3. Munsie MJ, Michalska AE, O'Brien CM, Trounson AO, Pera MF, Mountford PS. Isolation of pluripotent embryonic stem cells from reprogrammed adult mouse somatic cell nuclei. *Curr Biol* 2000;**10**:989–92.
4. Wakayama T, Tabar V, Rodriguez I, Perry AC, Studer L, Mombaerts P. Differentiation of embryonic stem cell lines generated from adult somatic cells by nuclear transfer. *Science* 2001;**292**:740–3.
5. Wakayama S, Jakt ML, Suzuki M, Araki R, Hikichi T, Kishigami S, et al. Equivalency of nuclear transfer-derived embryonic stem cells to those derived from fertilized mouse blastocysts. *Stem Cells* 2006;**24**:2023–33.
6. Stadtfeld M, Hochedlinger K. Induced pluripotency: history, mechanisms, and applications. *Genes Dev* 2010;**24**:2239–63.
7. Mombaerts P. Therapeutic cloning in the mouse. *Proc Natl Acad Sci USA* 2003;**100** (Suppl. 1):11924–5.
8. McGrath J, Solter D. Nuclear transplantation in mouse embryos. *J Exp Zool* 1983;**228**:355–62.
9. Ectors F, Ectors FJ, Delval A, Thonon F, Beckers JF. Cloning using nuclear transfer in bovine species: initial results. *Bull Mem Acad R Med Belg* 1993;**148**:262–9; Discussion 269–270.
10. Willadsen SM. Nuclear transplantation in sheep embryos. *Nature* 1986;**320**:63–5.
11. Li J, Greco V, Guasch G, Fuchs E, Mombaerts P. Mice cloned from skin cells. *Proc Natl Acad Sci USA* 2007;**104**:2738–43.
12. Sparman ML, Tachibana M, Mitalipov SM. Cloning of non-human primates: the road "less traveled by" *Int J Dev Biol* 2010;**54**:1671–8.
13. Jullien J, Pasque V, Halley-Stott RP, Miyamoto K, Gurdon JB. Mechanisms of nuclear reprogramming by eggs and oocytes: a deterministic process? *Nat Rev Mol Cell Biol* 2011;**12**:453–9.
14. Kim DS, Lee JS, Leem JW, Huh YJ, Kim JY, Kim HS, et al. Robust enhancement of neural differentiation from human ES and iPS cells regardless of their innate difference in differentiation propensity. *Stem Cell Rev* 2010;**6**:270–81.
15. Tada M, Takahama Y, Abe K, Nakatsuji N, Tada T. Nuclear reprogramming of somatic cells by in vitro hybridization with ES cells. *Curr Biol* 2001;**11**:1553–8.

16. Pells S, Di Domenico AI, Gallagher EJ, McWhir J. Multipotentiality of neuronal cells after spontaneous fusion with embryonic stem cells and nuclear reprogramming in vitro. *Cloning Stem Cells* 2002;**4**:331–8.

17. Cowan CA, Atienza J, Melton DA, Eggan K. Nuclear reprogramming of somatic cells after fusion with human embryonic stem cells. *Science* 2005;**309**:1369–73.

18. Tada M, Tada T, Lefebvre L, Barton SC, Surani MA. Embryonic germ cells induce epigenetic reprogramming of somatic nucleus in hybrid cells. *EMBO J* 1997;**16**:6510–20.

19. Yu J, Vodyanik MA, He P, Slukvin II, Thomson JA. Human embryonic stem cells reprogram myeloid precursors following cell-cell fusion. *Stem Cells* 2006;**24**:168–76.

20. Simonsson S, Gurdon J. DNA demethylation is necessary for the epigenetic reprogramming of somatic cell nuclei. *Nat Cell Biol* 2004;**6**:984–90.

21. Byrne JA, Simonsson S, Western PS, Gurdon JB. Nuclei of adult mammalian somatic cells are directly reprogrammed to oct-4 stem cell gene expression by amphibian oocytes. *Curr Biol* 2003;**13**:1206–13.

22. Ying QL, Nichols J, Evans EP, Smith AG. Changing potency by spontaneous fusion. *Nature* 2002;**416**:545–8.

23. Terada N, Hamazaki T, Oka M, Hoki M, Mastalerz DM, Nakano Y, et al. Bone marrow cells adopt the phenotype of other cells by spontaneous cell fusion. *Nature* 2002;**416**:542–5.

24. Flasza M, Shering AF, Smith K, Andrews PW, Talley P, Johnson PA. Reprogramming in interspecies embryonal carcinoma-somatic cell hybrids induces expression of pluripotency and differentiation markers. *Cloning Stem Cells* 2003;**5**:339–54.

25. Do JT, Scholer HR. Cell fusion-induced reprogramming. *Methods Mol Biol* 2010;**636**:179–90.

26. Takahashi K, Yamanaka S. Induction of pluripotent stem cells from mouse embryonic and adult fibroblast cultures by defined factors. *Cell* 2006;**126**:663–76.

27. Takahashi K, Tanabe K, Ohnuki M, Narita M, Ichisaka T, Tomoda K, et al. Induction of pluripotent stem cells from adult human fibroblasts by defined factors. *Cell* 2007;**131**:861–72.

28. Yu J, Vodyanik MA, Smuga-Otto K, Antosiewicz-Bourget J, Frane JL, Tian S, et al. Induced pluripotent stem cell lines derived from human somatic cells. *Science* 2007;**318**:1917–20.

29. Aasen T, Raya A, Barrero MJ, Garreta E, Consiglio A, Gonzalez F, et al. Efficient and rapid generation of induced pluripotent stem cells from human keratinocytes. *Nat Biotechnol* 2008;**26**:1276–84.

30. Maherali N, Hochedlinger K. Induced pluripotency of mouse and human somatic cells. *Cold Spring Harb Symp Quant Biol* 2008;**73**:157–62.

31. Aoi T, Yae K, Nakagawa M, Ichisaka T, Okita K, Takahashi K, et al. Generation of pluripotent stem cells from adult mouse liver and stomach cells. *Science* 2008;**321**:699–702.

32. Stadtfeld M, Nagaya M, Utikal J, Weir G, Hochedlinger K. Induced pluripotent stem cells generated without viral integration. *Science* 2008;**322**:945–9.

33. Kim JB, Zaehres H, Arauzo-Bravo MJ, Scholer HR. Generation of induced pluripotent stem cells from neural stem cells. *Nat Protoc* 2009;**4**:1464–70.

34. Eminli S, Utikal J, Arnold K, Jaenisch R, Hochedlinger K. Reprogramming of neural progenitor cells into induced pluripotent stem cells in the absence of exogenous Sox2 expression. *Stem Cells* 2008;**26**:2467–74.

35. Brown ME, Rondon E, Rajesh D, Mack A, Lewis R, Feng X, et al. Derivation of induced pluripotent stem cells from human peripheral blood T lymphocytes. *PLoS One* 2010;**5**: e11373.

36. Hanna J, Markoulaki S, Schorderet P, Carey BW, Beard C, Wernig M, et al. Direct reprogramming of terminally differentiated mature B lymphocytes to pluripotency. *Cell* 2008;**133**:250–64.

37. Loh YH, Agarwal S, Park IH, Urbach A, Huo H, Heffner GC, et al. Generation of induced pluripotent stem cells from human blood. *Blood* 2009;**113**:5476–9.

38. Ye Z, Zhan H, Mali P, Dowey S, Williams DM, Jang YY, et al. Human-induced pluripotent stem cells from blood cells of healthy donors and patients with acquired blood disorders. *Blood* 2009;**114**:5473–80.

39. Staerk J, Dawlaty MM, Gao Q, Maetzel D, Hanna J, Sommer CA, et al. Reprogramming of human peripheral blood cells to induced pluripotent stem cells. *Cell Stem Cell* 2010;**7**:20–4.

40. Haase A, Olmer R, Schwanke K, Wunderlich S, Merkert S, Hess C, et al. Generation of induced pluripotent stem cells from human cord blood. *Cell Stem Cell* 2009;**5**:434–41.

41. Giorgetti A, Montserrat N, Aasen T, Gonzalez F, Rodriguez-Piza I, Vassena R, et al. Generation of induced pluripotent stem cells from human cord blood using OCT4 and SOX2. *Cell Stem Cell* 2009;**5**:353–7.

42. Stadtfeld M, Brennand K, Hochedlinger K. Reprogramming of pancreatic beta cells into induced pluripotent stem cells. *Curr Biol* 2008;**18**:890–4.

43. Utikal J, Maherali N, Kulalert W, Hochedlinger K. Sox2 is dispensable for the reprogramming of melanocytes and melanoma cells into induced pluripotent stem cells. *J Cell Sci* 2009;**122**: 3502–10.

44. Zhao HX, Li Y, Jin HF, Xie L, Liu C, Jiang F, et al. Rapid and efficient reprogramming of human amnion-derived cells into pluripotency by three factors OCT4/SOX2/NANOG. *Differentiation* 2010;**80**:123–9.

45. Gekas C, Graf T. Induced pluripotent stem cell-derived human platelets: one step closer to the clinic. *J Exp Med* 2010;**207**:2781–4.

46. Ruiz S, Brennand K, Panopoulos AD, Herrerias A, Gage FH, Izpisua-Belmonte JC. High-efficient generation of induced pluripotent stem cells from human astrocytes. *PLoS One* 2010;**5**:e15526.

47. Sun N, Panetta NJ, Gupta DM, Wilson KD, Lee A, Jia F, et al. Feeder-free derivation of induced pluripotent stem cells from adult human adipose stem cells. *Proc Natl Acad Sci USA* 2009;**106**:15720–5.

48. Aoki T, Ohnishi H, Oda Y, Tadokoro M, Sasao M, Kato H, et al. Generation of induced pluripotent stem cells from human adipose-derived stem cells without c-MYC. *Tissue Eng Part A* 2010;**16**:2197–206.

49. Meissner A, Wernig M, Jaenisch R. Direct reprogramming of genetically unmodified fibroblasts into pluripotent stem cells. *Nat Biotechnol* 2007;**25**:1177–81.

50. Okita K, Ichisaka T, Yamanaka S. Generation of germline-competent induced pluripotent stem cells. *Nature* 2007;**448**:313–7.

51. Huangfu D, Osafune K, Maehr R, Guo W, Eijkelenboom A, Chen S, et al. Induction of pluripotent stem cells from primary human fibroblasts with only Oct4 and Sox2. *Nat Biotechnol* 2008;**26**:1269–75.

52. Kim JB, Greber B, Arauzo-Bravo MJ, Meyer J, Park KI, Zaehres H, et al. Direct reprogramming of human neural stem cells by OCT4. *Nature* 2009;**461**:649–53.

53. Feng B, Jiang J, Kraus P, Ng JH, Heng JC, Chan YS, et al. Reprogramming of fibroblasts into induced pluripotent stem cells with orphan nuclear receptor Esrrb. *Nat Cell Biol* 2009;**11**: 197–203.

54. Stadtfeld M, Maherali N, Breault DT, Hochedlinger K. Defining molecular cornerstones during fibroblast to iPS cell reprogramming in mouse. *Cell Stem Cell* 2008;**2**:230–40.

55. Lai MI, Wendy-Yeo WY, Ramasamy R, Nordin N, Rosli R, Veerakumarasivam A, et al. Advancements in reprogramming strategies for the generation of induced pluripotent stem cells. *J Assist Reprod Genet* 2011;**28**:291–301.

56. Maherali N, Sridharan R, Xie W, Utikal J, Eminli S, Arnold K, et al. Directly reprogrammed fibroblasts show global epigenetic remodeling and widespread tissue contribution. *Cell Stem Cell* 2007;**1**:55–70.

57. Takahashi K, Okita K, Nakagawa M, Yamanaka S. Induction of pluripotent stem cells from fibroblast cultures. *Nat Protoc* 2007;**2**:3081–9.

58. Yamanaka S. Strategies and new developments in the generation of patient-specific pluripotent stem cells. *Cell Stem Cell* 2007;**1**:39–49.

59. Park IH, Arora N, Huo H, Maherali N, Ahfeldt T, Shimamura A, et al. Disease-specific induced pluripotent stem cells. *Cell* 2008;**134**:877–86.

60. Park TS, Galic Z, Conway AE, Lindgren A, van Handel BJ, Magnusson M, et al. Derivation of primordial germ cells from human embryonic and induced pluripotent stem cells is significantly improved by coculture with human fetal gonadal cells. *Stem Cells* 2009;**27**:783–95.

61. Nelson TJ, Martinez-Fernandez A, Yamada S, Mael AA, Terzic A, Ikeda Y. Induced pluripotent reprogramming from promiscuous human stemness related factors. *Clin Transl Sci* 2009;**2**:118–26.

62. Brambrink T, Foreman R, Welstead GG, Lengner CJ, Wernig M, Suh H, et al. Sequential expression of pluripotency markers during direct reprogramming of mouse somatic cells. *Cell Stem Cell* 2008;**2**:151–9.

63. Hockemeyer D, Soldner F, Cook EG, Gao Q, Mitalipova M, Jaenisch R. A drug-inducible system for direct reprogramming of human somatic cells to pluripotency. *Cell Stem Cell* 2008;**3**:346–53.

64. Maherali N, Ahfeldt T, Rigamonti A, Utikal J, Cowan C, Hochedlinger K. A high-efficiency system for the generation and study of human induced pluripotent stem cells. *Cell Stem Cell* 2008;**3**:340–5.

65. Wernig M, Zhao JP, Pruszak J, Hedlund E, Fu D, Soldner F, et al. Neurons derived from reprogrammed fibroblasts functionally integrate into the fetal brain and improve symptoms of rats with Parkinson's disease. *Proc Natl Acad Sci USA* 2008;**105**:5856–61.

66. Woltjen K, Michael IP, Mohseni P, Desai R, Mileikovsky M, Hamalainen R, et al. piggyBac transposition reprograms fibroblasts to induced pluripotent stem cells. *Nature* 2009;**458**: 766–70.

67. Soldner F, Hockemeyer D, Beard C, Gao Q, Bell GW, Cook EG, et al. Parkinson's disease patient-derived induced pluripotent stem cells free of viral reprogramming factors. *Cell* 2009;**136**:964–77.

68. Sommer CA, Sommer AG, Longmire TA, Christodoulou C, Thomas DD, Gostissa M, et al. Excision of reprogramming transgenes improves the differentiation potential of iPS cells generated with a single excisable vector. *Stem Cells* 2010;**28**:64–74.

69. Umehara H, Kimura T, Ohtsuka S, Nakamura T, Kitajima K, Ikawa M, et al. Efficient derivation of embryonic stem cells by inhibition of glycogen synthase kinase-3. *Stem Cells* 2007;**25**:2705–11.

70. Okita K, Nakagawa M, Hyenjong H, Ichisaka T, Yamanaka S. Generation of mouse induced pluripotent stem cells without viral vectors. *Science* 2008;**322**:949–53.

71. Gonzalez F, Barragan Monasterio M, Tiscornia G, Montserrat Pulido N, Vassena R, Batlle Morera L, et al. Generation of mouse-induced pluripotent stem cells by transient expression of a single nonviral polycistronic vector. *Proc Natl Acad Sci U S A* 2009;**106**:8918–22.

72. Yu J, Hu K, Smuga-Otto K, Tian S, Stewart R, Slukvin II, et al. Human induced pluripotent stem cells free of vector and transgene sequences. *Science* 2009;**324**:797–801.

73. Zhou W, Freed CR. Adenoviral gene delivery can reprogram human fibroblasts to induced pluripotent stem cells. *Stem Cells* 2009;**27**:2667–74.

74. Fusaki N, Ban H, Nishiyama A, Saeki K, Hasegawa M. Efficient induction of transgene-free human pluripotent stem cells using a vector based on Sendai virus, an RNA virus that does not integrate into the host genome. *Proc Jpn Acad Ser B Phys Biol Sci* 2009;**85**: 348–62.

75. Freimuth P, Philipson L, Carson SD. The coxsackievirus and adenovirus receptor. *Curr Top Microbiol Immunol* 2008;**323**:67–87.
76. Yamanaka S. A fresh look at iPS cells. *Cell* 2009;**137**:13–7.
77. Kaji K, Norrby K, Paca A, Mileikovsky M, Mohseni P, Woltjen K. Virus-free induction of pluripotency and subsequent excision of reprogramming factors. *Nature* 2009;**458**:771–5.
78. Chang DF, Tsai SC, Wang XC, Xia P, Senadheera D, Lutzko C. Molecular characterization of the human NANOG protein. *Stem Cells* 2009;**27**:812–21.
79. Elick TA, Bauser CA, Fraser MJ. Excision of the piggyBac transposable element in vitro is a precise event that is enhanced by the expression of its encoded transposase. *Genetica* 1996;**98**:33–41.
80. Fraser MJ, Ciszczon T, Elick T, Bauser C. Precise excision of TTAA-specific lepidopteran transposons piggyBac (IFP2) and tagalong (TFP3) from the baculovirus genome in cell lines from two species of Lepidoptera. *Insect Mol Biol* 1996;**5**:141–51.
81. Yusa K, Rad R, Takeda J, Bradley A. Generation of transgene-free induced pluripotent mouse stem cells by the piggyBac transposon. *Nat Methods* 2009;**6**:363–9.
82. Jia F, Wilson KD, Sun N, Gupta DM, Huang M, Li Z, et al. A nonviral minicircle vector for deriving human iPS cells. *Nat Methods* 2010;**7**:197–9.
83. Ryan MD, Drew J. Foot-and-mouth disease virus 2A oligopeptide mediated cleavage of an artificial polyprotein. *EMBO J* 1994;**13**:928–33.
84. Chen ZY, He CY, Ehrhardt A, Kay MA. Minicircle DNA vectors devoid of bacterial DNA result in persistent and high-level transgene expression in vivo. *Mol Ther* 2003;**8**:495–500.
85. Chen ZY, He CY, Kay MA. Improved production and purification of minicircle DNA vector free of plasmid bacterial sequences and capable of persistent transgene expression in vivo. *Hum Gene Ther* 2005;**16**:126–31.
86. Zhou H, Wu S, Joo JY, Zhu S, Han DW, Lin T, et al. Generation of induced pluripotent stem cells using recombinant proteins. *Cell Stem Cell* 2009;**4**:381–4.
87. Kim D, Kim CH, Moon JI, Chung YG, Chang MY, Han BS, et al. Generation of human induced pluripotent stem cells by direct delivery of reprogramming proteins. *Cell Stem Cell* 2009;**4**:472–6.
88. Warren L, Manos PD, Ahfeldt T, Loh YH, Li H, Lau F, et al. Highly efficient reprogramming to pluripotency and directed differentiation of human cells with synthetic modified mRNA. *Cell Stem Cell* 2010;**7**:618–30.
89. Hannon GJ. RNA interference. *Nature* 2002;**418**:244–51.
90. Yakubov E, Rechavi G, Rozenblatt S, Givol D. Reprogramming of human fibroblasts to pluripotent stem cells using mRNA of four transcription factors. *Biochem Biophys Res Commun* 2010;**394**:189–93.
91. Yu J, Chau KF, Vodyanik MA, Jiang J, Jiang Y. Efficient feeder-free episomal reprogramming with small molecules. *PLoS One* 2011;**6**:e17557.
92. McGann CJ, Odelberg SJ, Keating MT. Mammalian myotube dedifferentiation induced by newt regeneration extract. *Proc Natl Acad Sci USA* 2001;**98**:13699–704.
93. Miyamoto K, Yamashita T, Tsukiyama T, Kitamura N, Minami N, Yamada M, et al. Reversible membrane permeabilization of mammalian cells treated with digitonin and its use for inducing nuclear reprogramming by Xenopus egg extracts. *Cloning Stem Cells* 2008;**10**:535–42.
94. Taranger CK, Noer A, Sorensen AL, Hakelien AM, Boquest AC, Collas P. Induction of dedifferentiation, genomewide transcriptional programming, and epigenetic reprogramming by extracts of carcinoma and embryonic stem cells. *Mol Biol Cell* 2005;**16**:5719–35.
95. Bru T, Clarke C, McGrew MJ, Sang HM, Wilmut I, Blow JJ. Rapid induction of pluripotency genes after exposure of human somatic cells to mouse ES cell extracts. *Exp Cell Res* 2008;**314**:2634–42.

96. Xu YN, Guan N, Wang ZD, Shan ZY, Shen JL, Zhang QH, et al. ES cell extract-induced expression of pluripotent factors in somatic cells. *Anat Rec (Hoboken)* 2009;**292**:1229–34.
97. Rajasingh J, Lambers E, Hamada H, Bord E, Thorne T, Goukassian I, et al. Cell-free embryonic stem cell extract-mediated derivation of multipotent stem cells from NIH3T3 fibroblasts for functional and anatomical ischemic tissue repair. *Circ Res* 2008;**102**:e107–17.
98. Cho HJ, Lee CS, Kwon YW, Paek JS, Lee SH, Hur J, et al. Induction of pluripotent stem cells from adult somatic cells by protein-based reprogramming without genetic manipulation. *Blood* 2010;**116**:386–95.
99. Ding S, Schultz PG. A role for chemistry in stem cell biology. *Nat Biotechnol* 2004;**22**:833–40.
100. Anastasia L, Pelissero G, Venerando B, Tettamanti G. Cell reprogramming: expectations and challenges for chemistry in stem cell biology and regenerative medicine. *Cell Death Differ* 2010;**17**:1230–7.
101. Bain G, Gottlieb DI. Expression of retinoid X receptors in P19 embryonal carcinoma cells and embryonic stem cells. *Biochem Biophys Res Commun* 1994;**200**:1252–6.
102. Sorm F, Vesely J. The activity of a new antimetabolite, 5-azacytidine, against lymphoid leukaemia in AK mice. *Neoplasma* 1964;**11**:123–30.
103. Christman JK. 5-Azacytidine and 5-aza-2'-deoxycytidine as inhibitors of DNA methylation: mechanistic studies and their implications for cancer therapy. *Oncogene* 2002;**21**:5483–95.
104. Li W, Wei W, Zhu S, Zhu J, Shi Y, Lin T, et al. Generation of rat and human induced pluripotent stem cells by combining genetic reprogramming and chemical inhibitors. *Cell Stem Cell* 2009;**4**:16–9.
105. Shi Y, Do JT, Desponts C, Hahm HS, Scholer HR, Ding S. A combined chemical and genetic approach for the generation of induced pluripotent stem cells. *Cell Stem Cell* 2008;**2**:525–8.
106. Lyssiotis CA, Foreman RK, Staerk J, Garcia M, Mathur D, Markoulaki S, et al. Reprogramming of murine fibroblasts to induced pluripotent stem cells with chemical complementation of Klf4. *Proc Natl Acad Sci USA* 2009;**106**:8912–7.
107. Huangfu D, Maehr R, Guo W, Eijkelenboom A, Snitow M, Chen AE, et al. Induction of pluripotent stem cells by defined factors is greatly improved by small-molecule compounds. *Nat Biotechnol* 2008;**26**:795–7.
108. Kubicek S, O'Sullivan RJ, August EM, Hickey ER, Zhang Q, Teodoro ML, et al. Reversal of H3K9me2 by a small-molecule inhibitor for the G9a histone methyltransferase. *Mol Cell* 2007;**25**:473–81.
109. Mikkelsen TS, Hanna J, Zhang X, Ku M, Wernig M, Schorderet P, et al. Dissecting direct reprogramming through integrative genomic analysis. *Nature* 2008;**454**:49–55.
110. Burlacu A. Can 5-azacytidine convert the adult stem cells into cardiomyocytes? A brief overview. *Arch Physiol Biochem* 2006;**112**:260–4.
111. Ruau D, Ensenat-Waser R, Dinger TC, Vallabhapurapu DS, Rolletschek A, Hacker C, et al. Pluripotency associated genes are reactivated by chromatin-modifying agents in neurosphere cells. *Stem Cells* 2008;**26**:920–6.
112. Araki H, Yoshinaga K, Boccuni P, Zhao Y, Hoffman R, Mahmud N. Chromatin-modifying agents permit human hematopoietic stem cells to undergo multiple cell divisions while retaining their repopulating potential. *Blood* 2007;**109**:3570–8.
113. Rajasingh J, Thangavel J, Siddiqui MR, Gomes I, Gao XP, Kishore R, et al. Improvement of cardiac function in mouse myocardial infarction after transplantation of epigenetically-modified bone marrow progenitor cells. *PLoS One* 2011;**6**:e22550.
114. Ambros V. The functions of animal microRNAs. *Nature* 2004;**431**:350–5.
115. Bartel DP. MicroRNAs: genomics, biogenesis, mechanism, and function. *Cell* 2004;**116**:281–97.
116. Rana TM. Illuminating the silence: understanding the structure and function of small RNAs. *Nat Rev Mol Cell Biol* 2007;**8**:23–36.

117. Kim VN, Han J, Siomi MC. Biogenesis of small RNAs in animals. *Nat Rev Mol Cell Biol* 2009;**10**:126–39.
118. Kanellopoulou C, Muljo SA, Kung AL, Ganesan S, Drapkin R, Jenuwein T, et al. Dicer-deficient mouse embryonic stem cells are defective in differentiation and centromeric silencing. *Genes Dev* 2005;**19**:489–501.
119. Murchison EP, Partridge JF, Tam OH, Cheloufi S, Hannon GJ. Characterization of Dicer-deficient murine embryonic stem cells. *Proc Natl Acad Sci USA* 2005;**102**:12135–40.
120. Barroso-delJesus A, Romero-Lopez C, Lucena-Aguilar G, Melen GJ, Sanchez L, Ligero G, et al. Embryonic stem cell-specific miR302-367 cluster: human gene structure and functional characterization of its core promoter. *Mol Cell Biol* 2008;**28**:6609–19.
121. Xu N, Papagiannakopoulos T, Pan G, Thomson JA, Kosik KS. MicroRNA-145 regulates OCT4, SOX2, and KLF4 and represses pluripotency in human embryonic stem cells. *Cell* 2009;**137**:647–58.
122. Wang Y, Keys DN, Au-Young JK, Chen C. MicroRNAs in embryonic stem cells. *J Cell Physiol* 2009;**218**:251–5.
123. Makeyev EV, Zhang J, Carrasco MA, Maniatis T. The MicroRNA miR-124 promotes neuronal differentiation by triggering brain-specific alternative pre-mRNA splicing. *Mol Cell* 2007;**27**:435–48.
124. Latronico MV, Condorelli G. MicroRNAs and cardiac pathology. *Nat Rev Cardiol* 2009;**6**:419–29.
125. Judson RL, Babiarz JE, Venere M, Blelloch R. Embryonic stem cell-specific microRNAs promote induced pluripotency. *Nat Biotechnol* 2009;**27**:459–61.
126. Banito A, Rashid ST, Acosta JC, Li S, Pereira CF, Geti I, et al. Senescence impairs successful reprogramming to pluripotent stem cells. *Genes Dev* 2009;**23**:2134–9.
127. Li Z, Yang CS, Nakashima K, Rana TM. Small RNA-mediated regulation of iPS cell generation. *EMBO J* 2011;**30**:823–34.
128. Takahashi J. Stem cell therapy for Parkinson's disease. *Expert Rev Neurother* 2007;**7**:667–75.
129. Park IH, Zhao R, West JA, Yabuuchi A, Huo H, Ince TA, et al. Reprogramming of human somatic cells to pluripotency with defined factors. *Nature* 2008;**451**:141–6.
130. Liu H, Zhu F, Yong J, Zhang P, Hou P, Li H, et al. Generation of induced pluripotent stem cells from adult rhesus monkey fibroblasts. *Cell Stem Cell* 2008;**3**:587–90.
131. Han X, Han J, Ding F, Cao S, Lim SS, Dai Y, et al. Generation of induced pluripotent stem cells from bovine embryonic fibroblast cells. *Cell Res* 2011;**21**:1509–12.
132. Kim JB, Zaehres H, Wu G, Gentile L, Ko K, Sebastiano V, et al. Pluripotent stem cells induced from adult neural stem cells by reprogramming with two factors. *Nature* 2008;**454**:646–50.
133. Eminli S, Foudi A, Stadtfeld M, Maherali N, Ahfeldt T, Mostoslavsky G, et al. Differentiation stage determines potential of hematopoietic cells for reprogramming into induced pluripotent stem cells. *Nat Genet* 2009;**41**:968–76.
134. Hu K, Yu J, Suknuntha K, Tian S, Montgomery K, Choi KD, et al. Efficient generation of transgene-free induced pluripotent stem cells from normal and neoplastic bone marrow and cord blood mononuclear cells. *Blood* 2011;**117**:e109–19.
135. Shi Y, Desponts C, Do JT, Hahm HS, Scholer HR, Ding S. Induction of pluripotent stem cells from mouse embryonic fibroblasts by Oct4 and Klf4 with small-molecule compounds. *Cell Stem Cell* 2008;**3**:568–74.
136. Li W, Zhao XY, Wan HF, Zhang Y, Liu L, Lv Z, et al. iPS cells generated without c-Myc have active Dlk1-Dio3 region and are capable of producing full-term mice through tetraploid complementation. *Cell Res* 2011;**21**:550–3.
137. Polo JM, Liu S, Figueroa ME, Kulalert W, Eminli S, Tan KY, et al. Cell type of origin influences the molecular and functional properties of mouse induced pluripotent stem cells. *Nat Biotechnol* 2010;**28**:848–55.

138. Tat PA, Sumer H, Pralong D, Verma PJ. The efficiency of cell fusion-based reprogramming is affected by the somatic cell type and the in vitro age of somatic cells. *Cell Reprogram* 2011;**13**:331–44.

139. Jenuwein T, Allis CD. Translating the histone code. *Science* 2001;**293**:1074–80.

140. Scheper W, Copray S. The molecular mechanism of induced pluripotency: a two-stage switch. *Stem Cell Rev* 2009;**5**:204–23.

141. Wu X, Li Y, Xue L, Wang L, Yue Y, Li K, et al. Multiple histone site epigenetic modifications in nuclear transfer and in vitro fertilized bovine embryos. *Zygote* 2011;**19**:31–45.

142. Mali P, Chou BK, Yen J, Ye Z, Zou J, Dowey S, et al. Butyrate greatly enhances derivation of human induced pluripotent stem cells by promoting epigenetic remodeling and the expression of pluripotency-associated genes. *Stem Cells* 2010;**28**:713–20.

143. Trojer P, Reinberg D. Histone lysine demethylases and their impact on epigenetics. *Cell* 2006;**125**:213–7.

144. Klose RJ, Bird AP. Genomic DNA methylation: the mark and its mediators. *Trends Biochem Sci* 2006;**31**:89–97.

145. Bui HT, Wakayama S, Kishigami S, Park KK, Kim JH, Thuan NV, et al. Effect of trichostatin A on chromatin remodeling, histone modifications, DNA replication, and transcriptional activity in cloned mouse embryos. *Biol Reprod* 2010;**83**:454–63.

146. Bird A. DNA methylation de novo. *Science* 1999;**286**:2287–8.

147. Shoemaker R, Wang W, Zhang K. Mediators and dynamics of DNA methylation. *Wiley Interdiscip Rev Syst Biol Med* 2011;**3**:281–98.

148. Han J, Sachdev PS, Sidhu KS. A combined epigenetic and non-genetic approach for reprogramming human somatic cells. *PLoS One* 2010;**5**:e12297.

149. Ng HH, Jeppesen P, Bird A. Active repression of methylated genes by the chromosomal protein MBD1. *Mol Cell Biol* 2000;**20**:1394–406.

150. Cheng X, Gadue P. Liver regeneration from induced pluripotent stem cells. *Mol Ther* 2010;**18**:2044–5.

151. Lyko F. DNA methylation learns to fly. *Trends Genet* 2001;**17**:169–72.

152. Curradi M, Izzo A, Badaracco G, Landsberger N. Molecular mechanisms of gene silencing mediated by DNA methylation. *Mol Cell Biol* 2002;**22**:3157–73.

153. Si J, Boumber YA, Shu J, Qin T, Ahmed S, He R, et al. Chromatin remodeling is required for gene reactivation after decitabine-mediated DNA hypomethylation. *Cancer Res* 2010;**70**:6968–77.

154. Annunziato AT, Hansen JC. Role of histone acetylation in the assembly and modulation of chromatin structures. *Gene Expr* 2000;**9**:37–61.

155. Strahl BD, Allis CD. The language of covalent histone modifications. *Nature* 2000;**403**:41–5.

156. Shi YJ, Matson C, Lan F, Iwase S, Baba T, Shi Y. Regulation of LSD1 histone demethylase activity by its associated factors. *Mol Cell* 2005;**19**:857–64.

157. Lan F, Nottke AC, Shi Y. Mechanisms involved in the regulation of histone lysine demethylases. *Curr Opin Cell Biol* 2008;**20**:316–25.

158. Bernstein BE, Mikkelsen TS, Xie X, Kamal M, Huebert DJ, Cuff J, et al. A bivalent chromatin structure marks key developmental genes in embryonic stem cells. *Cell* 2006;**125**:315–26.

159. Azuara V, Perry P, Sauer S, Spivakov M, Jorgensen HF, John RM, et al. Chromatin signatures of pluripotent cell lines. *Nat Cell Biol* 2006;**8**:532–8.

160. Loh YH, Zhang W, Chen X, George J, Ng HH. Jmjd1a and Jmjd2c histone H3 Lys 9 demethylases regulate self-renewal in embryonic stem cells. *Genes Dev* 2007;**21**:2545–57.

161. Doi A, Park IH, Wen B, Murakami P, Aryee MJ, Irizarry R, et al. Differential methylation of tissue- and cancer-specific CpG island shores distinguishes human induced pluripotent stem cells, embryonic stem cells and fibroblasts. *Nat Genet* 2009;**41**:1350–3.

162. Deng J, Shoemaker R, Xie B, Gore A, LeProust EM, Antosiewicz-Bourget J, et al. Targeted bisulfite sequencing reveals changes in DNA methylation associated with nuclear reprogramming. *Nat Biotechnol* 2009;**27**:353–60.
163. Okita K, Yamanaka S. Induced pluripotent stem cells: opportunities and challenges. *Philos Trans R Soc Lond B Biol Sci* 2011;**366**:2198–207.
164. Suzuki Jr. J, Therrien J, Filion F, Lefebvre R, Goff AK, Perecin F, et al. Loss of methylation at H19 DMD is associated with biallelic expression and reduced development in cattle derived by somatic cell nuclear transfer. *Biol Reprod* 2011;**84**:947–56.
165. Hai T, Hao J, Wang L, Jouneau A, Zhou Q. Pluripotency maintenance in mouse somatic cell nuclear transfer embryos and its improvement by treatment with the histone deacetylase inhibitor TSA. *Cell Reprogram* 2011;**13**:47–56.
166. Dey D, Evans GR. Generation of induced pluripotent stem (iPS) cells by nuclear reprogramming. *Stem Cells Int* 2011;**2011**:619583.
167. Araki R, Jincho Y, Hoki Y, Nakamura M, Tamura C, Ando S, et al. Conversion of ancestral fibroblasts to induced pluripotent stem cells. *Stem Cells* 2010;**28**:213–20.
168. Samavarchi-Tehrani P, Golipour A, David L, Sung HK, Beyer TA, Datti A, et al. Functional genomics reveals a BMP-driven mesenchymal-to-epithelial transition in the initiation of somatic cell reprogramming. *Cell Stem Cell* 2010;**7**:64–77.
169. Li R, Liang J, Ni S, Zhou T, Qing X, Li H, et al. A mesenchymal-to-epithelial transition initiates and is required for the nuclear reprogramming of mouse fibroblasts. *Cell Stem Cell* 2010;**7**:51–63.
170. Boyer LA, Mathur D, Jaenisch R. Molecular control of pluripotency. *Curr Opin Genet Dev* 2006;**16**:455–62.
171. Boyer LA, Plath K, Zeitlinger J, Brambrink T, Medeiros LA, Lee TI, et al. Polycomb complexes repress developmental regulators in murine embryonic stem cells. *Nature* 2006;**441**:349–53.
172. Chan EM, Ratanasirintrawoot S, Park IH, Manos PD, Loh YH, Huo H, et al. Live cell imaging distinguishes bona fide human iPS cells from partially reprogrammed cells. *Nat Biotechnol* 2009;**27**:1033–7.
173. Sanges D, Lluis F, Cosma MP. Cell-fusion-mediated reprogramming: pluripotency or transdifferentiation? Implications for regenerative medicine. *Adv Exp Med Biol* 2011;**713**:137–59.
174. Prigione A, Adjaye J. Modulation of mitochondrial biogenesis and bioenergetic metabolism upon in vitro and in vivo differentiation of human ES and iPS cells. *Int J Dev Biol* 2010;**54**:1729–41.
175. Bock C, Kiskinis E, Verstappen G, Gu H, Boulting G, Smith ZD, et al. Reference maps of human ES and iPS cell variation enable high-throughput characterization of pluripotent cell lines. *Cell* 2011;**144**:439–52.
176. Lister R, Pelizzola M, Kida YS, Hawkins RD, Nery JR, Hon G, et al. Hotspots of aberrant epigenomic reprogramming in human induced pluripotent stem cells. *Nature* 2011;**471**:68–73.
177. Ohi Y, Qin H, Hong C, Blouin L, Polo JM, Guo T, et al. Incomplete DNA methylation underlies a transcriptional memory of somatic cells in human iPS cells. *Nat Cell Biol* 2011;**13**:541–9.
178. Han JW, Yoon YS. Induced pluripotent stem cells: emerging techniques for nuclear reprogramming. *Antioxid Redox Signal* 2011;**15**:1799–820.
179. Hanna J, Cheng AW, Saha K, Kim J, Lengner CJ, Soldner F, et al. Human embryonic stem cells with biological and epigenetic characteristics similar to those of mouse ESCs. *Proc Natl Acad Sci USA* 2010;**107**:9222–7.

180. Marson A, Foreman R, Chevalier B, Bilodeau S, Kahn M, Young RA, et al. Wnt signaling promotes reprogramming of somatic cells to pluripotency. *Cell Stem Cell* 2008;**3**:132–5.
181. Li W, Ding S. Generation of novel rat and human pluripotent stem cells by reprogramming and chemical approaches. *Methods Mol Biol* 2010;**636**:293–300.
182. Nussbaum J, Minami E, Laflamme MA, Virag JA, Ware CB, Masino A, et al. Transplantation of undifferentiated murine embryonic stem cells in the heart: teratoma formation and immune response. *FASEB J* 2007;**21**:1345–57.
183. Davis RL, Weintraub H, Lassar AB. Expression of a single transfected cDNA converts fibroblasts to myoblasts. *Cell* 1987;**51**:987–1000.
184. Ieda M, Fu JD, Delgado-Olguin P, Vedantham V, Hayashi Y, Bruneau BG, et al. Direct reprogramming of fibroblasts into functional cardiomyocytes by defined factors. *Cell* 2010;**142**:375–86.
185. Efe JA, Hilcove S, Kim J, Zhou H, Ouyang K, Wang G, et al. Conversion of mouse fibroblasts into cardiomyocytes using a direct reprogramming strategy. *Nat Cell Biol* 2011;**13**:215–22.
186. Outani H, Okada M, Hiramatsu K, Yoshikawa H, Tsumaki N. Induction of chondrogenic cells from dermal fibroblast culture by defined factors does not involve a pluripotent state. *Biochem Biophys Res Commun* 2011;**411**:607–12.
187. Kim J, Efe JA, Zhu S, Talantova M, Yuan X, Wang S, et al. Direct reprogramming of mouse fibroblasts to neural progenitors. *Proc Natl Acad Sci USA* 2011;**108**:7838–43.
188. Pasha Z, Haider H, Ashraf M. Efficient non-viral reprogramming of myoblasts to stemness with a single small molecule to generate cardiac progenitor cells. *PLoS One* 2011;**6**:e23667.
189. Vierbuchen T, Ostermeier A, Pang ZP, Kokubu Y, Sudhof TC, Wernig M. Direct conversion of fibroblasts to functional neurons by defined factors. *Nature* 2010;**463**:1035–41.
190. Takeuchi JK, Bruneau BG. Directed transdifferentiation of mouse mesoderm to heart tissue by defined factors. *Nature* 2009;**459**:708–11.
191. Zhou Q, Melton DA. Extreme makeover: converting one cell into another. *Cell Stem Cell* 2008;**3**:382–8.
192. Laiosa CV, Stadtfeld M, Xie H, de Andres-Aguayo L, Graf T. Reprogramming of committed T cell progenitors to macrophages and dendritic cells by C/EBP alpha and PU.1 transcription factors. *Immunity* 2006;**25**:731–44.
193. Szabo E, Rampalli S, Risueno RM, Schnerch A, Mitchell R, Fiebig-Comyn A, et al. Direct conversion of human fibroblasts to multilineage blood progenitors. *Nature* 2010;**468**:521–6.
194. Chambers SM, Studer L. Cell fate plug and play: direct reprogramming and induced pluripotency. *Cell* 2011;**145**:827–30.
195. Zhao T, Zhang ZN, Rong Z, Xu Y. Immunogenicity of induced pluripotent stem cells. *Nature* 2011;**474**:212–5.
196. Dhodapkar KM, Feldman D, Matthews P, Radfar S, Pickering R, Turkula S, et al. Natural immunity to pluripotency antigen OCT4 in humans. *Proc Natl Acad Sci USA* 2010;**107**: 8718–23.
197. Chen YT, Venditti CA, Theiler G, Stevenson BJ, Iseli C, Gure AO, et al. Identification of CT46/HORMAD1, an immunogenic cancer/testis antigen encoding a putative meiosis-related protein. *Cancer Immun* 2005;**5**:9.
198. Okita K, Nagata N, Yamanaka S. Immunogenicity of induced pluripotent stem cells. *Circ Res* 2011;**109**:720–1.

Induction of Somatic Cell Reprogramming Using the MicroRNA miR-302

Karen Kelley and
Shi-Lung Lin

Division of Regenerative Medicine, WJWU
& LYNN Institute for Stem Cell Research,
Santa Fe Springs, California, USA

Since the discovery of pluripotent stem cells, scientists have envisioned their use in regenerative medicine. Unfortunately, such application of embryonic pluripotent stem cells has been impeded by ethical concerns as well as other obstacles. In light of this, the scientific community has begun to explore somatic cell reprogramming (SCR) as a means of producing induced pluripotent stem cells (iPSCs) from somatic cells. Although still far from being clinically applicable, SCR has become a hot research topic, with many groups working to understand its underlying mechanism. The standard method for inducing SCR is achieved by forced expression of four transcription factors defined by Yamanaka and Yu *et al.* Regrettably, iPSCs produced by the four-factor method tend to be tumorigenic, making them unsafe for clinical application. Recently, a new method has been identified to generate iPSCs through forced expression of an embryonic stem cell (ESC)-enriched microRNA, miR-302. This method holds a distinct advantage over the four-factor method because it can reprogram somatic cells to tumor-free iPSCs. Also, these miR-302-induced iPSCs, termed "mirPSCs," demonstrate a clear mechanism, which explains the process of reprogramming as a response to global DNA demethylation—the first sign

Progress in Molecular Biology
and Translational Science, Vol. 111
http://dx.doi.org/10.1016/B978-0-12-398459-3.00004-6

83

of SCR. Nevertheless, miR-302-induced reprogramming is dose-dependent, and microRNA (miRNA) concentration must be within a specific range for the reprogramming to occur. In addition, excessive overexpression of miR-302 in mirPS cells must not occur; otherwise, they will undergo early senescence. mirPSCs represent a new source of pluripotent stem cells without the tumorigenicity traditionally attributed to iPSCs. Looking forward, the next challenge lies with surmounting senescence, an obstacle that often limits stem cell expansion and prevents researchers from growing the large quantities of iPSCs needed for therapeutic use.

I. Introduction

A stem cell possesses two functions: self-renewal and differentiation into other cell types. During early embryogenesis, stem cells are activated, but as the body develops and cells differentiate, there is a gradual decrease in the number of stems cells. Embryonic stem cells (ESCs) are pluripotent and have the ability to differentiate into three germ layers: ectoderm, mesoderm, and endoderm. Adult stem cells maintain some capacity to differentiate, but they are tissue-specific and multipotent rather than pluripotent. Stem cell research has been a hot topic because pluripotent stem cells hold great potential in disease treatment and regenerative medicine. Studies using ESCs have already revealed much about human development. However, in addition to the moral and ethical issues surrounding the use of ESCs in research, there are also technical challenges to their application in clinical therapy. First, the cell production process is inefficient and relies heavily on nonhuman components to support the growth of ESCs, frequently resulting in foreign antigen contamination. Second, immune rejection is probable because even a trace amount of foreign antigens can contaminate donor cells. Third, tumor formation has always deterred the application of these ESCs. Finally, we have not yet uncovered the mechanism underlying tumor prevention during early embryonic cell development.

The recent discovery of somatic cell reprogramming (SCR) has given us the opportunity to move away from using ESCs by providing methods to generate a new source of pluripotent stem cells: induced pluripotent stem cells (iPS cells or iPSCs). Similar to ESCs, iPSCs are capable of self-renewal and differentiating into ectoderm, mesoderm, and endoderm. Based on the current knowledge, SCR can be achieved through somatic cell nuclear transfer, forced expression of four previously defined transcription factors (either Oct4–Sox2–Klf4–c-Myc or Oct4–Sox2–Nanog–Lin28), or ectopic expression of the human ESC-specific microRNA (miRNA), namely, miR-302.

SCNT is the process of generating pluripotent stem cells by replacing the nucleus in a donor oocyte with that of a somatic cell.[1,2] SCNT has been used in cloning but is not widely used for iPSC generation because of the enormous amount of stress to the donor and acceptor cells, the low success rate, and moral objections to the use of human oocytes. Currently, the standard method of generating iPSCs involves the coexpression of four defined Yamanaka (Oct4–Sox2–Klf4–c-Myc) or Thomson (Oct4–Sox2–Nanog–Lin28) factors to convert somatic cells into pluripotent stem cells. The reprogramming success rate is slim, varying from 0.002% to 0.2%.[3–5] Although the exact mechanism of SCR remains unclear, both octamer-binding transcription factor 4 (*Oct4*) and sex determining region Y-box 2 (*Sox2*) appear to be critical components. Even though these reprogrammed cells raise less ethical concerns than SCNT, they still pose a risk for use in regenerative medicine, as cells produced by this method are potentially tumorigenic.

A new attempt to improve the previous four-factor method involves the forced expression of the ESC-specific miRNA, that is, miR-302, in somatic cells. We have previously found that miR-302 is capable of reprogramming human cancer cells into a normal ESC-like state.[6] Further studies showed that miR-302, when expressed at a level that is greater than human ESCs, suppresses >70% of both cyclin-dependent kinase 2 and 4/6 (CDK2 and CDK4/6) pathways at the G1 cell cycle checkpoint and reduces cell proliferation.[7] This approach for iPSC generation is safer and more efficient than induction by the four previously defined factors. Most importantly, miR-302 iPSCs (mirPSCs) overcome the tumorigenicity problem that has plagued four-factor iPSCs. Unfortunately, mirPSCs are not free of problems, as they are slow to propagate because of a restricted cell cycle rate and propensity to undergo senescence when the cellular miR-302 concentration is twice the level in human ESCs. Hence, finding a balance between tumorigenicity and senescence is important to advance the application of iPSCs toward therapeutic uses.

The mechanism behind iPSC generation is still unclear, but cells produced by any of these processes undergo global DNA demethylation in order to reset gene expression to an ESC-like profile and consequently achieve full reprogramming. Additionally, all the reprogramming methods known so far require the activation of both Oct4 and Sox2, which are the transcription activators of miR-302. It is conceivable that the relationship between Oct4–Sox2 and miR-302 may provide important insights into the mechanism underlying SCR. Given that iPSCs are still surrounded by obstacles such as tumorigenicity and senescence, further studies need to be carried out to understand the SCR mechanism before iPSCs can be used safely in medical applications.

II. Mechanism of Reprogramming

A. MicroRNA and Reprogramming

miRNA was first found in *Caenorhabditis elegans* 1993.[8] miRNAs are typically ~22 nt long and produced from short hairpin RNAs (shRNAs) that serve as precursors to the mature miRNAs. The primary precursors of miR-NAs, known as pri-miRNAs, are transcribed by RNA polymerase II (Pol II), or in some cases by Pol III. They are then further processed into hairpin-like miRNA precursors (pre-miRNAs) by the endoribonuclease Drosha before being exported out of the nucleus in a Ran-GTP and exportin-5-dependent manner, and finally cleaved into mature miRNAs in the cytoplasm by RNaseIII Dicers.[9-14] After forming RNA-induced silencing complexes with argonaute proteins and Dicers, mature miRNAs bind to complementary sequences located in the 3'-untranslated regions (3'-UTR) of their target messenger RNAs (mRNAs) and hence suppress protein translation. One miRNA sequence is capable of targeting several mRNA targets because of nonstandard or mismatched pairing that often occurs between the miRNA and its targets. The importance of miRNA in stem cell development and maintenance is demonstrated by Dicer1$^{-/-}$ mice, which not only lack Oct4 expression but also die before axis formation.[15]

miR-302 is the most abundant miRNA found in human ESCs and iPSCs, but its expression is absent in differentiated cells.[16,17] In the human zygote, miR-302 serves as a major gene silencer and regulator of pluripotency. The entire miR-302 cluster, along with another miRNA miR-367, is found in the intronic region of *La ribonucleoprotein domain family member 7* (*LARP7/PIP7S*), a gene positioned at the 4q25 locus of human chromosome 4; this region is commonly associated with longevity.[18] The miR-302 familial cluster consists of four sense miRNAs (b, c, a, d) and three known antisense miRNAs (b*, c*, a*).[16] The complete role of miR-302 is yet uncharacterized, however it has been shown to be involved in the maintenance of normal ESC status, and—if ectopically expressed in differentiated somatic cells—can reprogram them to iPSCs. Although murine ESCs express the homologs miR-291/294/295 rather than miR-302, recent studies show that murine iPSCs produced by the four-factor method also display abundant miR-302 expression, implicating a role for miR-302 in SCR.[19]

In addition to being a powerful tumor suppressor in humans, miR-302 has been found to target and orchestrate the translational suppression of several key epigenetic regulators, including AOF1/2 (also named LSD2/1 or KDM1B/1), DNMT1, MECP1-p66, MECP2, and MBD2 during the initiation of SCR.[7,20] The total targets of miR-302 include over 600 cellular genes, many of which are involved in differentiation and developmental signaling, and their suppression leads to disruptions in the RAS–MAPK, TGFß–SMAD, and LEFTY pathways, strongly suggesting that miR-302 may play a focal role in maintaining stem cell populations by inhibiting differentiation. For reprogramming somatic cells, we have shown that miR-302 is the most crucial factor for initiating global DNA demethylation and epigenetic modification in both iPSCs and SCNT cells.[20] It is noted that global demethylation promotes Oct4–Nanog overexpression in mouse embryos and mouse–human fused heterokaryons.[21,22] Furthermore, miR-302 can directly silence nuclear receptor subfamily 2, group F, number 2 (NR2F2), a transcriptional repressor of Oct4, to activate Oct4 transcription,[23] which explains how miR-302 activates Oct4 expression in human ESCs.

miR-302 silences its targets by binding to complementary mRNA sequences in a nonstringent manner, increasing the number of targets available to a single miRNA sequence. As a consequence of mismatch accommodation during binding, the degree of silencing is directly related to the concentration of miR-302. Since miR-302b, c, a, and d share an identical seed sequence, more than one miR-302 familial member may redundantly regulate the same target to enhance the gene silencing effect (Fig. 1B). Although miR-367 has a slightly different seed sequence, it shares many gene targets with miR-302 and likely may increase the reprogramming efficiency of iPSC formation.

B. Mechanism of miR-302 Biogenesis

There are several minor differences in the expression patterns of mirPSCs and ESCs. First, the miR-302 cluster we designed is processed from the artificial intron *SpRNAi*, whose expression is driven by a constitutive cytomegaloviral promoter or a tetracycline-inducible (Tet-On) promoter. The use of an inducible promoter is advantageous for studying gene targeting as it is related to miR-302 concentration. By this approach, we have determined the minimum miR-302 concentration required for inducing SCR to be approximately 1.1–1.3 times the miR-302 level found in human ESCs.[6,20] After reprogramming, the ectopic miR-302 expression further stimulates the cells' intrinsic miR-302–367 expression via Oct4–Sox2 activation. Through microarray analysis of mirPSCs, we have observed expression of all miR-302 familial members except the antisense strand to miR-302b, termed "miR-302b*." Therefore, miR-302 function characterized both here and in our previous studies describes the combinational activity and interaction of miR-302a, a*, b, c, c*, and d but not miR-302b*. Interestingly, there is evidence that miR-302b* may play a critical role in regulating the entire miR-302 cluster expression, as the miRNA-targeting program PICTAR-VERT (http://pictar.mdc-berlin.de/) predicts Dicer1 to be a target of miR-302b*. Dicer1 is responsible for the processing of pre-miRNAs into mature miRNAs; hence, decreasing Dicer1 may result in lower concentrations of mature miR-302 as well as other

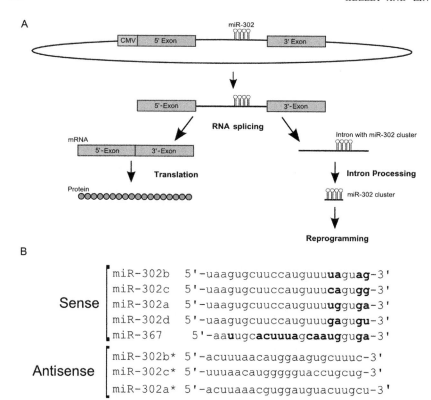

FIG. 1. The miR-302 family and biogenesis. (A) Schematic mechanism of miR-302 biogenesis. The miR-302 cluster is located in the intronic region between the two exons of its encoding gene. The advantage of this intronic expression system is that it uses the cell's intrinsic RNA-splicing and miRNA-processing machineries for miR-302 biogenesis, preventing the risk of RNA accumulation. (B) Sequence alignment of mature miR-302–367 familial members. The first 17 ribonucleotides in miR-302b, a, c, and d are identical, including the seed sequences, while the homolog miR-367 shares some similarities in the seed sequence and target genes. Asterisks denote the miR-302 antisense sequences. Conserved bases are depicted in red. (For interpretation of the references to color in this figure legend, the reader is referred to the online version of this chapter.)

ESC-specific miRNAs. Owing to the absence of miR-302b*, our method is exempted from this regulatory mechanism, so as to provide sufficient miR-302 for inducing SCR.

RNA accumulation is a problem with direct (exonic) expression of miRNA or miRNA-like shRNA due to the cytotoxicity induced by oversaturation of the natural miRNA biogenesis system.[24] Native miR-302 bypasses this cytotoxic pathway because it is expressed through an intronic biogenesis mechanism, in which miR-302 is located in an intron and coexpressed with the intron-containing gene (Fig. 1A; Refs. 6,10,25). After transcription, the intron is released from the gene transcript by RNA splicing and further processed into precursor miR-302 (pre-miR-302). Given that the miRNA-encoding intron contains nonsense stop codons recognized by the intracellular nonsense-mediated decay (NMD) system,[26,27] NMD preserves pre-miR-302 hairpins and degrades all other nonhairpin structures. As a result, this intronic biogenesis mechanism has an evolutionary advantage in safety control to prevent RNA oversaturation.

C. miR-302 Induces Global Demethylation

Global DNA demethylation is the first sign that cells have begun the process of SCR to attain ESC-like pluripotency.[28] Until recently, SCR has been accomplished using four-factor induction, which presents no link with global demethylation. In contrast, as more information regarding miR-302's functions and its targeted genes becomes available, a connection between miR-302 and global demethylation has become apparent.

The first step of miR-302-induced SCR involves global demethylation and histone modification. Global demethylation likely occurs in a passive mechanism triggered by decreased expression of several key epigenetic regulators. Based on current bioinformatics and experimental data, these epigenetic regulators include DNA (cytosine-5-)-methyltransferase 1 (DNMT1), both members of the AOF/LSD/KDM family, MBD2, MECP1-p66, MECP2, and HDAC2.[6,20] DNMT1 functions to maintain the epigenome by methylating CpGs on newly synthesized DNA during S-phase of the cell cycle, and its suppression is necessary for the passive global demethylation during reprogramming. Deficiency of DNMT1 function following zygotic cleavage events during early embryogenesis results in global demethylation in ESCs.[29–32] The miRNA-target prediction program provided by the European Bioinformatics Institute EMBL-EBI has shown that DNMT1 is a direct target of miR-302. In addition, the DNMT1 level is also indirectly affected by repression of other miR-302 targets such as lysine-specific histone demethylase 1 (AOF2/LSD1/KDM1). Silencing AOF2 prevents DNMT1-mediated maintenance of genomic DNA methylation, as AOF2 is required for stabilizing the DNMT1 activity.[20] This kind of redundant targeting DNMT1 for silencing ensures its low level during reprogramming.

AOF1 and AOF2 regulate gene transcription by demethylating histone 3 on lysine 4 (H3K4), which silences transcription.[33–35] Studies using knockout mice have demonstrated that AOF1 is essential for de novo DNA methylation during oogenesis, while AOF2 knockout mice are embryonic-lethal due to a lack of DNA methylation—therefore no differentiation occurs.[33,36] Deficiency of either AOF1 or AOF2 elevates H3K4 methylation (H3K4me2/3) and is sufficient to cause global DNA demethylation. Studies using the inhibitor tranylcypromine to disrupt AOF2 function in embryonal carcinoma cells also showed significant elevation of H3K4me2/3 modification and Oct4 activation.[34,35] On the other hand, miR-302 further promotes DNA demethylation via targeting methyl CpG-binding proteins 1 and 2 (MECP1 and MECP2) and HDAC2, a member of the histone deacetylase family. While HDAC2 is a weak target of miR-302, recent research suggests that decreased HDAC2 levels promotes reprogramming efficiency in mirPSCs,[37] further emphasizing the need for high miR-302 concentrations in order to achieve iPSC generation. It has been measured that miR-302-induced Oct4–Sox2–Nanog expression occurs when over 80% of the DNMT1, AOF1/2, and MECP1/2 expression are concurrently silenced and the level of HDAC2 is reduced to approximately 50%.[20]

Induction of SCR is highly dependent on miR-302 concentration. Using inducible miR-302 expression in human cells, we have determined three critical miR-302 concentrations.[20] First, when expressed at the same level as that found in human ESCs, miR-302 is sufficient to mediate changes in cell morphology but not induce SCR. Second, when increased to 1.3-fold above the ESC level, miR-302 can activate the expression of the core reprogramming factors Oct4–Sox2–Nanog and thus initiate SCR. Last, at 1.5–1.7-folds greater than the ESC level, the miR-302 concentration is optimal for reprogramming human cells to iPSCs and forming embryoid bodies. However, when the level exceeds two times the ESC level, miR-302 may cause severe cell cycle arrest at the G0/G1 phase transition, which ultimately hinders iPSC formation. In contrast with humans, murine ESCs do not express high levels of miR-302, but they do express the miR-302 homologs miR-291/294/295. The reason for a different mechanism in mice remains elusive, but four-factor-induced mouse iPSCs are suspected to undergo the same mechanism as human iPSCs as indicated by miR-302 levels that are 30 times higher than miR-291/294/295.[19]

Mature miR-302 is localized to the cytoplasm while DNA demethylation occurs in the nucleus. Hence, miR-302 must utilize an intermediate to achieve global DNA demethylation. Experiments using SCNT have shown that reprogramming occurs at a high efficiency (>93%) when a somatic nucleus was transferred into the mirPSC cytoplasm.[20] Global demethylation of the transferred nucleus was observed only a few days post-SCNT. In contrast, transfer of a mirPSC nucleus into a somatic cytoplasm fails to form iPSCs. This finding demonstrates that the factor initiating SCR is located in the cytoplasm rather than in the nucleus, and these results are consistent with the previous SCNT experiments using oocyte cytoplasm. Even though oocytes do not contain an abundance of miR-302, they are rich in the miR-302 homologs miR-200c and miR-371–373. Unlike miR-302, the previously defined transcription factors Oct4–Sox2–Klf4–c-Myc and Oct4–Sox2–Nanog–Lin28 are all present in the nucleus. Thus, if these transcription factors were responsible for global demethylation, one would expect to see SCR to occur from the transfer of a reprogrammed nucleus; however, this is not the case. One alternative explanation behind the lack of iPSC formation from the hybrids of mirPSC nucleus and somatic cytoplasm may involve an unidentified inhibitor against reprogramming. Even so, no reports have substantiated such a theory and the addition of miR-302 into the above hybrids makes iPSC formation possible, indicating the crucial role of miR-302 in global demethylation and SCR.[20]

D. miR-302 Activates Oct4 Expression

A positive feed-forward mechanism exists between miR-302 expression and Oct4–Sox2–Nanog activation (Fig. 2). Ectopic expression of miR-302 to levels of approximately 1.1–1.3 million copies/cell induces strong Oct4–Sox2–Nanog coexpression, suggesting that mirPSCs and four-factor iPSCs may share a similar reprogramming mechanism.[6,20] Expressions of Lin28, as well as other human ESC markers, are observed 1–3 days later than Oct4–Sox2–Nanog coexpression, further indicating that miR-302-induced SCR may function through the same iPSC generation pathway using previously identified Oct4–Sox2–Nanog–Lin28 factors.[20] Oct4 and Sox2, the most critical reprogramming factors in both of the Yamanaka's and Thomson's iPSC generation methods, have also been found to stimulate miR-302 expression.[38] As a result, the mutual stimulation of miR-302 and Oct4–Sox2–Nanog forms a positive feed-forward regulation loop to maintain the stem cell status of the reprogrammed iPSCs. This finding explains why miR-302, Oct4, Sox2, and Nanog are all essential markers for both human ESCs and iPSCs.

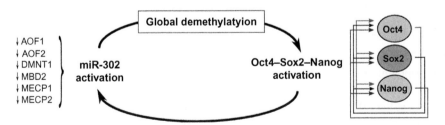

FIG. 2. Diagram of the positive regulatory network between miR-302 and Oct4–Sox2–Nanog. miR-302 expression leads to global DNA demethylation and Oct4–Sox2–Nanog activation. Global DNA demethylation is achieved through cosuppression of several important epigenetic regulators AOF1/2, DNMT1, MECP1/2, and HDAC2. In somatic cells, the promoters of *Oct4* and *Nanog* are highly methylated to prevent gene transcription. After DNA demethylation, Oct4–Sox2–Nanog are activated and further stimulate the expression of other ESC-specific genes as well as miR-302 to maintain the process of reprogramming. (For color version of this figure, the reader is referred to the online version of this chapter.)

Previous studies have identified NR2F2 and germ cell nuclear factor (GCNF) as transcriptional repressors of Oct4.[23,39] Based on the online miRNA-targeting prediction programs, both NR2F2 and GCNF are direct targets of miR-302, which silences these transcriptional repressors to activate Oct4 expression in iPSCs. As mentioned previously, Oct4–Sox2–Nanog are key members of a feed-forward regulatory network necessary for the maintenance of stem cell status[20,40]; Oct4 activation in conjunction with global demethylation can lead to increases of all three transcription factors, which also have been defined as core reprogramming factors in iPSCs. Subsequently, as miR-302 level declines during differentiation, NR2F2 and GCNF are inversely increased and are able to suppress Oct4 transcription by methylating the Oct4 promoter; corresponding levels of Sox2 and Nanog will decrease as a response to Oct4 suppression. Interestingly, Oct4–Sox2–Nanog and Tcf3 can bind to the promoters of miR-302 and many other ESC-specific miRNAs to stimulate their expressions.[38] Therefore, all these involved factors are part of a positive regulatory loop, with expression of miR-302 leading to increases in Oct4–Sox2–Nanog and vice versa (Fig. 2). Our studies suggest that global demethylation is attributed to miR-302 function rather than Oct4–Sox2–Nanog coexpression. Global demethylation in mouse embryos and mouse–human fused heterokaryons also shows elevated Oct4 and Nanog expressions.[21,22] In general, formation of mirPSC colonies takes a much shorter time (1–2 weeks) than that of four-factor iPSCs (2–3 weeks). It is conceivable that four-factor iPSC formation takes longer since Oct4–Sox2–Nanog must first function through miR-302-induced global demethylation in order to continue the reprogramming process.

III. Role of miR-302 in Early Embryogenesis

A. MicroRNA and Zygotic Development

Two types of miRNAs are responsible for establishing and maintaining pluripotent stem cells *in vivo*: one is inherited from oocytes (some maternal miRNAs) and the other, such as miR-302, is produced by zygotes. As shown in Fig. 3, maternal miRNAs are abundantly expressed in oocytes and begin to decline after fertilization. Following the decrease of maternal miRNAs, miR-302 expression rises dramatically after the first zygotic cell division. Zygotic demethylation occurs in chorus, establishing a colony of reprogrammed pluripotent stem cells at the 2- to 8-cell stage of embryonic development. Before the morula stage, these zygotic cells divide slowly (20–24 h/cycle) and maintain their stemness as miR-302 begins to decline around the 16- to 32-cell stage. At this point, remethylation of the genomic DNA occurs, ESC-specific transcription decreases, and differentiation begins.

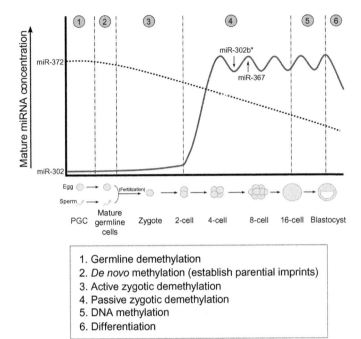

FIG. 3. Role of miRNA regulation during early embryogenesis. The chart on the right describes the state of genomic DNA methylation that coincides with each step of embryonic development. *In vivo*, from the PGC stage to the one-cell zygote, pluripotency is preserved by maternal miRNAs, such as miR-200c and miR-371–373 (indicated as miR-372). However, maternal miRNA levels decline after fertilization and miR-302 expression rapidly increases following the first zygotic cell division, leading to a maternal–zygotic transition in maintenance of stem cell pluripotency. miR-200c, miR-371–373, and miR-302 are homologues that share numerous target genes in common. Once the critical miR-302 concentration is reached, this level is maintained by miR-302b° to prevent its overexpression. During the 16- to 32-cell stage, miR-302 expression starts to decline, allowing remethylation of the genomic DNA to set up primary somatic gene expression patterns for later differentiation. (For color version of this figure, the reader is referred to the online version of this chapter.)

As pluripotency is preserved from germline to zygotic development, many miR-302-like homologues may have a similar capacity to induce SCR. However, many of them may also play a role in stem cell tumorigenicity. Recent studies have implicated that miR-372 is an oncogenic miRNA (onco-miR) that promotes tumorigenesis in primary human cells.[41] Four-factor iPSCs supplemented with miR-372 significantly increase both reprogramming efficiency and tumorigenicity, resulting in colonies prone to form tumors.[42] On the other hand, we have shown that miR-302 is a tumor suppressor in humans and mirPSCs proliferate at a relatively slow cell cycle rate similar to that of the early zygote at 2- to 8-cell stage (20–24 h/cycle).[6,7] Even more, miR-302 can directly reprogram cancer/tumor cells into a normal ESC-like pluripotent state. Based on these differences, careful selection of a tumor-free miRNA for iPSC generation is critical for therapy development.

B. Comparison Between SCR-Mediated Global Demethylation and Zygotic Demethylation

Global demethylation is required for both zygotic development and SCR, but the processes are not identical. In nature, global demethylation occurs during two periods of embryonic development. First, germline demethylation erases parental imprints during primordial germ cell (PGC) migration into the gonads approximately embryonic days E10.5–E13.5.[30,32,43,44] The second event, zygotic demethylation, occurs during the 2- to 8-cell stage in mammalian zygotes and requires simultaneous downregulation of MECP1 and 2, along with methyl-CpG-binding domain 2 (MBD2)—all targets of miR-302. Although both processes involve global demethylation, parental imprints in the epigenome are erased and reestablished only during germline demethylation.[45,46] Similar to the process in zygotes, global demethylation in iPSCs occurs with parental imprints intact, as SCR involves a forced demethylation mechanism that does not go through the germline stage. Somatic cells are converted directly to ESC-like pluripotent cells, bypassing germline demethylation (Fig. 4). As a result, SCR is missing some germline components necessary for full reprogramming of the cell genome as in zygotic demethylation, the consequences of which remain to be determined. Notably, cells reprogrammed by SCNT display transcriptomic and epigenomic patterns that are more similar to ECS than iPSCs produced using the four-factor methods.[47] Hence, SCR may result in certain defects in iPSCs because of the lack of germline components.

Zygotic demethylation at the 2- to 8-cell stage occurs in a passive manner. Zygotic expression of DNMT1 is low, and maternal DNMT1 (inherited from the oocyte) is excluded from the nucleus by an unknown mechanism.[48–50] Similarly, miR-302-induced SCR also requires a low DNMT1 environment for iPSC generation. Previous studies have shown that while AOF1/2, MECP1/2, DNMT1, and HDAC2 reached maximum depression 3 days after forced miR-302 expression in iPSCs, elevation in the levels of Oct4, Sox2, and Nanog was not observed until the fifth day.[7] The cell cycle was at its slowest for the first 3 days, with almost no division detected during this time. Then, one or two divisions were observed during the fourth and fifth days, in accordance with a moderate increase in the cell cycle rate. This phenomenon of attenuated cell divisions is identical to the natural passive process of zygotic demethylation, suggesting that deficiency of germline components in SCR does not affect this passive demethylation mechanism.

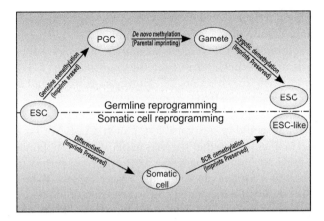

FIG. 4. Comparison of germline, zygotic, and somatic cell reprogramming. *In vivo*, germline reprogramming initiates with global DNA demethylation and erasure of parental imprints; these imprints are later reestablished during PGC to gamete transition through a process known as *de novo* methylation. Next, after fertilization, zygotic reprogramming occurs to induce the second global demethylation event, which removes only certain DNA methylation sites related to pluripotency and differentiation but not parental imprints, and hence activates ESC-specific gene expression. In parallel to zygotic reprogramming, SCR is a forced demethylation mechanism that occurs in differentiated somatic cells and does not erase parental imprints. Thus, iPSCs so obtained are similar but not identical to ESCs due to lack of germline reprogramming. (For color version of this figure, the reader is referred to the online version of this chapter.)

Currently, most of the evidence points to a passive demethylation mechanism for SCR, even though an active mechanism has not been ruled out. If passive demethylation is the sole mechanism of SCR, it would result in two hemimethylated cells per iPSC colony, the existence of which has never been found. Whether these hemimethylated cells exist, or simply unobserved because they undergo apoptosis or are demethylated by an active demethylation mechanism, remains unknown. Active demethylation does occur in embryos and is coordinated by maternal components prior to the first zygotic cleavage.[30,32,51] However, both miR-302 and four-factor iPSCs undergo a forced reprogramming process that likely bypasses this active mechanism. Several theories have been described, involving the 5-methylcytosine (5mC)/ 5-hydroxymethylcytosine (5hmC) deamination pathway and the mismatch/ excision repair pathway to remove methyl groups from DNA, but the exact mechanism is still unclear.[52–54]

We recently discovered that miR-302-targeted AOF2 silencing leads to elevated levels of activation-induced cytidine deaminase (AID).[20] Knockdown of AID using siRNA can partially prevent miR-302-induced demethylation of the Oct4 and Nanog promoters, implying that an active demethylation mechanism may be involved in iPSC formation. Previous studies have identified a role for AID in global demethylation and Oct4–Nanog activation in both early murine embryos as well as mouse–human fused heterokaryons.[21,22] AID is expressed in oocytes, PGCs, early embryos, and blood B cells, and its function is to deaminate 5mC, resulting in thymine–guanine (T–G) mispairing due to cytosine (C) to T conversion.[55,56] However, as we do not observe any increase of T–G mispairing in iPSCs, some repairing mechanisms may be active in SCR. Recent researchers have proposed that a mammalian version of the base excision repair could be responsible for such a task, yet no enzyme has been identified in mammals. Other theories implicate that Tet familial enzymes may convert 5mC to 5hmC, which can be further converted to C by the spontaneous loss of a formaldehyde group.[57] Nevertheless, there are several questions in this theory. First, spontaneous conversion of 5hmC to C is inefficient. Second, Oct4 regulates Tet expression but Oct4 activation occurs after global demethylation; also, depletion of Tet has no effect on the expression of Oct4, Sox2, and Nanog in ESCs.[53] Finally, if Tet is depleted in the 2-cell stage of the mouse embryo, abnormalities are not seen until much later than zygotic demethylation, when the trophectoderm has developmental and signaling deficiencies.[58] Until these questions are addressed, whether active demethylation occurs during SCR remains to be determined.

IV. Dual Role of miR-302: Reprogramming Effector and Tumor Suppressor

A. Pluripotency Versus Tumorigenicity: Mechanism

Regenerative medicine holds the key to future therapies for curing genetic disorders, physical abnormalities, injures, cancers/tumors, and aging. Supply and safety of the stem cells are two of the most important bottlenecks in modern regenerative medicine. Previous efforts have succeeded in isolating various pluripotent/multipotent stem cells from embryos, umbilical cords/amniotic fluids, and reproductive organs; however, none of these stem cells can be expanded to large enough quantities for clinical use. On the other hand, tumor formation is still a problem for ESCs and four-factor iPSCs. Oncogenic factors such as c-Myc and Klf4 are frequently used to boost the survival and proliferative rates of iPSCs, creating an inevitable problem of tumorigenicity that hinders the therapeutic usefulness of these cells. Because of their fast cell proliferation (12–16 h/cycle) and easy teratoma transformation, many researchers consider these four-factor iPSCs to be cancer stem cells. Following the recent discovery of miR-302-induced SCR, using a forced reprogramming mechanism mimicking the natural zygotic development, we now have a clear view of the regulatory network necessary for stem cell generation while also preventing tumor formation from the stem cell.[7,20] With this advantage, we can generate tumor-free iPSCs derived from the bountiful pool of somatic cells.

The pluripotency of mirPSCs has been thoroughly evaluated using various standard stem cell assays. Analyses using microarrays and western blots revealed that mirPSCs/iPSCs strongly express Oct4, Sox2, Nanog, Lin28, and many other ESC gene markers, sharing over 92% similarity to human ESCs. Bisulfite DNA sequencing assays have further demonstrated that global demethylation profiles in mirPSCs/iPSCs highly resemble the human ESC-specific pattern, particularly in the promoter regions of core reprogramming factor genes, *Oct4*, *Nanog*, *Sox2*, and *UTF1*. Notably, teratoma formation assays have indicated that mirPSCs are highly pluripotent but not tumorigenic, as they preferentially form teratoma-like tissue cysts in the uteruses and peritoneal cavities but rarely in other tissues of pseudopregnant female immuno-compromised SCID-beige mice.[6,20] Unlike tumor-prone embryonal teratomas, these tissue cysts contain various but relatively organized tissue regions that are derived from all three embryonic germ layers (ectoderm, mesoderm, and definitive endoderm). The presence of partial homeobox (HOX) gradient expression is also frequently observed in these tissue cysts. When xenografted into normal male mice, mirPSCs can be easily differentiated and assimilated by the surrounding tissues, indicating a potential application for regenerative medicine.[7] Taken together, these findings suggest that miR-302 reprograms human somatic cells to ESC-like iPSCs with a high degree of pluripotency without tumorigenicity.

Normal and cancerous cells respond differently to miR-302-mediated tumor suppression effect. In humans, miR-302 triggers massive reprogrammed cell death (apoptosis) in fast proliferative tumor/cancer cells, whereas slow-growing normal cells can tolerate this inhibitory effect on cell proliferation.[6,7] It is conceivable that tumor/cancer cells may not survive in such a slow cell cycle state because of their high metabolism and rapid consumption rates. This result provides a very beneficial advantage in preventing the formation of tumorigenic iPSCs. In this tumor inhibition mechanism (Fig. 5), miR-302 not only suppresses both of the cyclin E–CDK2 and cyclin D–CDK4/6 pathways to block >70% of the G1–S cell cycle transition but also silences BMI1 polycomb ring finger oncogene (BMI1) to increase the expression of two cell cycle inhibitors, namely, p16Ink4a and p14/p19Arf.[7] 16Ink4a functions to inhibit cyclin D-dependent CDK4/6 activities via phosphorylation of retinoblastoma protein Rb and subsequently prevents Rb from releasing E2F-dependent transcription required for S-phase entry.[59,60] In addition, p14/p19Arf prevents HDM2, an Mdm2 p53 binding protein homolog, from binding to p53 and permits the p53-dependent transcription responsible for G1 arrest or apoptosis.[61] As a result, through simultaneous

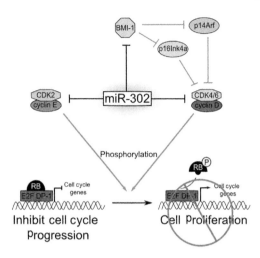

FIG. 5. Mechanism of miR-302-mediated tumor suppression in human iPSCs. miR-302 not only concurrently suppresses G1-phase checkpoint regulators CDK2, cyclin D, and BMI-1 but also indirectly activates p16Ink4a and p14/p19Arf to quench most (>70%) of the cell cycle activities during SCR. E2F is also a predicted target of miR-302. Relative quiescence at the G0/G1 state may prevent possible random growth and/or tumor-like transformation of the reprogrammed iPSCs, leading to a more accurate and safer reprogramming process, by which premature cell differentiation and tumorigenicity are both inhibited. (For color version of this figure, the reader is referred to the online version of this chapter.)

activation of multiple cell cycle attenuation checkpoints and tumor suppression pathways, miR-302 induces iPSC formation while preventing any possible stem cell tumorigenicity.

miR-302-mediated tumor suppression is a dose-dependent effect, which occurs only when its cellular concentration reaches ≥ 1.3-folds higher than the level in human ESCs, as determined by northern blot analysis.[7] This miR-302 concentration is also the optimal concentration for reprogramming.[6,20] Using an inducible miR-302 expression system in iPSCs, we found that at levels equal to or lower than the miR-302 level in human ESCs, miR-302 silences large tumor suppressor homolog 2 (LATS2) but not CDK2 to promote cell proliferation; while at a higher concentration (beyond the ESC level), both CDK2 and LATS2 are targeted for silencing and hence cell cycle is attenuated at the G1-phase checkpoint.[7] Given that LATS2 inhibits cell cycle via blocking the cyclin E-CDK2 pathway during the G1–S phase transit,[62] the silencing of CDK2 by miR-302 can compensate this LATS2 function while LATS2 is silenced. Moreover, this CDK2 silencing effect can also counteract the suppressive effect of miR-367 on CDKN1C (p57, Kip2), which is a cell cycle inhibitor against both CDK2 and CDK4, subsequently leading to a reduced cell cycle rate. Therefore, by synchronizing its best reprogramming and tumor suppression effects at the same concentration, miR-302 is able to generate tumor-free iPSCs suitable for application in regenerative medicine.

B. Cell Cycle Regulation: iPSCs Versus Zygotes

Through dose-dependent gene silencing, miR-302 can fine-tune or even alter the function of its target genes at different rates or during different stages of development. It is conceivable that early human embryonic cells before the morula stage may express higher miR-302 than blastocyst-derived ESCs. The 2- to 8-cell stage zygotic cells grow at a relatively slow cell cycle rate (20–24 h/cycle) than that of blastocyst-derived ESCs (15–16 h/cycle).[63,64] Even though both zygotes and ESCs are known to display a short G1 phase, the G1 phase of a 2- to 8-cell stage zygotic cell is still significantly longer than that of ESCs by 4 ± 1 h. Nevertheless, iPSC generation was reported to involve either a cell-cycle-dependent (with Klf4) or cell-cycle-independent (with Nanog) reprogramming process,[65] in which the former cannot be explained by the tumor suppression function of miR-302, indicating that iPSCs generated with Klf4 and c-Myc may bypass the tumor suppression function of miR-302. In fact, this hypothesis also overlooks the fact that Klf4 is an upstream transcription factor of Nanog.[66,67] Klf4 induces not only Nanog but also many other oncogenes, which promote rapid cell proliferation as a result unrelated to the reprogramming function of Nanog.[66–69] Therefore, even though a fast cell cycle can increase iPSC number, it is not essential for SCR initiation to form iPSCs.

When induced by high miR-302 concentration, mirPSCs form three-dimensional colonies with a relatively slow cell cycle rate of approximately 20–24 h/cycle—identical to the rate of natural embryonic cells at the 2- to 8-cell stage.[20] On the other hand, the Yamanaka-factor iPSCs exhibit faster cell cycle rate of 12–16 h/cycle, comparable to that of some cancer cell lines, such as melanoma Colo-829 and prostate cancer LNCaP cells. Early zygotic embryos before the morula stage typically have a longer cell cycle (20–24 h/cycle) than late blastocyst-derived ESCs (15–16 h/cycle). Also, similar to murine ESCs, mirPSCs are sensitive to growth stimulation by leukemia inhibitory factor (LIF) (Fig. 6). Murine ESCs are considered to be at an earlier stage of development (embryonic stage E3–E3.5) than the human ESCs (E5–E6), where LIF sensitivity is typical.[70,71] Taken together, these findings suggest that mirPSCs may represent an earlier stage of embryonic development than blastocyst-derived ESCs.

V. Balancing Stem Cell Tumorigenicity and Senescence

In light of the miR-302's dual function in reprogramming and tumor suppression, mirPSCs appear to be an ideal candidate to replace ESCs for clinical applications. iPSCs have been reported to exhibit problems including

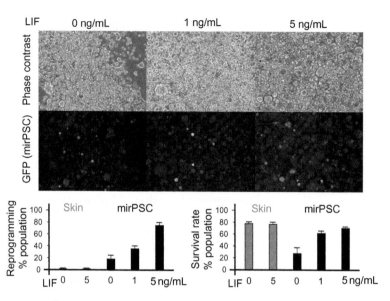

FIG. 6. Similar properties shared by human mirPSCs and mouse ESCs at an earlier embryonic stage. Human mirPSCs form three-dimensional cell colonies (red fluorescent) and are sensitive to the growth stimulation of LIF. Upper panels show that LIF stimulation significantly increases both the reprogramming efficiency and survival rates of mirPSCs in a dose-dependent matter, while lower charts indicate the relative levels of changes corresponding to the LIF concentration. In human HEK-293 cells, the reprogramming rates after 5 ng/mL LIF stimulation can reach over 83% of the miR-302-transfected cell population. (For interpretation of the references to color in this figure legend, the reader is referred to the online version of this chapter.)

early senescence and limited expansion.[52,72] However, mirPSCs still pose challenges, as miR-302 indirectly increases the expression of p16Ink4a and p14/p19Arf by silencing BMI1 (Fig. 5). BMI1, an oncogenic cancer stem cell marker, promotes G1–S cell cycle transition through inhibition of p16Ink4a and p14Arf tumor suppressor activities.[73] Recently, we found that excessive expression of miR-302 beyond 1.5–1.7-folds of the level found in ESCs leads to increased senescence proportional to miR-302 concentration. Normal adult cells undergo a limited number of divisions before reaching a quiescent state of replicative senescence. Cells that escape replicative senescence often become immortalized; thus replicative senescence is a cellular defense mechanism against tumor formation. As miR-302 increases the expression of both p16Ink4a and p14/p19Arf, two of the major regulators in replicative senescence, miR-302 overexpression may exacerbate early senescence in iPSCs.

Previous studies have identified several genes encoded in the Ink4a/Arf locus, such as p16Ink4a and p14/p19Arf, which pose a barrier to the success of iPSC reprogramming.[72,74–76] In particular, after serial passaging of iPSCs, both p16Ink4a-RB and p14/19Arf-p53 activities are gradually enhanced and finally commit to the onset of early senescence.[72,76] In murine iPSCs, p19Arf rather than p16Ink4a is the main factor in activating senescence, whereas in humans p16Inka plays a larger role than p14Arf.[72,74] Owing to forced miR-302 elevation, mirPSCs often present a higher degree of early senescence than four-factor iPSCs and under the same feeder-free conditions; four-factor iPSCs can be cultivated over 30–50 passages, whereas mirPSCs reach only a maximum of 26 passages.

Senescence is most commonly associated with the shortening of the telomeres after cells have undergone many divisions. After multiple divisions, telomeres can become so short that genes near the ends of chromosomes are subject to damage. Surprisingly, telomere shortening does not explain senescence in mirPSCs. Using telomeric repeat amplification protocol and telomerase activity ELISA, we have found that mirPSCs possess normal or slightly longer telomeres—discounting telomere shortening as a cause for early senescence. The reason for this is likely mirPSCs having high activities of human telomerase reverse transcriptase (hTERT), the enzyme responsible for maintaining telomere length. Reports have shown that decreasing both AOF2 and HDAC2 increases hTERT expression.[77,78] Both AOF2 and HDAC2 are valid targets of miR-302 and suppressed during mirPSC formation,[20] allowing hTERT expression to be elevated. Thus, the senescence of mirPSCs results from increased p16Ink4a-RB and/or p14/19Arf-p53 activities rather than loss of telomere length or hTERT.

The balance of tumorigenicity and senescence is critical to determining the safety of iPSCs. In mirPSCs, both mechanisms are controlled by p16Ink4a and p14/p19Arf activities, whereas in four-factor iPSCs overexpression of Oct4, Sox2, and Klf4 were found to repress the expression of p16Ink4a and p14/19Arf, leading to rapid cell proliferation.[74] Conceivably, the counteracting effects of miR-302 and Oct4-Sox2-Klf4 on the p16Ink4a and p14/19Arf expression may be the key in reaching the balance between tumorigenicity and senescence in iPSCs. To this, our identification of the positive feed-forward regulation between miR-302 and Oct4–Sox2–Nanog provides the first insight into this important mechanism (Fig. 2). In this reprogramming cycle, miR-302 and Oct4–Sox2–Nanog are able to mutually induce each other to constantly maintain the ESC-like pluripotent state of iPSCs. Therefore, at a certain point in the cycle, the interaction between miR-302 and Oct4–Sox2–Nanog/Klf4 may result in a stable iPSC state suitable for stem cell self-renewal without the problem of either tumor formation or early senescence. Based on this concept, identifying and using the correct combination of miR-302 and Oct4–Sox2–Nanog/Klf4 for generating both tumor- and senescence-free iPSCs will be the next challenge.

VI. Conclusion

Through deciphering the roles of miR-302 target genes in epigenetic modification and cell cycle regulation, our studies reveal that miR-302 has two parallel functions that occur simultaneously: reprogramming and tumor suppression. These two functions are important aspects of both ESCs and iPSCs and are controlled by distinct molecular pathways. To this, our studies have paved the fundamental groundwork for understanding the occurrence of SCR while preventing stem cell tumorigenicity. Since we have uncovered the involvement of at least nine miR-302 target genes thus far, understanding the roles of remaining genes targeted by miR-302 may provide further insights into the detailed mechanism underlying other aspects of SCR.

Reprogramming is a stepwise process that starts with global demethylation. During the first step of SCR, miR-302 forces global demethylation, bypassing germline demethylation. This process is achieved through cosuppression of AOF1/2, DNMT1, MECP1/2, and partially HDAC2, leading to genome-wide passive demethylation as newly synthesized DNAs remain unmethylated. Resupplementation of either AOF2 or DNMT1 can interfere with miR-302-induced reprogramming process, suggesting that concurrent silencing of these proteins is crucial for SCR. Because of the lack of germline demethylation, iPSCs may present several drawbacks. First, the consequences of retaining parental imprints during SCR are yet uncharacterized, but may prevent the formation of a universally compatible iPSCs for immediate use in clinical applications. It often takes several weeks or months to obtain iPSCs from patients. Second, iPSCs omit certain germline elements required for normal zygotic development, such as paternal protamines and maternal transcripts. Loss of these germline elements may cause developmental abnormalities in iPSCs. Last, due to the lack of normal induction from fertilization, iPSCs often require compensatory stimulation from oncogene or onco-miR activities (i.e., c-Myc and Klf4 or miR-200c/miR-367) to increase their survival rate, which may contribute to tumorigenicity.

After global demethylation, many genes are accessible to transcription machinery, allowing the expression of ESC-specific transcription factors. These include Oct4–Sox2–Nanog–Lin28, expression of which contributes to reprogramming by simultaneously stimulating miR-302 as well as one another. This feed-forward regulatory loop is an important aspect of miR-302-induced SCR and helps in maintaining stem cell pluripotency. The mechanism underlying miR-302-induced SCR also sheds light on the four-factor reprogramming mechanism. While miR-302 induces activation of Oct4–Sox2–Nanog–Lin28, four-factor induction in turn activates miR-302, which coordinates genome-wide DNA demethylation. Until now, the mechanism of demethylation during ESC formation is not fully understood. Even though our research focus is on miR-302, there are other miR-302 homologues present in the oocyte cytoplasm that may provide similar reprogramming effects, but their functions remain largely unknown. Conventional four-factor iPSC generation requires oncogenic stimulation from Klf4 and/or c-Myc, which often result in tumor-prone iPSCs. Recent identification of tumor-free mirPSCs overcomes this problem in that miR-302 not only induces reprogramming but also inhibits tumorigenicity by suppressing both of the CDK2- and CDK4/6-dependent cell cycle pathways during SCR. However, other problems exit in these tumor-free mirPSCs. miR-302 silences BMI1 to promote p16Ink4a and p14/19Arf activities, which causes early senescence and limits life span. Given that a balance between tumorigenicity and senescence has been found in the feed-forward regulation cycle of miR-302 and Oct4–Sox2–Nanog/Klf4 (Fig. 2), an effort to pinpoint the balanced position will be very rewarding.

Until now, most of the work done in stem cell research was focused on protein–protein interactions while the mechanism underlying iPSC generation was largely overlooked. Using the current pool of established data, this review proposes a novel SCR mechanism that is controlled by a single miRNA, instead of proteins. However, there are still many concerns to be addressed regarding the processes involved in iPSC generation *in vitro*, as well at ESC formation *in vivo*. We expect to improve our understanding of these processes through well-thought-out experimental designs and deductive reasoning. The aim has been to use stem cells to advance the field of medicine, and by providing insights into the function of miR-302 during reprogramming, we hope to achieve this goal in the near future.

REFERENCES

1. Wakayama T, Perry AC, Zuccotti M, Johnson KR, Yanagimachi R. Full-term development of mice from enucleated oocytes injected with cumulus cell nuclei. *Nature* 1998;**394**:369–74.
2. Wilmut I, Schnieke AE, McWhir J, Kind AJ, Campbell KH. Viable offspring derived from fetal and adult mammalian cells. *Nature* 1997;**385**:810–3.
3. Takahashi K, Yamanaka S. Induction of pluripotent stem cells from mouse embryonic and adult fibroblast cultures by defined factors. *Cell* 2006;**126**:663–76.
4. Wernig M, Meissner A, Foreman R, Brambrink T, Ku M, Hochedlinger K, et al. In vitro reprogramming of fibroblasts into a pluripotent ES-cell-like state. *Nature* 2007;**448**:318–24.
5. Yu J, Vodyanik MA, Smuga-Otto K, Antosiewicz-Bourget J, Frane JL, Tian S, et al. Induced pluripotent stem cell lines derived from human somatic cells. *Science* 2007;**318**:1917–20.
6. Lin SL, Chang D, Chang-Lin S, Lin CH, Wu DTS, Chen DT, et al. MiR-302 reprograms human skin cancer cells into a pluripotent ES-cell-like state. *RNA* 2008;**14**:2115–24.
7. Lin SL, Chang D, Ying SY, Leu D, Wu DTS. MicroRNA miR-302 inhibits the tumorigenecity of human pluripotent stem cells by coordinate suppression of CDK2 and CDK4/6 cell cycle pathways. *Cancer Res* 2010;**70**:9473–82.
8. Lee R, Feinbaum R, Ambros V. The C. elegans heterochronic gene lin-4 encodes small RNAs with antisense complementarity to lin-14. *Cell* 1993;**75**:843–54.

9. Danin-Kreiselman M, Lee CY, Chanfreau G. RNAse III-mediated degradation of unspliced pre-mRNAs and lariat introns. *Mol Cell* 2003;**11**:1279–89.
10. Lin SL, Chang D, Wu DY, Ying SY. A novel RNA splicing-mediated gene silencing mechanism potential for genome evolution. *Biochem Biophys Res Commun* 2003;**310**:754–60.
11. Lee YS, Nakahara K, Pham JW, Kim K, He Z, Sontheimer EJ, et al. Distinct roles for Drosophila Dicer-1 and Dicer-2 in the siRNA/miRNA silencing pathways. *Cell* 2004;**117**:69–81.
12. Lee Y, Ahn C, Han J, Choi H, Kim J, Yim J, et al. The nuclear RNase III Drosha initiates microRNA processing. *Nature* 2003;**425**:415–9.
13. Lund E, Guttinger S, Calado A, Dahlberg JE, Kutay U. Nuclear export of microRNA precursors. *Science* 2004;**303**:95–8.
14. Yi R, Qin Y, Macara IG, Cullen BR. Exportin-5 mediates the nuclear export of pre-microRNAs and short hairpin RNAs. *Genes Dev* 2003;**17**:3011–6.
15. Bernstein E, Duncan EM, Masui O, Gil J, Heard E, Allis CD. Mouse polycomb proteins bind differentially to methylated histone H3 and RNA and are enriched in facultative heterochromatin. *Mol Cell Biol* 2006;**26**:2560–9.
16. Suh MR, Lee Y, Kim JY, Kim SK, Moon SH, Lee JY, et al. Human embryonic stem cells express a unique set of microRNAs. *Dev Biol* 2004;**270**:488–98.
17. Wilson KD, Venkatasubrahmanyam S, Jia F, Sun N, Butte AJ, Wu JC. MicroRNA profiling of human-induced pluripotent stem cells. *Stem Cells Dev* 2009;**18**:749–58.
18. Puca AA, Daly MJ, Brewster SJ, Matise TC, Barrett J, Shea-Drinkwater M, et al. A genome-wide scan for linkage to human exceptional longevity identifies a locus on chromosome 4. *Proc Natl Acad Sci USA* 2000;**98**:10505–8.
19. Li Z, Yang C, Nakashima K, Rana T. Small RNA-mediated regulation of iPS cell generation. *EMBO J* 2011;**30**:823–34.
20. Lin SL, Chang D, Lin CH, Ying SY, Leu D, Wu DTS. Regulation of somatic cell reprogramming through inducible mir-302 expression. *Nucleic Acids Res* 2011;**39**:1054–65.
21. Bhutani N, Brady JJ, Damian M, Sacco A, Corbel SY, Blau HM. Reprogramming towards pluripotency requires AID-dependent DNA demethylation. *Nature* 2010;**463**:1042–7.
22. Popp C, Dean W, Feng S, Cokus SJ, Andrews S, Pellegrini M, et al. Genome-wide erasure of DNA methylation in mouse primordial germ cells is affected by AID deficiency. *Nature* 2010;**463**:1101–5.
23. Rosa A, Brivanlou A. A regulatory circuitry comprised of miR-302 and transcription factors OCT4 and NR2F2 regulates human embryonic stem cell differentiation. *EMBO J* 2011;**30**:237–48.
24. Grimm D, Streetz KL, Jopling CL, Storm TA, Pandey K, Davis CR, et al. Fatality in mice due to oversaturation of cellular microRNA/short hairpin RNA pathways. *Nature* 2006;**441**:537–41.
25. Barroso-delJesus A, Romero-López C, Lucena-Aguilar G, Melen GJ, Sanchez L, Ligero G, et al. Embryonic stem cell-specific miR302-367 cluster: human gene structure and functional characterization of its core promoter. *Mol Cell Biol* 2008;**28**:6609–19.
26. Lewis BP, Green RE, Brenner SE. Evidence for the widespread coupling of alternative splicing and nonsense-mediated mRNA decay in humans. *Proc Natl Acad Sci USA* 2003;**100**:189–92.
27. Zhang G, Taneja KL, Singer RH, Green MR. Localization of pre-mRNA splicing in mammalian nuclei. *Nature* 1994;**372**:809–12.
28. Simonsson S, Gurdon J. DNA demethylation is necessary for the epigenetic reprogramming of somatic cell nuclei. *Nat Cell Biol* 2004;**6**:984–90.
29. Hirasawa R, Chiba H, Kaneda M, Tajima S, Li E, Jaenisch R, et al. Maternal and zygotic Dnmt1 are necessary and sufficient for the maintenance of DNA methylation imprints during preimplantation development. *Genes Dev* 2008;**22**:1607–16.
30. Mayer W, Niveleau A, Walter J, Fundele R, Haaf T. Demethylation of the zygotic paternal genome. *Nature* 2000;**403**:501–2.

31. Monk M, Boubelik M, Lehnert S. Temporal and regional changes in DNA methylation in the embryonic, extraembryonic and germ cell lineages during mouse embryo development. *Development* 1987;**99**:371–82.
32. Santos F, Hendrich B, Reik W, Dean W. Dynamic reprogramming of DNA methylation in the early mouse embryo. *Dev Biol* 2002;**241**:172–82.
33. Ciccone DN, Su H, Hevi S, Gay F, Lei H, Bajko J, et al. KDM1B is a histone H3K4 demethylase required to establish maternal genomic imprints. *Nature* 2009;**461**:415–8.
34. Lee MG, Wynder C, Cooch N, Shiekhattar R. An essential role for CoREST in nucleosomal histone 3 lysine 4 demethylation. *Nature* 2005;**437**:432–5.
35. Lee MG, Wynder C, Schmidt DM, McCafferty DG, Shiekhattar R. Histone H3 lysine 4 demethylation is a target of nonselective antidepressive medications. *Chem Biol* 2006;**13**: 563–7.
36. Wang J, Hevi S, Kurash JK, Lei H, Gay F, Bajko J, et al. The lysine demethylase LSD1 (KDM1) is required for maintenance of global DNA methylation. *Nat Genet* 2008;**41**:125–9.
37. Anokye-Danso F, Trivedi C, Juhr D, Gupta M, Cui Z, Tian Y, et al. Highly efficient miRNA-mediated reprogramming of mouse and human somatic cells to pluripotency. *Cell Stem Cell* 2011;**8**:376–88.
38. Marson A, Levine SS, Cole MF, Frampton GM, Brambrink T, Johnstone S, et al. Connecting microRNA genes to the core transcriptional regulatory circuitry of embryonic stem cells. *Cell* 2008;**134**:521–33.
39. Fuhrmann G, Chung A, Jackson K, Hummelke G, Baniahmad A, Sutter J, et al. Mouse germline restriction of Oct4 expression by germ cell nuclear factor. *Dev Cell* 2001;**1**:377–87.
40. Young RA. Control of the embryonic stem cell state. *Cell* 2011;**144**:940–54.
41. Voorhoeve PM, le Sage C, Schrier M, Gillis AJM, Stoop H. A genetic screen implicates miRNA-372 and miRNA-373 as oncogenes in testicular germ cell tumors. *Cell* 2006;**124**: 1169–81.
42. Subramanyam D, Lamouille S, Judson R, Liu J, Bucay N, Derynck R, et al. Multiple targets of miR-302 and miR-372 promote reprogramming of human fibroblasts to induced pluripotent stem cells. *Nat Biotechnol* 2011;**29**:443–8.
43. Hajkova P, Erhardt S, Lane N, Haaf T, El-Maarri O, Reik W, et al. Epigenetic reprogramming in mouse primordial germ cells. *Mech Dev* 2002;**117**:15–23.
44. Szabó PE, Mann JR. Biallelic expression of imprinted genes in the mouse germ line: implications for erasure, establishment, and mechanisms of genomic imprinting. *Genes Dev* 1995;**9**:1857–68.
45. Stöger R, Kubicka P, Liu CG, Kafri T, Razin A, Cedar H, et al. Maternal-specific methylation of the imprinted mouse Igf2r locus identifies the expressed locus as carrying the imprinting signal. *Cell* 1993;**73**:61–71.
46. Tremblay KD, Saam JR, Ingram RS, Tilghman SM, Bartolomei MS. A paternal-specific methylation imprint marks the alleles of the mouse H19 gene. *Nat Genet* 1995;**9**:407–13.
47. Kim K, Doi A, Wen B, Ng K, Zhao R, Cahan P, et al. Epigenetic memory in induced pluripotent stem cells. *Nature* 2010;**467**:285–90.
48. Bestor TH. The DNA methyltransferases of mammals. *Hum Mol Genet* 2000;**9**:2395–402.
49. Carlson LL, Page AW, Bestor TH. Properties and localization of DNA methyltransferase in preimplantation mouse embryos: implications for genomic imprinting. *Genes Dev* 1992;**6**: 2536–41.
50. Vassena R, Dee Schramm R, Latham KE. Species-dependent expression patterns of DNA methyltransferase genes in mammalian oocytes and preimplantation embryos. *Mol Reprod Dev* 2005;**72**:430–6.
51. Oswald J, Engemann S, Lane N, Mayer W, Olek A, Fundele R, et al. Active demethylation of the paternal genome in the mouse zygote. *Curr Biol* 2000;**10**:475–8.

52. Feng Q, Lu SJ, Klimanskaya I, Gomes I, Kim D, Chung Y, et al. Hemangioblastic derivatives from human induced pluripotent stem cells exhibit limited expansion and early senescence. *Stem Cell* 2010;**28**:704–12.

53. Koh KP, Yabuuchi A, Rao S, Huang Y, Cunniff K, Nardone J, et al. Tet1 and tet2 regulate 5-hydroxymethylcytosine production and cell lineage specification in mouse embryonic stem cells. *Cell Stem Cell* 2011;**8**:200–13.

54. Weaver JR, Susiarjo M, Bartolomei MS. Imprinting and epigenetic changes in the early embryo. *Mamm Genome* 2009;**20**:532–43.

55. Conticello SG, Langlois MA, Yang Z, Neuberger MS. DNA deamination in immunity: AID in the context of its APOBEC relatives. *Adv Immunol* 2007;**94**:37–73.

56. Morgan HD, Dean W, Coker HA, Reik W, Petersen-Mahrt SK. Activation-induced cytidine deaminase deaminates 5-methylcytosine in DNA and is expressed in pluripotent tissues: implications for epigenetic reprogramming. *J Biol Chem* 2004;**279**:52353–60.

57. Privat E, Sowers LC. Photochemical deamination and demethylation of 5-methylcytosine. *Chem Res Toxicol* 1996;**9**:745–50.

58. Ito S, D'Alessio AC, Taranova OV, Hong K, Sowers LC, Zhang Y. Role of Tet proteins in 5mC to 5hmC conversion, ES-cell self-renewal and inner cell mass specification. *Nature* 2010;**466**: 1129–33.

59. Parry D, Bates S, Mann DJ, Peters G. Lack of cyclin D-Cdk complexes in Rb-negative cells correlated with high levels of p16INK4/MTS1 tumor suppressor gene product. *EMBO J* 1995;**14**:503–11.

60. Quelle DE, Zindy F, Ashmun RA, Sherr CJ. Alternative reading frames of the NK4a tumor suppressor gene encode two unrelated proteins capable of inducing cell cycle arrest. *Cell* 1995;**83**:993–1000.

61. Kamijo T, Zindy F, Roussel MF, Quelle DE, Downing JR, Ashmun RA, et al. Tumor suppression at the mouse INK4a locus mediated by the alternative reading frame product p19ARF. *Cell* 1997;**91**:649–59.

62. Li Y, Pei J, Xia H, Ke H, Wang H, Tao W. Lats2, a putative tumor suppressor, inhibits G1/S transition. *Oncogene* 2003;**22**:4398–405.

63. Becker KA, Ghule PN, Therrien JA, Lian JB, Stein JL, van Wijnen AJ, et al. Self-renewal of human embryonic stem cells is supported by a shortened G1 cell cycle phase. *J Cell Physiol* 2006;**209**:883–93.

64. Cowan CA, Klimanskaya I, McMahon J, Atienza J, Witmyer J, Zucker JP, et al. Derivation of embryonic stem-cell lines from human blastocysts. *N Engl J Med* 2004;**350**:1353–6.

65. Hanna J, Saha K, Pando B, van Zon J, Lengner CJ, Creyghton MP, et al. Direct cell reprogramming is a stochastic process amenable to acceleration. *Nature* 2009;**462**:595–601.

66. Jiang J, Chan YS, Loh YH, Cai J, Tong GQ, Lim CA, et al. A core Klf circuitry regulates self-renewal of embryonic stem cells. *Nat Cell Biol* 2008;**10**:353–60.

67. Rowland BD, Bernards R, Peeper DS. The KLF4 tumor suppressor is a transcriptional repressor of p53 that acts as a context-dependent oncogene. *Nat Cell Biol* 2005;**7**:1074–82.

68. Kim J, Chu J, Shen X, Wang J, Orkin SH. An extended transcriptional network for pluripotency of embryonic stem cells. *Cell* 2008;**132**:1049–61.

69. Nandan MO, Yang VW. The role of Krüppel-like factors in the reprogramming of somatic cells to induced pluripotent stem cells. *Histol Histopathol* 2009;**24**:1343–55.

70. Brons IG, Smithers LE, Trotter MW, Rugg-Gunn P, Sun B, de Sousa Chuva, et al. Derivation of pluripotent epiblast stem cells from mammalian embryos. *Nature* 2007;**448**:191–5.

71. Tesar PJ, Chenoweth JG, Brook FA, Davies TJ, Evans EP, Mack DL, et al. New cell lines from mouse epiblast share defining features with human embryonic stem cells. *Nature* 2007;**448**: 196–9.

72. Banito A, Rashid ST, Acosta JC, Li S, Pereira CF, Geti I, et al. Senescence impairs successful reprogramming to pluripotent stem cells. *Genes Dev* 2009;**23**:2134–9.
73. Jacobs JJ, Kieboom K, Marino S, DePinho RA, van Lohuizen M. The oncogene and Polycomb-group gene bmi-1 regulates cell proliferation and senescence through the ink4a locus. *Nature* 1999;**397**:164–8.
74. Li H, Collado M, Villasante A, Strati K, Ortega S, Cañamero M, et al. The Ink4/Arf locus is a barrier for iPS cell reprogramming. *Nature* 2009;**460**:1136–9.
75. Marión RM, Strati K, Li H, Murga M, Blanco R, Ortega S, et al. A p53-mediated DNA damage response limits reprogramming to ensure iPS cell genomic integrity. *Nature* 2009;**460**:1149–53.
76. Utikal J, Polo JM, Stadtfeld M, Maherali N, Kulalert W, Walsh RM, et al. Immortalization eliminates a roadblock during cellular reprogramming into iPS cells. *Nature* 2009;**460**:1145–8.
77. Won J, Yim J, Kim TK. Sp1 and Sp3 recruit histone deacetylase to repress transcription of human telomerase reverse transcriptase (hTERT) promoter in normal human somatic cells. *J Biol Chem* 2002;**277**:38230–8.
78. Zhu O, Liu C, Ge Z, Fang X, Zhang X, Straat K, et al. Lysine-specific demethylase 1 (LSD1) is required for the transcriptional repression of the telomerase reverse transcriptase (hTERT) gene. *PLoS One* 2008;**3**:e1446.

From Ontogenesis to Regeneration: Learning how to Instruct Adult Cardiac Progenitor Cells

Isotta Chimenti,[*] Elvira Forte,[†] Francesco Angelini,[*] Alessandro Giacomello,[†] and Elisa Messina[†]

[*]Department of Medical Surgical Sciences and Biotechnologies, "Sapienza" University of Rome, Latina, Italy

[†]Department of Molecular Medicine, Pasteur Institute, Cenci-Bolognetti Foundation, "Sapienza" University of Rome, Rome, Italy

Since the first observations over two centuries ago by Lazzaro Spallanzani on the extraordinary regenerative capacity of urodeles, many attempts have been made to understand the reasons why such ability has been largely lost in metazoa and whether or how it can be restored, even partially. In this context, important clues can be derived from the systematic analysis of the relevant distinctions among species and of the pathways involved in embryonic development, which might be induced and/or recapitulated in adult tissues. This chapter provides an overview on regeneration and its mechanisms, starting with the lesson learned from lower vertebrates, and will then focus on recent advancements and novel insights concerning regeneration in the adult mammalian heart, including the discovery of resident cardiac progenitor cells (CPCs). Subsequently, it explores all the important pathways involved in regulating differentiation during development and embryogenesis, and that might potentially provide important clues on how to activate and/or modulate regenerative processes in the adult myocardium, including the potential activation of endogenous CPCs. Furthermore the importance of the stem cell niche is discussed, and how it is possible to create *in vitro* a microenvironment and culture system to provide adult CPCs with the ideal conditions promoting their regenerative ability. Finally, the state of clinical translation of cardiac cell therapy is presented. Overall, this chapter provides a new perspective on how to approach cardiac regeneration, taking advantage of important lessons from development and optimizing biotechnological tools to obtain the ideal conditions for cell-based cardiac regenerative therapy.

I. Regeneration: From Urodeles and Teleosts to Mammals

Living organisms are self-sufficient systems, able to maximize the exploitation of available energy and also to efficiently provide for their own maintenance and, possibly, repair. The ability to regenerate damaged tissues or even amputated organs is a skill widespread in the animal kingdom. Metazoa are able to reconstruct complex body structures and considerable parts of their bodies, but among vertebrates only urodeles, amphibians, and teleost fishes have exceptional regenerative potential, being able to efficiently rebuild a wide variety of tissues and organs, including whole limbs, tail, retina, jaw, and heart. Salamanders and zebrafish are surely among the most striking examples of heart regeneration occurring (as in the other organs) through the formation of the so-called blastema at the wound surface.[1] This is a zone of mesenchymal proliferation that gradually redifferentiates in all damaged tissues by precise axial identity, through a considerably autonomous morphogenesis. This environment, which is created on the lesion by migration of adjacent epithelial cells, can promote cell dedifferentiation even on implanted heterologous cells.[2] It is a very special tissue, able to generate new progenitors and to start a new morphogenetic process, as occurs during development. There seems to be an intrinsic difference in responsiveness to dedifferentiation stimuli for vertebrates with high regenerative capacities compared to mammalian cells, and this difference may lie both in the ability to produce different signals and in the responsiveness to these signals. In particular, a secondary enzymatic activity of a proteolytic product of thrombin (one of the final effectors of the coagulation cascade) has been identified in the serum of blastema-forming urodeles. Such ligand, generated by the cleavage of serum components, can initiate a cascade able to switch on the resumption of the cell cycle, promoting directly or indirectly the phosphorylation of the retinoblastoma (Rb) protein[3] (a cell cycle inhibitor which is hyperphosphorylated in myotubes of newt after exposure to high concentrations of serum.[4]) Even though this proteolytic product can be also generated in mammalian sera, it has no activity on mammalian myotubes. So there appears to be a close link between the very early stages of wound healing, namely coagulation, and the following induction of cells to exit from a quiescent state to proliferate and redifferentiate via the inactivation of one of the key cell cycle negative regulators.

However, as well discussed by Ausoni et al.,[1] while the coagulation path is aimed at achieving heart repair and regeneration in urodeles, conversely, it leads to rapid hemostasis and fibrosis in mammals, which is in fact the physiopathologic trait of the human heart's inability to self-regenerate after an acute or chronic damage.

Indeed, the distribution of regenerative ability in the phylogenetic tree suggests that it may be an ancestral trait suppressed, or at least discouraged, by evolution.[1,5] One may speculate that the more complex the physiology and biology of an organism, the more risky and relatively less beneficial it would be to rebuild damaged parts: the time required to repair a complex structure would prolong too much the period in which the animal is very vulnerable, and then, beyond a certain level of evolution, regeneration does not appear to be an advantageous trait, considering the long-term risk of oncogenic transformation or the immediate occurrence of catastrophic events in case of acute vascular damages. In fact, high regenerative abilities in mammals are observed in a limited number of tissues, such as liver and epidermis, and they are largely limited by an irreversible program of differentiation.

Taking into account all these considerations, this chapter focuses on the issues currently debated in the field of cardiac regenerative medicine and reports the state of the art concerning the origin of cardiac progenitors as well as their molecular and functional properties useful for the assessment and improvement of their clinical translation potential. Particular attention is given to the important role of comparative and developmental biology studies in the troublesome task of "rebuilding" the heart. In fact, important insights into the genetic pathways and mechanisms that control myocardial differentiation during ontogenesis are strongly contributing to the advancements in this field, both in basic and translational research.

II. To the "Heart" of the Problem

The dogma of the human heart as a postmitotic organ has been challenged since the late 1990s by the discovery of myocytes and immature cells in mitosis, both in physiological and pathological conditions.[6] In this regard, data have been collected since then, by means of multiple approaches. Sometimes inconsistent results have been obtained, but since the physiologic turnover seems to be very low, it is not surprising that different strategies might have different success rates in detecting such an elusive process. Despite this, in recent years it has become more and more evident that new cardiomyocytes are physiologically generated over the life of adult mammals. Recent experiments on genetic fate mapping in mice[7] and ^{14}C isotope dating in the nucleus of cardiomyocytes in humans[8] have unequivocally demonstrated a slow but detectable turnover in the postnatal heart cell population in mammals. Other studies with alternative approaches have drawn different conclusions, suggesting much higher turnover rates[9]; therefore, further data is needed to settle such controversies. Furthermore, since 2003, several populations of resident cardiac stem cells (CSCs) and/or cardiac progenitor cells (CPCs) have been identified and/or isolated from the adult mammalian heart. Some of these populations are able to significantly proliferate *ex vivo* and to efficiently give rise to new cardiomyocytes and vascular cells when reinjected into animal models of heart failure (HF).[10] Despite this, the mammalian heart is clearly unable to regenerate or repair itself after significant injuries, such as ischemic events, and that is the obvious reason why regenerative medicine offers great promise in this field for the understanding of the biology of (non)regeneration, together with searching for novel approaches to overcome the heart's physiological limits.

One of the most interesting models of heart regeneration is that of the zebrafish, which has been shown to be able to efficiently regenerate up to 20% of the ventricular mass.[11] Such efficient cardiac muscle regeneration provides a model for understanding how natural heart regeneration may be blocked or enhanced in different species. By genetic tracing of cardiomyocyte lineage in the adult fish, it has been established that regenerated heart muscle cells after significant ventricle resection are derived from the proliferation of previously differentiated cardiomyocytes. Such proliferating cardiomyocytes undergo limited dedifferentiation, manifested by disassembly of sarcomeric structures as well as detachment and activation of cell cycle promoters. Such data provided the first direct evidence that stem/progenitor cells are not significantly responsible for the regenerative process.[12]

Using genetic fate-mapping approaches as well, a population of cardiomyocytes from the subepicardial ventricular layer has been identified, which becomes activated after resection of the ventricular apex. They express the embryonic cardiogenesis gene Gata4 soon after trauma, which localizes into proliferating cardiomyocytes at the injury site, and contribute prominently to cardiac muscle regeneration. Between 2 and 4 weeks post-injury, electrical conduction is reestablished between the existing and regenerated cardiomyocytes.[13]

The adult mammalian heart shows several morphological differences compared to the adult zebrafish or newt heart. The overall structure in mammals is more complex, based on four chambers and two separated blood circulations (pulmonary and systemic). In contrast, the adult fish has only two chambers and one single blood circulation. At the cellular level, zebrafish's cardiomyocytes, despite being terminally differentiated, are small, mononucleated, and able to proliferate, while differentiated cardiomyocytes in mammals withdraw from the cell cycle shortly after birth, and, in many species, become binucleated or multinucleated gradually in the postnatal life. Nevertheless, the embryonic mammalian heart is much more similar to the adult zebrafish heart, since it contains mononucleated diploid cardiomyocytes still able to proliferate through disassembling of the contractile structures, as also clearly demonstrated by the ability to compensate for the loss of cardiac tissue during embryogenesis.[14]

With these premises, Porrello et al.[15] tested whether the early postnatal murine heart could regenerate like that of the zebrafish and newt. They removed the apex of the left ventricle (approximately 15% of ventricular myocardium) of 1-day-old neonatal mice in hypothermia, thus preventing excessive blood loss during surgery. By 3 weeks, the lesion was completely healed and no functional differences were detectable compared to sham-operated controls. Regeneration progressed together with activation of epicardial genes, widespread cell proliferation in the whole heart, and disorganization of sarcomeric structures. No signs of fibrosis or hypertrophy were observed in the regenerating apex, further indicating a regenerating rather then repairing process. Such regenerative capacity is already lost by postnatal day 7, notably corresponding to the timing for binucleation and withdrawal from the cell cycle for mouse cardiomyocytes.

Based on the latter three studies cited, from a mechanistic point of view, heart regeneration in zebrafish seems to be very similar to the regenerative process observed for the early postnatal murine heart. In both organisms, cardiac cells detached and disassembled their sarcomeric structures, which, as pointed out before, is a required process before cell division. Therefore, the main regenerative process seems to be activation, dedifferentiation, and cell cycle reentry of preexisting cardiomyocytes. There is one study, though, in contrast with this hypothesis, based on genetic fate mapping,[7] which could not detect a physiologic cardiomyocyte turnover during postnatal life in mice, although the incomplete efficiency of the transgenic reporter might be responsible for the inability to detect a very slow rate of cardiomyocyte turnover. Certainly, a deeper understanding of the biology and pathways of regeneration in zebrafish or amphibians will provide key insights on how to improve and/or tune certain processes in mammals as well, such as cell cycle reentry and progenitors activation.

III. Lessons from Development

The vast majority of clinically relevant cardiovascular diseases result from the death of cardiac cells, which are not promptly replaced given the poor regenerative capacity of the heart, or from defects in cardiogenesis, which lead to a wide range of congenital heart anomalies, including major life-threatening defects (such as cardiac arrhythmias or failure to separate systemic and pulmonary circulatory systems). In order to reverse the process of heart disease progression and to regenerate lost tissue, a deep knowledge of the events occurring during cardiac development is undoubtedly useful and required.

The identification of tissue-specific enhancer/promoters and the advances in conditional gene targeting have allowed *in vivo* tracking of specific cell types at almost any stage of cardiovascular development.[16] This approach enables the identification of stem cells and related progenitors in the context of the intact heart *in situ*, without the need for direct injection or transplantation[17–19] and has contributed significantly to our understanding of cell lineage diversification within the heart.

Furthermore, during differentiation of embryonic stem cells (ESCs), cardiovascular cells are generated spontaneously through a biological process that recapitulates the cellular and molecular events normally occurring during embryonic development.[20,21] This observation has created great enthusiasm among developmental biologists, outlining the possibility to study the complex cellular and molecular mechanisms governing cardiovascular differentiation and cardiovascular diseases through a reductionist *in vitro* approach, as well as to generate cardiovascular cells for cellular therapy and drug or toxicity screening.[21,22]

Until recently, the formation of cardiac smooth muscle and endothelial lineages in the heart has been ascribed to distinct populations of embryonic progenitors. However, the existence of a common progenitor, for both muscle and nonmuscle lineages of the heart, is now well accepted (Fig. 1). The earliest precursors for heart-forming cells generate in the vertebrate mesoderm at the onset of gastrulation, when a transition from the expression of the T-box transcription factor Brachyury T (Bry) to expressing mesoderm posterior (Mesp)1 occurs.[23,24]

Gastrulation is a critical phase in embryonic development which takes place within 5–7 days after fertilization, depending on the species. Future mesoderm and definitive endoderm cells delaminate from the ectoderm along a furrow—the primitive streak—to form distinct germ layers. In addiction, cell fates are defined and the organization of a general body plan, which serves as a blueprint of the embryonic body, is outlined. The heart is the first functional organ formed during organogenesis. Cells destined to become cardiac mesoderm cells are fated, induced, and specified before and during gastrulation. The first precardiac cells to ingress the primitive streak are Mesp1 positive. Cell lineage, loss of function, and chimeric analyses have shown that all cells of the future heart, as well as cells of the main vessels, derive from cells that have expressed Mesp1 at one point during embryonic development.[24,25] In addition to being the earliest marker of cardiovascular development, Mesp1

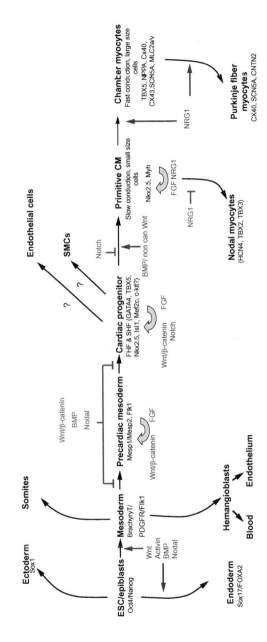

FIG. 1. Cardiogenic lineage differentiation hierarchy. A schematic overview of the signaling pathways known to positively (green) or negatively (red) regulate the progression through the different stages of cardiac differentiation during embryogenesis. (For interpretation of the references to color in this figure legend, the reader is referred to the online version of this chapter.)

also plays a very important role during the earliest step of cardiovascular differentiation: in fact, the combined deletion of Mesp1 and Mesp2 leads to the absence of mesoderm and cardiac specification.[23]

Many reports have documented the critical roles of intercellular signaling pathways in cellular specification (Fig. 1). BMPs, Wnts, and the TGFβ family member Nodal play evolutionarily conserved roles in inducing mesoderm and endoderm; thus, they are important to promote cardiogenic differentiation and, for this reason, are widely used to initiate cardiogenesis in ESC cultures (for example, see Refs. 26–28). However, they exert an inhibitory effect as gastrulation ensues when uncommitted Flk1[+]/MesP1[+] progenitors adopt a cardiac fate. In this context, extracellular inhibitors of BMPs (Noggin, Cerberus-1), Wnts (Dkk1), and Nodal (Cerberus-1) are evolutionarily conserved inducers of cardiac mesoderm,[29,30] committing multipotent progenitors (Flk1[+]/MesP1[+]) to a cardiac fate, as marked by the heart-forming transcription factors Nkx2.5, Isl1, Tbx1, Tbx5, and Mef2c. Nodal, BMP, and Wnt are broadly expressed in the embryo and do not block differentiation of vascular endothelial and smooth muscle cells, which arise from the same progenitor pool; thus, Cer1 and Dkk1 (and possibly other antagonists) effectively distinguish areas of potency for heart versus systemic vasculature.[26] BMPs are needed subsequently for differentiation of committed precursors[31]; therefore, the exposure to these factors must be tightly regulated.

Once the heart field is established, proliferation and fate of competent progenitors are controlled by signals such as FGFs and Notch. In particular, Notch suppresses cardiogenesis through RBPj, as shown in ESC cultures and Xenopus embryos.[32–34] Since Notch can sustain proliferation of progenitor cells by blocking their differentiation, it is not surprising that the absence of Notch has been linked to precocious differentiation and exhaustion of resident progenitors or precursors, such as skeletal muscle satellite cells[35] and epithelial precursors of the developing pancreas.[36] In summary, Notch plays an important gatekeeper role, and manipulation of Notch signaling could help expand the cardiopoietic progenitor pool in the adult.

Not all cardiac progenitors develop identically. Mesp1[+] cardiac mesodermal cells segregate into two cardiogenic heart fields which lie contiguously along their entire length[37] (Fig. 1); so it is likely that their specification occurs by the same molecules at the same time. However, differentiation of the second field is delayed relative to the first. In fact, the first population of cells migrating to the heart-forming region, referred to as the first or primary heart field (FHF), originates bilaterally in the anterior splanchnic mesoderm and gives rise to a group of cardiovascular precursors that form a crescent-shaped epithelium (cardiac crescent) at the cranial and craniolateral parts of the embryo, and that will ultimately contribute to both atria and to all cardiomyocytes of the left ventricle. No unique molecular marker has been identified for this field, although Tbx5 expression has been associated with FHF derivatives.[38] Progenitors in the cardiac crescent become organized into a two-layered epithelium within the splanchnopleure, and then the two cardiogenic areas fold around the primitive foregut, forming the primary heart tube. Mesenchymal cells arising from this delamination give rise to the endothelial cell lining of the heart, thus forming an endocardial tube surrounded by the myocardial epithelium. Nkx2.5,[39] Flt1,[40] Mef2b/c,[41] Gata4,[42,43] and dHand/eHand[44] are all associated with cardiogenesis and are expressed in the crescent-shaped heart primordium, myocardium, and/or endocardium.

The inflow region of the linear heart tube is located caudally, and the outflow region cranially. Subsequently, this primary structure undergoes a complex progression termed cardiac looping, in which it adopts a spiral shape with its outer surface sweeping rightward, the inflow portion of the heart, including the common atrium, being forced dorsally and cranially so that it is now above the developing ventricles.[45] As the linear heart tube elongates and undergoes rightward looping, cardiovascular progenitors from the second or anterior heart field (SHF) migrate into the heart tube, contributing cells that will form the main parts of the atria, the right ventricle, and the outflow tract of the myocardium.[46–48] The SHF appears in the extracrescent tissue and is largely derived from a population of cells located dorsally and in front of the heart tube, in the pharyngeal and splanchnic mesoderm. The delayed differentiation of the SHF is probably due to a higher and sustained exposure to Wnts, produced by the neural plate, compared to the FHF cells located more laterally. In fact, Wnt/β-catenin signaling maintains SHF cells in a proliferative state until they are needed to build the right ventricle and outflow tract.[49–52] Lineage tracing experiments have demonstrated that most of the SHF-derived cells can be traced back to multipotent heart progenitors that express the LIM-homeodomain transcription factor Isl1.[19,47] A subset of undifferentiated Isl1$^+$ cells remains embedded in the embryonic heart after its formation, and a few cells are still detectable shortly after birth.[18] Given that the most common cause of human HF is the loss of functional muscle cells in the left ventricular wall, it would be particularly relevant to determine whether FHF progenitors also persist after birth. Random labeling of cardiac precursors during embryonic development has revealed the existence of rare clones that contributed to both FHF and SHF lineages and could represent a common cardiovascular progenitor for both heart fields.[53] Lineage tracing experiments have led to the identification of such a common progenitor, which can be marked and isolated based on the expression of Brachyury (Bry, a mesodermal marker) and Flk1 (cell surface receptor that encodes VGFR2).[54] The Flk1$^+$ population appears to develop as cells exit the primitive streak and begin to migrate to form the cardiac crescent.[55] Studying ESCs differentiation, two different waves of Bry$^+$/Flk1$^+$ cells have been identified: the first one contributed to hemagioblasts and the second one to a population of multipotent progenitors with cardiomyocyte, vascular smooth muscle, and endothelial potential, which displayed molecular profiles consistent with cells of the primary and secondary heart fields. In fact, some colonies became positive for the SHF marker Isl1, whereas others were negative for Isl1 but expressed the FHF-associated factor Tbx5.

During the cardiac looping stage, delamination of the endocardium generates endocardial cushions (ECs) in the atrioventricular (AV) canal, which are precursors of the tricuspid and mitral valves. ECs will also give rise to the aorticopulmonary septum, which divides the outflow tract into the aorta and pulmonary artery, and to the aortic and pulmonary valves. Other features of this developmental stage are the formation of trabeculae, the spongiform layer of myocytes along the inner surface of the ventricles, and the interventricular septum.

Early cardiomyocytes formed in the heart tube are poorly striated, with few organized sarcomeres and a composition of voltage-gated channels consistent with an immature electrophysiological phenotype. As the looping heart tube continues into a four-chambered structure, there is a marked expansion in the thickness of the ventricular chamber wall due to the expansion of the compact zone, which contains the cells that represent the most proliferative cardiac myocytes in the mammalian heart.[56] A subset of these cells migrates out toward the interior of the chamber to form trabecular, fingerlike structures. Trabeculation increases the myocardial surface for proper oxygenation in the absence of an intact coronary circulation. These cells are more differentiated, showing an increased sarcomeric organization and a lower mitotic rate.[45,57] It is particularly important to understand the signals regulating these discrete steps of cardiomyocyte differentiation, in order to drive ESC/CPC-derived cardiomyocytes to a fully mature phenotype and prevent arrhythmias, which could derive from the implantation of immature cardiomyocytes. During the remodeling phase of heart development, division of the heart chambers by septation is completed, and distinct left and right ventricles and left and right atria are evident. This is achieved by further spiraling of the heart tube such that the outflow region becomes wedged between the developing ventricles on the ventral side and the inflow region spans the ventricles dorsally.[45] Muscular interatrial and interventricular septae fuse with the nonmuscular AV septum, which is derived from the endocardial cushions of the AV canal, thus completing the separation of the chambers.

Two additional sources contribute to heart formation: the cardiac neural crest (NC) and the proepicardium. Cardiac NC cells take part in the remodeling of the aortic arch arteries and the septation of the common outflow tract into the aortic and pulmonary arteries.[58,59] Lineage-tracing experiments suggest that NC cells also contribute to the conduction system and the epicardium, but this remains controversial.[60–63] Recent reports show that nestin$^+$ stem cells of NC origin reside in the adult heart and take part in *de novo* blood vessel formation and reparative fibrosis after ischemic injury.[64]

During the looping stage, cells of the proepicardium (a transient cauliflower structure from splachnopleuric mesoderm) attach to the exterior surface of the heart and spread out over the entire organ in a single epithelial cell layer called the epicardium.[65] Shortly thereafter, the epicardial epithelium generates a mesenchymal stem cell population named epicardial-derived progenitor cells, or EPDCs. EPDCs invade the cardiac tissue and differentiate into interstitial fibroblasts, perivascular fibroblasts, and smooth muscle cells of the developing coronary blood vessels.[66,67] Cell-lineage tracing experiments using Cre-recombinase technology have shown the potential contribution of the epicardium to a small fraction of ventricular myocytes, suggesting a more substantial role of EPDCs in heart tissue formation than previously thought.[68,69] In addition to contributing to the cellular makeup of the heart, the epicardium and myocardium mutually engage in both paracrine and direct cellular interactions that are required for the growth and development of each compartment. Consistently, during zebrafish regeneration, a widespread activation of the epicardium occurs, which contributes directly to neovascularization and produces important factors for the reactivation of a subset of subepicardial Gata4-positive cardiomyocytes within the ventricular wall.[12,13] A reactivation of the epicardium has been observed also in adult mammalian hearts in response to myocardial infarction (MI) or treatments with specific growth factors (Thymosin β4,[70–72] prokineticins[73]) and it contributes mainly to vascular cell types, even though a subpopulation of c-kit$^+$/CD34$^+$ epicardial progenitors has been identified[74] that can give rise also to Gata4$^+$/Nkx2.5$^+$ cardiac progenitors. It has been also recently shown that a small percentage of cardiomyocytes can be obtained from Wt1$^+$ cells in infarcted hearts of mice preventively primed with Thymosin β4.[75] Overall, such pathways involved in the epicardium/myocardium cross talk also provide important clues on how to turn on and/or encourage specific genetic programs toward cardiomyogenesis.

As previously mentioned, a key signaling pathway in the regulation of growth and differentiation of all cardiovascular cell lineages is the Notch signaling pathway.[76,77] In adult animals, Notch pathway regulates heart regeneration in zebrafish[78] and has been implicated in the injury-repair response of the mammalian heart.[79,80] In this respect, an interesting approach for the identification of progenitor cells activated in response to injury consists in the use of a transgenic Notch reporter mouse.[81] In addition to scattered endothelial and interstitial cells, "Notch-activated progenitors" were unexpectedly highly enriched in the adult epicardium, confirming the importance of this fibrous mesothelial covering in cardiac homeostasis. Notch-activated epicardial-derived cells are multipotent stromal cells derived from the epicardium by

EMT, and behave as a "tissue-repair module": they contribute to fibrosis by default, but are capable of expressing muscle genes if appropriately instructed, thus providing a platform for developing new cardioregenerative strategies and drugs.

Despite the poor regenerative capacity of the adult mammalian heart, several populations of endogenous CPCs have been isolated in recent years based on the expression of surface markers, such as c-kit[82] and sca1,[83] or based on functional properties, such as the ability to efflux Hoescht dye[84] or to spontaneously migrate out of cardiac explants and form multicellular clusters called cardiospheres (CSs).[85] Their developmental origin is not clear[6]: they may be remnants of cardiogenesis[86] or circulating cells originating from bone marrow, or they may arise via mesangioblasts from the surrounding vasculature.[83,87] No consensus view has emerged on their capacity for self-renewal and potential for forming cardiomyocytes versus other lineages. Nonetheless, these progenitor cell populations express an overlapping set of cardiogenic markers, regardless of the criteria used for their isolation. In particular, they share the expression of many heart-specific transcription factors (Nkx2.5, MEF2c, Gata4) even in their basal undifferentiated state, resembling an atypical form of cardiac mesoderm intermediate, which is transiently expressed in embryos and ESC-derived embryoid bodies.[26]

Learning from all the lessons mentioned above about development and embryogenesis, it is possible to speculate that evolutionarily conserved signals driving heart development might somehow be forcibly activated and/or encouraged in the context of the adult mammalian heart to improve the regenerative capacity of endogenous resident CPCs to be used as a therapeutic product in autologous cell therapy.

To date, in comparison with the wealth of information concerning directed differentiation of mouse and human ESCs, little has been done to analyze the role of known positive and negative regulators of myogenesis on postnatal heart-derived cells.

Knowledge derived from developmental biology studies turned out to be important also for the identification of transcription factors capable of inducing direct transdifferentiation of murine fibroblasts into induced cardiomyocytes.[88] Using microarray analyses, transcription factors and epigenetic remodeling factors were identified, with greater expression in mouse cardiomyocytes than in cardiac fibroblasts at embryonic day 12.5. Among them, 13 factors that caused severe developmental cardiac defects and embryonic lethality when mutated in animal models were selected; additionally, Mesp1 was included, given its cardiac transdifferentiation effect in Xenopus.[89] Using an approach similar to the one adopted by Yamanaka *et al.*,[90] it has been shown that three transcription factors, Gata4/Mef2c/Tbx5, are sufficient to obtain functional cardiomyocytes from postnatal cardiac and dermal fibroblasts, which resemble neonatal cardiomyocytes in their global gene expression profile and electrophysiological features. However, additional epigenetic regulators, microRNAs, or signaling proteins may be identified to increase the efficiency and robustness of the reprogramming event.

In conclusion, a systematic analysis of the signaling and genetic networks that control embryonic cardiogenesis can provide important information on how to modulate the proliferation and differentiation of both embryonic and adult-derived CSCs/CPCs, to improve therapeutic regeneration. One of the main challenges is, in fact, the ability to reactivate differentiative embryonic pathways in adult cells to redirect and/or control their commitment and function. In this respect, a wide knowledge of specific expression markers defining stem cell activation and/or differentiation stages is very useful to follow cell fate and maturation, but do not necessarily define in adult cells the same functional properties as during embryo development, especially when considering single markers or univocal patterns. Therefore, apart from an accurate understanding of gene expression and molecular pathway profiling, effective tools are needed to specifically stimulate regenerative/ontological functions in adult progenitor cells. To achieve this goal, protocols are needed to mimic *in vitro* the appropriate tissue-like microenvironment for the cells.

IV. Understanding the Stem Cell "Niche" and Its Roles

Stem cells have the ability to renew themselves and to differentiate into distinct mature cell types. This attribute is very important during embryonic development and plays a crucial role in tissue regeneration. Most tissues of the adult organism maintain a population of putatively slow-cycling stem cells that maintain homeostasis and respond to injury when challenged. These cells are regulated and supported by the surrounding microenvironment, which constitutes a three-dimensional (3D) so-called niche that regulates cell fate decision, and in which the balance between self-renewal and differentiation is finely regulated.[91]

Stem cells within niches undergo either symmetric divisions, leading to amplification of the cell population, or asymmetric divisions, resulting in a constant number of multipotent stem cells and, at the same time, the production of progenitor cells, which are further recruited to give rise to a differentiated progeny outside the niche. There is no specific structure, and different cell types provide the appropriate microenvironment, ranging from the single-cell-formed niche in *Caenorhabditis elegans* to more complicated microenvironments, such as those regulating stem cell populations in neural, epidermal, hematopoietic, and intestinal niches of higher organisms, such as mammals.[92]

Despite this variety, all niches regulate stem or progenitor cell fate through the same biological mechanisms: secreted factors and cell–cell and cell–extracellular matrix (ECM) contacts.[93]

The importance of secreted factors came to light for the first time when studying hematopoiesis *in vitro*. Many studies have shown how proliferation, differentiation, and survival of progenitor cells depend on factors secreted from the surrounding cells. Even though several signal molecules are involved in the maintenance of the balance between self-renewal and differentiation (including Wnts, BMPs, FGFs, Notch, SCF, Ang-1, LIF, and the JAK–Stat pathway), only BMP and Wnts signals have emerged as common pathways for controlling stem cells fate from *Drosophila* to mammals.[94]

In addition to secreted factors, cell–cell and cell–ECM interactions are also necessary for the establishment of the niche and for stem cell anchorage. Several studies have demonstrated that ECM is able to bind soluble growth factors that act as activating or inhibitory signals for cells in proximity. E-cadherin-mediated interactions appear to be essential for the retention of undifferentiated cells in the niche; they contribute to the maintenance of stem cell self-renewal potential and, in *Drosophila*, control symmetric and asymmetric divisions. Integrin-mediated ECM adhesion, instead, is responsible for the orientation of the cell division axis.[93]

The mammalian heart contains three main cell lineages derived from Isl1/nkx2.5/flk-1 common cardiovascular progenitor cells: cardiac myocytes, smooth muscle cells, and endothelial cells. During heart development, both in humans and mice, these CPCs are detectable in many clusters, which are reminiscent of a stem cell niche. Although cardiac niches are still poorly defined, it has been shown that the Wnt/β-catenin pathway controls the specification, renewal, and differentiation of isl1 + progenitors through their maturation from ESCs, and that β-catenin mediates the proliferation of embryonic progenitors in both heart fields.[95]

The Wnt family consists of cysteine-rich glycoproteins of approximately 350–400 amino acids that contain an N-terminal signal peptide for secretion. When Wnt binds its seven-pass transmembrane receptor (Frizzled or Fzd) and LDL-receptor-related proteins 5 and 6 (LRP5 and LRP6), a signal cascade is activated, resulting in displacement of the multifunctional kinase GSK-3β from the APC/Axin/GSK-3β complex. Normally, this complex promotes the proteolytic degradation of β-catenin, but Wnt binding prevents this degradation, and β-catenin can then reach the nucleus and interact with transcription factors of the TCF/LEF family. This interaction promotes the displacement of Groucho–HDAC corepressors and induces the transcription of early differentiation markers such as Tbx5 and Gata4[96] (Fig. 2). Thus, the Wnt pathway has a role in maintaining SHF cells in a proliferative state before their necessary differentiation to contribute to the right ventricle and the outflow tract (as described

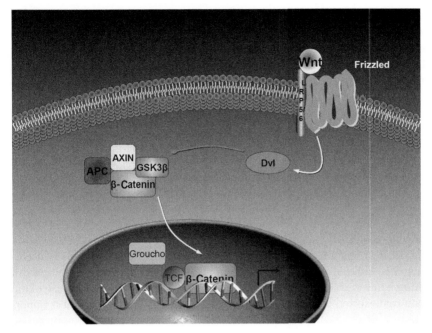

Fig. 2. Representation of the Wnt pathway, from receptor activation to gene expression modulation. (See Color Insert.)

earlier in this chapter) and provides an important example of a key developmental regulatory network that might be exploited to modulate genetic programs in adult tissues.

Cardiac niches in the adult mammalian heart have been described as interstitial structures, well distinguishable for their architectural organization, that harbor proliferating BrdU-positive cells.[97] Such niches contain CSCs and lineage-committed cells, which are connected to supporting cells represented by myocytes and fibroblasts. Gap and adherens junctions are detectable at the interface of early committed cells and supporting cells. Undifferentiated CSCs express alpha(4)-integrin, alpha(2)-chain of laminin, and fibronectin, and are able to divide symmetrically and asymmetrically.

Niche-like structures have been identified also in the subepicardium of adult mice using transmission electron microscopy.[98] Multiple cell lineages are present in these areas, such as adipocytes, fibroblasts, Schwann cells and nerve fibers, isolated smooth muscle cells, mast cells, macrophages, lymphocytes, interstitial Cajal-like cells (ICLCs), and CPCs. CPCs show typical features of immature cardiomyocytes, and in some cases are connected to each other to form columns surrounded by a basal lamina and embedded in a cellular network made by ICLCs. These latter cells and CPCs display complex intercellular communication through electron-dense nanostructures or through shed vesicles, supporting the hypothesis that ICLCs might act as supporting and regulatory nurse cells of the cardiac niches.

V. Recreating the Niche: The Importance of 3D Models

Undoubtedly, stem cell niches are highly complex structures that exhibit tissue-specific architecture and differ in their cellular and ECM components. For this reason, during the past years many studies have investigated and characterized tissue-specific niche microenvironments to understand the process that regulates cell maintenance and development, mainly to translate this knowledge into therapeutic applications. Cardiac niches, as all other adult niches, are regulatory structures, and this emphasizes the importance of creating a suitable *in vitro* biological system for adult and pluripotent stem cell cultures that is able to mimic all the appropriate niche features. Because of the main role exerted by cell–matrix contacts inside the niche *in vivo*, it is important to also re-create *in vitro* a sort of natural ECM with the appropriate structure.

Unfortunately, monolayer cell culture methods, such as microwell plates, tissue culture flasks, and Petri dishes, do not reflect the real physiology of tissues. As previously mentioned while describing the niche microenvironment, cells within a tissue undergo cell–cell and cell–ECM interactions that create an important 3D network through biochemical and mechanical signals. According to this evidence, the cellular environment is able to affect critical events in the cell cycle. Furthermore, the lack of a unique ECM microenvironment of each cell type can alter cell metabolism and reduce specific functionalities.

The use of a 3D cell culture model attempts to fulfill the above-mentioned requirements. In a 3D cell culture system, cells attach to one another and form natural cell-to-cell junctions. The ECM that they spontaneously secrete is the natural material to which the cells are attached. It is made of complex proteins in their native configuration, thus providing important biological instructions to the cells.[99] Under these conditions, cells can communicate and migrate as they naturally do *in vivo*. Gap junctions are much more prevalent in 3D than 2D models. The close proximity of cells also facilitates binding between surface adhesion molecules and surface receptors on adjacent cells.

The currently available 3D biological models include embryos and adult animals, organ slices, and cell cultures in ECM gels. Obviously, they provide tissue-specific information of different levels of complexity. Organ slices cultured on semiporous membranes or on 3D collagen gels encourage the natural cell disposition and differentiation, in a way similar to the normal organ, and they have found many applications in the study of the brain.[100]

Embryo culture analysis can provide precious information about cellular behavior in their normal environment, but the strict culture conditions required for mammalian embryos to maintain their normal cell physiology represent a strong technical limitation to their use.

Cell cultures represent a very standardized, but nevertheless artificial research model. From this perspective, cellular spheroids are straightforward 3D systems that exploit the natural tendency of many cell types to aggregate. It is possible to obtain spheroids from single cultures or cocultures of many cell types. They can be produced ranging from a simple semisuspension culture condition to a more complex rotating/wall vessel culture system. Spheroids have the not-so-obvious potential to aggregate in absence of an external scaffold and are considered, with increasing interest, as a promising vehicle for therapeutic biomedical studies.

Messina et al. were the first to describe a culture method to obtain spontaneous scaffold-free spheroids from resident adult CPCs, starting from human cardiac biopsies.[85] The cells that grow from biopsies include both resident CSCs and mesenchymal supporting cells. When these cells are collected and seeded on a poly-D-lysine coating, they grow as 3D floating colonies called CSs. CSs are able, once collected and seeded on fibronectin coating, to grow as monolayers, called cardiosphere-derived cells (CDCs),[101] which are able in the presence of a poly-D-lysine coating to form secondary CSs.[102]

CSs mimic in vitro in many aspects a niche microenvironment (Fig. 3). In fact, their architecture consists of a central core of c-kit$^+$ CSCs, surrounded by supporting CD105$^+$/c-kit$^-$ cells, and interlinked by ECM molecules, such as collagens IV, integrins, laminins, connexins, and vimentin.[85,103] Compared to those from a monolayer culture, cells of the CS have a lower proliferation rate, but the expression of stem cell markers is upregulated.[102–104]

In primary and secondary CSs, the proportion of c-kit$^+$ cells is higher compared to the same cells cultured as monolayers. Also, the expression of stem cell transcription factors such as Nanog and Sox2 is upregulated in 3D CSs,[103] as well as other factors such as IGF-1, HDAC-2, Dhh, Dll1, Dll3, and Tert, all of which are involved in maintaining a stemness state and managing reentry into the cell cycle.

Regarding the ECM and adhesion molecules such as laminin-β1, integrin-α2, and E-selectin, analogous profiles have been observed: indeed, mRNAs expression levels were increased in CSs. Since ECM and adhesion molecules are important for cell survival, these results consistently correlate to the better engraftment of stem cells in the form of CSs rather than CDCs when implanted in a damaged heart, as shown by Marban et al.[103,105] Moreover, to obtain an

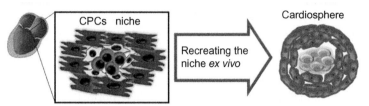

- CSCs + supporting cells (myocytes, fibroblasts, Cajal-like cells).

- Junctions (cadherins, connexins, integrins).

- Structural organization of cell types.

- Physiologic ECM.

- CPCs (c-kit$^+$) + supporting fibroblast-like cells (CD90+).

- Espression of connexin 43, cadherin 1, and integrin β2.

- BrdU+/c-kit$^+$ cell in the core; differentiation gradient.

- Production of ECM (collagens, fibronectin, laminins).

Fig. 3. The importance of recreating *in vitro* a niche-like microenvironment. Multiple features of the adult cardiac niche are reproduced, at least partially, in the 3D culture system of the cardiosphere, which seems to provide ideal biological, morphological, and functional stimuli for CPC culture and maturation. (See Color Insert.)

injectable suspension, collection of CSs is carried out without enzymatic digestion, while for monolayers it is necessary to use trypsinization, which may damage the ECM and adhesion molecules and undermine their functionality and following engraftment.

From this point of view, CSs, and in general all 3D cell culture forms, may represent a powerful way to increase stemness features and expression of ECM and adhesion molecules, resulting in a higher survival rate and increased regenerative potential for repair of damaged tissue.

VI. Cardiac Cell Therapy: The Era of Clinical Trials

HF and coronary heart disease are major risk factors for health and one of the greatest social costs for the Western World. Despite continuous advancements in drug and assisted therapy, conventional medical treatments for HF are not able to correct the related degenerative mechanism, that is, the loss and reduction in the number of heart muscle cells that respond to myocardium damage with hypertrophy and/or formation of scar tissue. Heart transplantation is currently the only treatment option for end-stage HF patients, but it is hampered by severe shortage of organs and immunological issues. Therefore, there is a pressing need to develop a therapy to overcome this deficiency and obtain efficient tissue repair, in addition to supporting the surviving healthy tissue.

In regenerative medicine, which by now has been successfully applied in many fields, cardiac cell therapy has recently become a realistic option that involves transplantation of cells with regenerative capabilities to restore (at least in part) damaged cardiac tissue as well as to improve blood supply and contractile function. Preclinical animal protocols and early clinical studies have clearly shown that adult stem cells of different origins may improve cardiac function following acute or chronic ischemia.[106,107]

Among the first regenerative candidates are ESCs, which possess infinite proliferative potential and are able to differentiate effectively in functional cardiomyocytes. Their use, however, presents major obstacles. First, ethical issues prevent the use of human ESCs in many countries. Second, they are allogeneic, and therefore imply immunosuppressive therapy. In addition, this type of cells, when undifferentiated, has a high and risky potential to form teratomas. All this has led through the years to alternative sources and has increased the interest in adult stem cells isolated from various tissues (bone, skeletal muscle, adipose tissue), such interest being related to the possibility they offer for autologous transplantation. However, many fierce disputes remain regarding the reproducibility of results, the contamination of samples with other stem cells, and the possibility that the beneficial effects and the expected plasticity observed in many animal models might actually be due to cell fusion and/or indirect mechanisms, especially paracrine.[108] Various cell types have been tested in animal models of MI, resulting in functional improvements, and adult stem cells, such as skeletal myoblasts and bone-marrow-derived stem cells, have moved quickly to clinical trials.

In the year 2000, the first clinical trials were started for cardiac cell therapy with bone-marrow-derived cells (hematopoietic, mesenchymal, mononuclear) or skeletal myoblasts in patients with acute MI, refractory angina, or chronic HF.[107,109]

For patients with acute MI, intracoronary infusion of mononuclear bone marrow cells, immediately after reopening the coronary artery involved, is the strategy that has been applied more widely. A recent meta-analysis of 13 randomized trials comprising a total of 811 patients reported a 2.99% increase in LVEF (left ventricular ejection fraction), a 4.74 mL reduction in LVESV (left ventricular end systolic volume), and 3.51% reduction in the lesion area.[110] However, several other large-scale trials, such as HEBE, REGENT, and SCAM, have not collected such encouraging data, often without detecting any increase in LVEF. Taken together, these data clearly show that the potential benefit of therapy with bone-marrow-derived stem cells shortly after an acute MI is still conflicting, and there is a concrete need for extended trials that incorporate the main findings of previous studies, concerning cell preparation as well as dose and timing of reinfusion, but focusing on clinically relevant endpoints, such as mortality, subsequent heart attacks, and HF.

The picture is different for patients with refractory angina who have exhausted conventional revascularization therapies, and for whom we can reasonably expect an increase in angiogenesis due to growth factors released by bone marrow cells, in particular the CD34+ fraction.[109] In this context, the most convincing evidence of cell therapy efficacy has come from randomized endoventricular catheter injections of CD34+ progenitors after mobilization induced by G-CSF. This study showed trends toward efficacy endpoints (angina frequency, nitroglycerin dosage, time of operation, Canadian Cardiovascular Society class) compared to placebo. A Phase IIb study is currently under way and will help confirm previous results in a larger patient population. Similar results were obtained in another study that included a similar group of patients with chronic ischemia refractory to medical treatment, who were treated with endocardial injection of unfractionated bone marrow mononuclear cells.

In patients with chronic HF, however, the results of the first approaches of cell therapy with bone marrow or skeletal cells were more heterogeneous: there was a reduction in symptoms associated with chronic insufficiency, but it was not possible to detect statistically significant differences of improved ventricular function compared to controls, with the exception of two studies in which much higher cell doses were used.[107,109] In addition, skeletal myoblast treatments often led to the development of potentially risky arrhythmias, seriously questioning the convenience of further attempts with this cell source.

Despite some initial enthusiastic disclosures, intracoronary infusion of mononuclear cells in patients with ischemic and nonischemic cardiomyopathy showed little or no efficacy, raising serious questions about the actual efficiency in cell engraftment long after an acute event, when recruitment and homing signals have probably ceased.[109]

Between 2008 and 2009, the first trials started to test the therapeutic potential of autologous resident CPCs in postinfarction patients with left ventricular dysfunction (see Table I and www.clinicaltrials.gov for details about each trial). Such approach is based on the strong rationale that the beneficial effects observed for noncardiac adult stem cells are transitory and have been shown to be indirect (i.e., paracrine) without real cardiac transdifferentiation and regeneration.[106–108] In fact, resident CPCs are autologous as well, but more importantly are intrinsically committed toward cardiac lineages, being able to efficiently differentiate *in vivo* in animal models of cell therapy into cardiomyocytes, smooth muscle, and endothelial cells.[10,85,101,103,105–107,117,118]

The CADUCEUS trial[113] consists of intracoronary infusion of CPCs isolated through the CS protocol.[85,101,114] Such therapy has succeeded in reducing scar size, increasing healthy heart muscle mass, improving regional

Ongoing Clinical Trials Employing Autologous Resident CPCs

Clinical trial	Trial number	Start date	Patients	Cell type	Procedure	Outcome	References
SCIPIO (Cardiac Stem Cell Infusion in Patients With Ischemic CardiOmyopathy)	NCT00474461	Feb 2009	Age: 18 to 75 years Diagnosis: post-MI LV dysfunction LVEF <40% before CABG	Autologous c-kit$^+$/lin$^-$ CPCs	1×10^6 autologous CPCs administered by intracoronary infusion at a mean of 113 days (SE 4) after surgery; controls were not given any treatment.	No adverse effects were reported. 1 year: LVEF ↑ by 12.3 ejection fraction units versus baseline (8 patients, $p = 0.0007$). Infarct size ↓ from 32.6 g (6.3) by 9.8 g (3.5; 30%) (MRI on 7 patients, $p = 0.04$).	111,112
CADUCEUS (CArdiosphere-Derived aUtologous StemCElls to Reverse ventricUlar dySfunction)	NCT00893360	May 2009	Age: 18–80 years Diagnosis: ischemic LV dysfunction and/or a recent MI, LVEF of ≥25 and ≤45	Autologous cardiosphere-derived cells (CDCs)	Intracoronary delivery of autologous CDCs. Group 1: 12.5×10^6 ($n = 4$); Group 2: 17.3×10^6 ($n = 1$); Group 3: 25×10^6 ($n = 12$). Control group ($n = 8$) usual medical management.	No adverse effects were reported 6 months: no significant change in LVEF Infarct size ↓ by 30–47%, significant ↑ viable cardiac mass (MRI).	113,114
ALCADIA (AutoLogous Human CArdiac-Derived Stem Cell to Treat Ischemic cArdiomyopathy)	NCT00981006	Apr 2010	Age: 20–80 years. Diagnosis: old MI due to coronary artery atherosclerotic disease LVEF of ≥15 and ≤35	Autologous human cardiac-derived stem cell (hCSC)	A bFGF gelatin sheet implantation and hCSC intramyocardial injection concomitant with CAGB (coronary artery bypass grafting).	Not yet available	115,116

LVEF, left ventricular ejection fraction; MI, myocardial infarction; SE, standard error; CABG, coronary artery bypass graft.

function, and attenuating adverse remodeling in cell-treated patients versus controls; follow-up has not been completed yet, though. A second trial, ALCADIA, is based on the infusion of autologous CPCs, isolated through a modified CS protocol, in combination with gelatin hydrogel incorporating basic FGF[115,116]; preliminary results of this trial have not been published yet. Finally, the ongoing SCIPIO trial,[111,112] by means of purified c-kit[+] resident cells, has achieved a significant recovery up to 12% on LVEF in cell-treated patients 1 year after cell infusion, also observing a reduction in infarct size.

Overall, these latest results with resident CPCs seem very promising, considering their intrinsic commitment toward cardiac lineages. Some flaws in using autologous CPCs have been pointed out, particularly related to the time required for their isolation and expansion; however, even allogeneic CPCs derived from excised transplant hearts have been demonstrated to be nonimmunogenic, safe, and with the same cardiac regenerative potency as their autologous counterpart, at least in rodents.[119]

Nevertheless, many hurdles and obstacles (e.g., low retention and engraftment efficiency, low cell survival and differentiation within the host tissue, induction of adequate paracrine and potency features, and efficient activation of endogenous regenerative potential) are still under careful analysis and need to be overcome, in order to optimize the efficiency and potency of cell therapy.[106] Being able to tune and modulate proliferation and differentiation of CPCs would be of extreme importance for maximizing their therapeutic effect. Thus, accurate insights are needed on the molecular pathways and gene expression profiles that regulate cardiac differentiation. As pointed out throughout this chapter, important lessons can be learnt from embryogenesis and organogenesis, where cell biology and function are finely controlled, and novel translational protocols for cell preparation and treatment might be derived from what we know about development.

VII. Concluding Remarks

As previously mentioned, the ability of the heart to regenerate has been gradually lost with evolution. The fibrotic response has emerged as the most efficient and rapid way to pad an injury occurring in the fast-beating, four-chambered mammalian heart, in order to prevent otherwise deadly myocardial ruptures. However, the loss of contractile cardiac muscle mass inevitably increases the hemodynamic burden on the remaining cardiac muscle. The cardiac chambers dilate and, eventually, cardiac contractile function declines, leading to HF. Current attempts to regenerate the heart include reactivation/enhancement of endogenous mechanisms of regeneration, and/or transplantation of cells capable of forming the three main cardiac cell types. The ideal cells to be injected should be easy to isolate and expand, nonimmunogenic or nontumorigenic. Much attention has been lately directed to the stem-cell-like cells obtained by reprogramming or direct transdifferentiation of adult somatic cells and to the surprisingly high number of endogenous progenitor cell populations identified in the mammalian adult heart. Regarding these latter, it is not clear yet what is their origin is and how the different populations described so far are related, but they all share the expression of an overlapping set of cardiogenic markers and have the potential to generate cardiomyocytes spontaneously or if appropriately instructed.[10] However, the isolation of cells based on the expression of single markers may not be sufficient to guarantee the necessary potency and features for optimal engraftment, survival, and correct differentiation *in vivo*.[120] Re-creating a condition resembling the natural tissue "niche" microenvironment seems to be mandatory to minimize artifacts due to *in vitro* culture conditions and to provide the cells with the appropriate physiological, biochemical, and mechanical signals. Indeed, it has already been shown that CPCs injected in the form of 3D cellular clusters (CSs) have higher resistance to oxidative stress and better engraftment, compared to cells dissociated or grown in monolayers.[103,105,121,122] Furthermore, it seems that the presence of stromal supportive cells may favor the maintenance of a "reservoir" of progenitor/stem cells in the core of the 3D structures,[103] and that a heterogeneous population, made of both stem and mesenchymal cells, has higher therapeutic potential than pure sorted pools.[120]

Moreover, to overcome the current obstacles to cell therapy, mostly due to low engraftment and differentiation efficiency, important lessons can be drawn from heart development during embryogenesis. Currently, much effort is directed toward learning from the multiple pathways finely regulating all the proliferative and differentiative switches involved in all the stages of organ development, to modulate the genetic programs of adult progenitor cells. Finding the ideal manipulation and/or pretreatment might efficiently increase the potency of candidate therapeutic cells, providing novel biotechnological tools for clinical translation. These multiple insights will also be useful for other applications such as tissue engineering[123] or pretreatment/induction of the host tissue by multiple pharmacological or gene targeting approaches.[106]

All these clues should guide future efforts toward the goal of switching on the right genetic programs in the ideal culture microenvironment to optimize the potency of cell products for heart regeneration.

ACKNOWLEDGMENT

This work was supported by the Pasteur Institute—Cenci Bolognetti Foundation.

References

1. Ausoni S, Sartore S. From fish to amphibians to mammals: in search of novel strategies to optimize cardiac regeneration. *J Cell Biol* 2009;**184**:357–64.
2. Lo DC, Allen F, Brockes JP. Reversal of muscle differentiation during urodele limb regeneration. *Proc Natl Acad Sci USA* 1993;**90**:7230–4.
3. Maden M. Regeneration: every clot has a thrombin lining. *Curr Biol* 2003;**13**:R517–8.
4. Brockes JP. Amphibian limb regeneration: rebuilding a complex structure. *Science* 1997;**276**:81–7.
5. Tanaka EM. Regeneration: if they can do it, why can't we? *Cell* 2003;**113**:559–62.
6. Chimenti I, Gaetani R, Barile L, Forte E, Ionta V, Angelini F, et al. Evidence for the existence of resident cardiac stem cells. In: Cohen IS, Glenn R, editors. *Regenerating the heart.* Totowa, New Jersey: Humana Press; 2011.
7. Hsieh PC, Segers VF, Davis ME, MacGillivray C, Gannon J, Molkentin JD, et al. Evidence from a genetic fate-mapping study that stem cells refresh adult mammalian cardiomyocytes after injury. *Nat Med* 2007;**13**:970–4.
8. Bergmann O, Bhardwaj RD, Bernard S, Zdunek S, Barnabe-Heider F, Walsh S, et al. Evidence for cardiomyocyte renewal in humans. *Science* 2009;**324**:98–102.
9. Kajstura J, Gurusamy N, Ogorek B, Goichberg P, Clavo-Rondon C, Hosoda T, et al. Myocyte turnover in the aging human heart. *Circ Res* 2010;**107**:1374–86.
10. Barile L, Messina E, Giacomello A, Marban E. Endogenous cardiac stem cells. *Prog Cardiovasc Dis* 2007;**50**:31–48.
11. Poss KD, Wilson LG, Keating MT. Heart regeneration in zebrafish. *Science* 2002;**298**:2188–90.
12. Jopling C, Sleep E, Raya M, Marti M, Raya A, Belmonte JC. Zebrafish heart regeneration occurs by cardiomyocyte dedifferentiation and proliferation. *Nature* 2010;**464**:606–9.
13. Kikuchi K, Holdway JE, Werdich AA, Anderson RM, Fang Y, Egnaczyk GF, et al. Primary contribution to zebrafish heart regeneration by gata4(+) cardiomyocytes. *Nature* 2010;**464**:601–5.
14. Drenckhahn JD, Schwarz QP, Gray S, Laskowski A, Kiriazis H, Ming Z, et al. Compensatory growth of healthy cardiac cells in the presence of diseased cells restores tissue homeostasis during heart development. *Dev Cell* 2008;**15**:521–33.
15. Porrello ER, Mahmoud AI, Simpson E, Hill JA, Richardson JA, Olson EN, et al. Transient regenerative potential of the neonatal mouse heart. *Science* 2011;**331**:1078–80.
16. Martin-Puig S, Wang Z, Chien KR. Lives of a heart cell: tracing the origins of cardiac progenitors. *Cell Stem Cell* 2008;**2**:320–31.
17. Dodou E, Verzi MP, Anderson JP, Xu SM, Black BL. Mef2c is a direct transcriptional target of ISL1 and GATA factors in the anterior heart field during mouse embryonic development. *Development* 2004;**131**:3931–42.
18. Laugwitz KL, Moretti A, Lam J, Gruber P, Chen Y, Woodard S, et al. Postnatal isl1+ cardioblasts enter fully differentiated cardiomyocyte lineages. *Nature* 2005;**433**:647–53.
19. Moretti A, Caron L, Nakano A, Lam JT, Bernshausen A, Chen Y, et al. Multipotent embryonic isl1+ progenitor cells lead to cardiac, smooth muscle, and endothelial cell diversification. *Cell* 2006;**127**:1151–65.
20. Kattman SJ, Adler ED, Keller GM. Specification of multipotential cardiovascular progenitor cells during embryonic stem cell differentiation and embryonic development. *Trends Cardiovasc Med* 2007;**17**:240–6.
21. Murry CE, Keller G. Differentiation of embryonic stem cells to clinically relevant populations: lessons from embryonic development. *Cell* 2008;**132**:661–80.

22. Bondue A, Tannler S, Chiapparo G, Chabab S, Ramialison M, Paulissen C, et al. Defining the earliest step of cardiovascular progenitor specification during embryonic stem cell differentiation. *J Cell Biol* 2011;**192**:751–65.

23. Kitajima S, Takagi A, Inoue T, Saga Y. MesP1 and MesP2 are essential for the development of cardiac mesoderm. *Development* 2000;**127**:3215–26.

24. Saga Y, Kitajima S, Miyagawa-Tomita S. Mesp1 expression is the earliest sign of cardiovascular development. *Trends Cardiovasc Med* 2000;**10**:345–52.

25. Saga Y, Miyagawa-Tomita S, Takagi A, Kitajima S, Miyazaki J, Inoue T. MesP1 is expressed in the heart precursor cells and required for the formation of a single heart tube. *Development* 1999;**126**:3437–47.

26. Mercola M, Ruiz-Lozano P, Schneider MD. Cardiac muscle regeneration: lessons from development. *Genes Dev* 2011;**25**:299–309.

27. Xu C, Police S, Hassanipour M, Gold JD. Cardiac bodies: a novel culture method for enrichment of cardiomyocytes derived from human embryonic stem cells. *Stem Cells Dev* 2006;**15**:631–9.

28. Yang L, Soonpaa MH, Adler ED, Roepke TK, Kattman SJ, Kennedy M, et al. Human cardiovascular progenitor cells develop from a KDR+ embryonic-stem-cell-derived population. *Nature* 2008;**453**:524–8.

29. Foley AC, Mercola M. Heart induction by Wnt antagonists depends on the homeodomain transcription factor Hex. *Genes Dev* 2005;**19**:387–96.

30. Yuasa S, Itabashi Y, Koshimizu U, Tanaka T, Sugimura K, Kinoshita M, et al. Transient inhibition of BMP signaling by Noggin induces cardiomyocyte differentiation of mouse embryonic stem cells. *Nat Biotechnol* 2005;**23**:607–11.

31. Schultheiss TM, Burch JB, Lassar AB. A role for bone morphogenetic proteins in the induction of cardiac myogenesis. *Genes Dev* 1997;**11**:451–62.

32. Rones MS, McLaughlin KA, Raffin M, Mercola M. Serrate and Notch specify cell fates in the heart field by suppressing cardiomyogenesis. *Development* 2000;**127**:3865–76.

33. Schroeder T, Fraser ST, Ogawa M, Nishikawa S, Oka C, Bornkamm GW, et al. Recombination signal sequence-binding protein Jkappa alters mesodermal cell fate decisions by suppressing cardiomyogenesis. *Proc Natl Acad Sci USA* 2003;**100**:4018–23.

34. Schroeder T, Meier-Stiegen F, Schwanbeck R, Eilken H, Nishikawa S, Hasler R, et al. Activated Notch1 alters differentiation of embryonic stem cells into mesodermal cell lineages at multiple stages of development. *Mech Dev* 2006;**123**:570–9.

35. Conboy IM, Rando TA. The regulation of Notch signaling controls satellite cell activation and cell fate determination in postnatal myogenesis. *Dev Cell* 2002;**3**:397–409.

36. Jensen J, Pedersen EE, Galante P, Hald J, Heller RS, Ishibashi M, et al. Control of endodermal endocrine development by Hes-1. *Nat Genet* 2000;**24**:36–44.

37. Abu-Issa R, Kirby ML. Patterning of the heart field in the chick. *Dev Biol* 2008;**319**:223–33.

38. Takeuchi JK, Ohgi M, Koshiba-Takeuchi K, Shiratori H, Sakaki I, Ogura K, et al. Tbx5 specifies the left/right ventricles and ventricular septum position during cardiogenesis. *Development* 2003;**130**:5953–64.

39. Lints TJ, Parsons LM, Hartley L, Lyons I, Harvey RP. Nkx-2.5: a novel murine homeobox gene expressed in early heart progenitor cells and their myogenic descendants. *Development* 1993;**119**:969.

40. Kaipainen A, Korhonen J, Pajusola K, Aprelikova O, Persico MG, Terman BI, et al. The related FLT4, FLT1, and KDR receptor tyrosine kinases show distinct expression patterns in human fetal endothelial cells. *J Exp Med* 1993;**178**:2077–88.

41. Molkentin JD, Firulli AB, Black BL, Martin JF, Hustad CM, Copeland N, et al. MEF2B is a potent transactivator expressed in early myogenic lineages. *Mol Cell Biol* 1996;**16**:3814–24.

42. Heikinheimo M, Scandrett JM, Wilson DB. Localization of transcription factor GATA-4 to regions of the mouse embryo involved in cardiac development. *Dev Biol* 1994;**164**:361–73.
43. Kuo CT, Morrisey EE, Anandappa R, Sigrist K, Lu MM, Parmacek MS, et al. GATA4 transcription factor is required for ventral morphogenesis and heart tube formation. *Genes Dev* 1997;**11**:1048–60.
44. Srivastava D, Cserjesi P, Olson EN. A subclass of bHLH proteins required for cardiac morphogenesis. *Science* 1995;**270**:1995–9.
45. Harvey RP. Patterning the vertebrate heart. *Nat Rev Genet* 2002;**3**:544–56.
46. Abu-Issa R, Waldo K, Kirby ML. Heart fields: one, two or more? *Dev Biol* 2004;**272**:281–5.
47. Cai CL, Liang X, Shi Y, Chu PH, Pfaff SL, Chen J, et al. Isl1 identifies a cardiac progenitor population that proliferates prior to differentiation and contributes a majority of cells to the heart. *Dev Cell* 2003;**5**:877–89.
48. Kelly RG, Brown NA, Buckingham ME. The arterial pole of the mouse heart forms from Fgf10-expressing cells in pharyngeal mesoderm. *Dev Cell* 2001;**1**:435–40.
49. Ai D, Fu X, Wang J, Lu MF, Chen L, Baldini A, et al. Canonical Wnt signaling functions in second heart field to promote right ventricular growth. *Proc Natl Acad Sci USA* 2007;**104**:9319–24.
50. Cohen ED, Wang Z, Lepore JJ, Lu MM, Taketo MM, Epstein DJ, et al. Wnt/beta-catenin signaling promotes expansion of Isl-1-positive cardiac progenitor cells through regulation of FGF signaling. *J Clin Invest* 2007;**117**:1794–804.
51. Klaus A, Saga Y, Taketo MM, Tzahor E, Birchmeier W. Distinct roles of Wnt/beta-catenin and Bmp signaling during early cardiogenesis. *Proc Natl Acad Sci USA* 2007;**104**:18531–6.
52. Lin L, Cui L, Zhou W, Dufort D, Zhang X, Cai CL, et al. Beta-catenin directly regulates Islet1 expression in cardiovascular progenitors and is required for multiple aspects of cardiogenesis. *Proc Natl Acad Sci USA* 2007;**104**:9313–8.
53. Meilhac SM, Esner M, Kelly RG, Nicolas JF, Buckingham ME. The clonal origin of myocardial cells in different regions of the embryonic mouse heart. *Dev Cell* 2004;**6**:685–98.
54. Kattman SJ, Huber TL, Keller GM. Multipotent flk-1+ cardiovascular progenitor cells give rise to the cardiomyocyte, endothelial, and vascular smooth muscle lineages. *Dev Cell* 2006;**11**:723–32.
55. Ema M, Takahashi S, Rossant J. Deletion of the selection cassette, but not cis-acting elements, in targeted Flk1-lacZ allele reveals Flk1 expression in multipotent mesodermal progenitors. *Blood* 2006;**107**:111–7.
56. Chien KR, Domian IJ, Parker KK. Cardiogenesis and the complex biology of regenerative cardiovascular medicine. *Science* 2008;**322**:1494–7.
57. Wessels A, Sedmera D. Developmental anatomy of the heart: a tale of mice and man. *Physiol Genomics* 2003;**15**:165–76.
58. Hutson MR, Kirby ML. Model systems for the study of heart development and disease. Cardiac neural crest and conotruncal malformations. *Semin Cell Dev Biol* 2007;**18**:101–10.
59. Snider P, Olaopa M, Firulli AB, Conway SJ. Cardiovascular development and the colonizing cardiac neural crest lineage. *ScientificWorldJournal* 2007;**7**:1090–113.
60. Kappetein AP, Gittenberger-de Groot AC, Zwinderman AH, Rohmer J, Poelmann RE, Huysmans HA. The neural crest as a possible pathogenetic factor in coarctation of the aorta and bicuspid aortic valve. *J Thorac Cardiovasc Surg* 1991;**102**:830–6.
61. Nakamura T, Colbert MC, Robbins J. Neural crest cells retain multipotential characteristics in the developing valves and label the cardiac conduction system. *Circ Res* 2006;**98**:1547–54.
62. Stoller JZ, Epstein JA. Cardiac neural crest. *Semin Cell Dev Biol* 2005;**16**:704–15.
63. Stottmann RW, Choi M, Mishina Y, Meyers EN, Klingensmith J. BMP receptor IA is required in mammalian neural crest cells for development of the cardiac outflow tract and ventricular myocardium. *Development* 2004;**131**:2205–18.

64. El-Helou V, Beguin PC, Assimakopoulos J, Clement R, Gosselin H, Brugada R, et al. The rat heart contains a neural stem cell population; role in sympathetic sprouting and angiogenesis. J Mol Cell Cardiol 2008;45:694–702.
65. Winter EM, Gittenberger-de Groot AC. Epicardium-derived cells in cardiogenesis and cardiac regeneration. Cell Mol Life Sci 2007;64:692–703.
66. Reese DE, Mikawa T, Bader DM. Development of the coronary vessel system. Circ Res 2002;91:761–8.
67. Wessels A, Perez-Pomares JM. The epicardium and epicardially derived cells (EPDCs) as cardiac stem cells. Anat Rec A Discov Mol Cell Evol Biol 2004;276:43–57.
68. Cai CL, Martin JC, Sun Y, Cui L, Wang L, Ouyang K, et al. A myocardial lineage derives from Tbx18 epicardial cells. Nature 2008;454:104–8.
69. Zhou B, Ma Q, Rajagopal S, Wu SM, Domian I, Rivera-Feliciano J, et al. Epicardial progenitors contribute to the cardiomyocyte lineage in the developing heart. Nature 2008;454:109–13.
70. Bock-Marquette I, Saxena A, White MD, Dimaio JM, Srivastava D. Thymosin beta4 activates integrin-linked kinase and promotes cardiac cell migration, survival and cardiac repair. Nature 2004;432:466–72.
71. Smart N, Risebro CA, Melville AA, Moses K, Schwartz RJ, Chien KR, et al. Thymosin beta4 induces adult epicardial progenitor mobilization and neovascularization. Nature 2007;445:177–82.
72. Bock-Marquette I, Shrivastava S, Pipes GC, Thatcher JE, Blystone A, Shelton JM, et al. Thymosin beta4 mediated PKC activation is essential to initiate the embryonic coronary developmental program and epicardial progenitor cell activation in adult mice in vivo. J Mol Cell Cardiol 2009;46:728–38.
73. Urayama K, Guilini C, Turkeri G, Takir S, Kurose H, Messaddeq N, et al. Prokineticin receptor-1 induces neovascularization and epicardial-derived progenitor cell differentiation. Arterioscler Thromb Vasc Biol 2008;28:841–9.
74. Limana F, Zacheo A, Mocini D, Mangoni A, Borsellino G, Diamantini A, et al. Identification of myocardial and vascular precursor cells in human and mouse epicardium. Circ Res 2007;101:1255–65.
75. Smart N, Bollini S, Dube KN, Vieira JM, Zhou B, Davidson S, et al. De novo cardiomyocytes from within the activated adult heart after injury. Nature 2011;474:640–4.
76. Nemir M, Pedrazzini T. Functional role of Notch signaling in the developing and postnatal heart. J Mol Cell Cardiol 2008;45:495–504.
77. Niessen K, Karsan A. Notch signaling in cardiac development. Circ Res 2008;102:1169–81.
78. Raya A, Koth CM, Buscher D, Kawakami Y, Itoh T, Raya RM, et al. Activation of Notch signaling pathway precedes heart regeneration in zebrafish. Proc Natl Acad Sci USA 2003;100 (Suppl 1):11889–95.
79. Croquelois A, Domenighetti AA, Nemir M, Lepore M, Rosenblatt-Velin N, Radtke F, et al. Control of the adaptive response of the heart to stress via the Notch1 receptor pathway. J Exp Med 2008;205:3173–85.
80. Gude NA, Emmanuel G, Wu W, Cottage CT, Fischer K, Quijada P, et al. Activation of Notch-mediated protective signaling in the myocardium. Circ Res 2008;102:1025–35.
81. Russell JL, Goetsch SC, Gaiano NR, Hill JA, Olson EN, Schneider JW. A dynamic notch injury response activates epicardium and contributes to fibrosis repair. Circ Res 2011;108:51–9.
82. Beltrami AP, Barlucchi L, Torella D, Baker M, Limana F, Chimenti S, et al. Adult cardiac stem cells are multipotent and support myocardial regeneration. Cell 2003;114:763–76.
83. Oh H, Bradfute SB, Gallardo TD, Nakamura T, Gaussin V, Mishina Y, et al. Cardiac progenitor cells from adult myocardium: homing, differentiation, and fusion after infarction. Proc Natl Acad Sci USA 2003;100:12313–8.

84. Hierlihy AM, Seale P, Lobe CG, Rudnicki MA, Megeney LA. The post-natal heart contains a myocardial stem cell population. *FEBS Lett* 2002;**530**:239–43.

85. Messina E, De Angelis L, Frati G, Morrone S, Chimenti S, Fiordaliso F, et al. Isolation and expansion of adult cardiac stem cells from human and murine heart. *Circ Res* 2004;**95**:911–21.

86. Bearzi C, Rota M, Hosoda T, Tillmanns J, Nascimbene A, De Angelis A, et al. Human cardiac stem cells. *Proc Natl Acad Sci USA* 2007;**104**:14068–73.

87. Dimmeler S, Zeiher AM, Schneider MD. Unchain my heart: the scientific foundations of cardiac repair. *J Clin Invest* 2005;**115**:572–83.

88. Ieda M, Fu JD, Delgado-Olguin P, Vedantham V, Hayashi Y, Bruneau BG, et al. Direct reprogramming of fibroblasts into functional cardiomyocytes by defined factors. *Cell* 2010;**142**:375–86.

89. David R, Brenner C, Stieber J, Schwarz F, Brunner S, Vollmer M, et al. MesP1 drives vertebrate cardiovascular differentiation through Dkk-1-mediated blockade of Wnt-signalling. *Nat Cell Biol* 2008;**10**:338–45.

90. Takahashi K, Yamanaka S. Induction of pluripotent stem cells from mouse embryonic and adult fibroblast cultures by defined factors. *Cell* 2006;**126**:663–76.

91. Li L, Xie T. Stem cell niche: structure and function. *Annu Rev Cell Dev Biol* 2005;**21**:605–31.

92. Walker MR, Patel KK, Stappenbeck TS. The stem cell niche. *J Pathol* 2009;**217**:169–80.

93. Vazin T, Schaffer DV. Engineering strategies to emulate the stem cell niche. *Trends Biotechnol* 2010;**28**:117–24.

94. Watt FM, Hogan BL. Out of Eden: stem cells and their niches. *Science* 2000;**287**:1427–30.

95. Qyang Y, Martin-Puig S, Chiravuri M, Chen S, Xu H, Bu L, et al. The renewal and differentiation of Isl1+ cardiovascular progenitors are controlled by a Wnt/beta-catenin pathway. *Cell Stem Cell* 2007;**1**:165–79.

96. Schlessinger K, Hall A, Tolwinski N. Wnt signaling pathways meet Rho GTPases. *Genes Dev* 2009;**23**:265–77.

97. Urbanek K, Cesselli D, Rota M, Nascimbene A, De Angelis A, Hosoda T, et al. Stem cell niches in the adult mouse heart. *Proc Natl Acad Sci USA* 2006;**103**:9226–31.

98. Popescu LM, Gherghiceanu M, Manole CG, Faussone-Pellegrini MS. Cardiac renewing: interstitial Cajal-like cells nurse cardiomyocyte progenitors in epicardial stem cell niches. *J Cell Mol Med* 2009;**13**:866–86.

99. Pampaloni F, Reynaud EG, Stelzer EH. The third dimension bridges the gap between cell culture and live tissue. *Nat Rev Mol Cell Biol* 2007;**8**:839–45.

100. Holopainen IE. Organotypic hippocampal slice cultures: a model system to study basic cellular and molecular mechanisms of neuronal cell death, neuroprotection, and synaptic plasticity. *Neurochem Res* 2005;**30**:1521–8.

101. Smith RR, Barile L, Cho HC, Leppo MK, Hare JM, Messina E, et al. Regenerative potential of cardiosphere-derived cells expanded from percutaneous endomyocardial biopsy specimens. *Circulation* 2007;**115**:896–908 [CIRCULATIONAHA.106.655209 [pii] 10.1161/CIRCULATIONAHA.106.655209].

102. Chimenti I, Smith RR, Li TS, Gerstenblith G, Messina E, Giacomello A, et al. Relative roles of direct regeneration versus paracrine effects of human cardiosphere-derived cells transplanted into infarcted mice. *Circ Res* 2010;**106**:971–80.

103. Li TS, Cheng K, Lee ST, Matsushita S, Davis D, Malliaras K, et al. Cardiospheres recapitulate a niche-like microenvironment rich in stemness and cell-matrix interactions, rationalizing their enhanced functional potency for myocardial repair. *Stem Cells* 2010;**28**:2088–98.

104. Gaetani R, Ledda M, Barile L, Chimenti I, De Carlo F, Forte E, et al. Differentiation of human adult cardiac stem cells exposed to extremely low-frequency electromagnetic fields. *Cardiovasc Res* 2009;**82**:411–20. doi:10.1093/cvr/cvp067 cvp067 [pii].

105. Lee ST, White AJ, Matsushita S, Malliaras K, Steenbergen C, Zhang Y, et al. Intramyocardial injection of autologous cardiospheres or cardiosphere-derived cells preserves function and

minimizes adverse ventricular remodeling in pigs with heart failure post-myocardial infarction. *J Am Coll Cardiol* 2011;**57**:455–65.

106. Forte E, Chimenti I, Barile L, Gaetani R, Angelini F, Ionta V, et al. Cardiac cell therapy: the next (re)generation. *Stem Cell Rev* 2011;**7**:1018–30.

107. Gaetani R, Barile L, Forte E, Chimenti I, Ionta V, Di Consiglio A, et al. New perspectives to repair a broken heart. *Cardiovasc Hematol Agents Med Chem* 2009;**7**:91–107.

108. Gnecchi M, Zhang Z, Ni A, Dzau VJ. Paracrine mechanisms in adult stem cell signaling and therapy. *Circ Res* 2008;**103**:1204–19.

109. Menasche P. Cardiac cell therapy: lessons from clinical trials. *J Mol Cell Cardiol* 2011;**50**:258–65.

110. Martin-Rendon E, Brunskill SJ, Hyde CJ, Stanworth SJ, Mathur A, Watt SM. Autologous bone marrow stem cells to treat acute myocardial infarction: a systematic review. *Eur Heart J* 2008;**29**:1807–18.

111. Linke A, Muller P, Nurzynska D, Casarsa C, Torella D, Nascimbene A, et al. Stem cells in the dog heart are self-renewing, clonogenic, and multipotent and regenerate infarcted myocardium, improving cardiac function. *Proc Natl Acad Sci USA* 2005;**102**:8966–71.

112. Bolli R, Chugh AR, D'Amario D, Loughran JH, Stoddard MF, Ikram S, et al. Cardiac stem cells in patients with ischaemic cardiomyopathy (SCIPIO): initial results of a randomised phase 1 trial. *Lancet* 2011;**378**:1847–57.

113. Makkar R, Smith RR, Cheng K, Malliaras K, Thomson LE, Berman D, et al. Intracoronary cardiosphere−derived cells for heart regeneration after myocardial infarction (CADUCEUS): a prospective, randomised phase 1 trial. *Lancet* 2012;**379**:895–904.

114. Johnston PV, Sasano T, Mills K, Evers R, Lee ST, Smith RR, et al. Engraftment, differentiation, and functional benefits of autologous cardiosphere-derived cells in porcine ischemic cardiomyopathy. *Circulation* 2009;**120**:1075–83 7 p following 83.

115. Takehara N, Tsutsumi Y, Tateishi K, Ogata T, Tanaka H, Ueyama T, et al. Controlled delivery of basic fibroblast growth factor promotes human cardiosphere-derived cell engraftment to enhance cardiac repair for chronic myocardial infarction. *J Am Coll Cardiol* 2008;**52**:1858–65.

116. Tateishi K, Ashihara E, Honsho S, Takehara N, Nomura T, Takahashi T, et al. Human cardiac stem cells exhibit mesenchymal features and are maintained through Akt/GSK-3beta signaling. *Biochem Biophys Res Commun* 2007;**352**:635–41.

117. Malliaras K, Marban E. Cardiac cell therapy: where we've been, where we are, and where we should be headed. *Br Med Bull* 2011;**98**:161–85.

118. Leri A, Kajstura J, Anversa P. Role of cardiac stem cells in cardiac pathophysiology: a paradigm shift in human myocardial biology. *Circ Res* 2011;**109**:941–61.

119. Malliaras K, Li TS, Luthringer D, Terrovitis J, Cheng K, Chakravarty T, et al. Safety and efficacy of allogeneic cell therapy in infarcted rats transplanted with mismatched cardiosphere-derived cells. *Circulation* 2012;**125**:100–12.

120. Smith RR, Chimenti I, Marban E. Unselected human cardiosphere-derived cells are functionally superior to c-Kit- or CD90-purified cardiosphere-derived cells. *Circulation* 2008;**118**: S_420.

121. Li TS, Cheng K, Malliaras K, Matsushita N, Sun B, Marban L, et al. Expansion of human cardiac stem cells in physiological oxygen improves cell production efficiency and potency for myocardial repair. *Cardiovasc Res* 2011;**89**:157–65.

122. Li TS, Marban E. Physiological levels of reactive oxygen species are required to maintain genomic stability in stem cells. *Stem Cells* 2010;**28**:1178–85.

123. Gaetani R, Rizzitelli G, Chimenti I, Barile L, Forte E, Ionta V, et al. Cardiospheres and tissue engineering for myocardial regeneration: potential for clinical application. *J Cell Mol Med* 2010;**14**:1071–7.

Roles of MicroRNAs and Myocardial Cell Differentiation

Tomohide Takaya,* Hitoo
Nishi,† Takahiro Horie,† Koh
Ono,† and Koji Hasegawa*

*Division of Translational Research,
National Hospital Organization Kyoto
Medical Center, Kyoto, Japan

†Department of Cardiovascular Medicine,
Graduate School of Medicine, Kyoto
University, Kyoto, Japan

As drug therapy is of limited efficacy in the treatment of heart diseases related to loss of cardiomyocytes, which have very poor division potential, regenerative medicine is expected to be a new strategy to address regenerative treatment in cardiac diseases. To achieve myocardial regeneration, elucidation of the mechanism of myocardial differentiation from stem cells is essential. Myocardial differentiation from embryonic pluripotent stem cells has been investigated worldwide, and remarkable developments such as establishment of induced pluripotent stem cells and transformation of somatic cells to cardiomyocytes have recently been made, markedly changing the strategy of regenerative medicine. At the same time, the close involvement of microRNA in the maintenance, proliferation, differentiation, and reprogramming of these stem cells has been revealed. In this report, microRNA is outlined, focusing on its role in myocardial differentiation.

Progress in Molecular Biology
and Translational Science, Vol. 111
http://dx.doi.org/10.1016/B978-0-12-398459-3.00006-X

139

I. Introduction

Cardiomyocytes proliferate and form the heart in the embryonic period, but proliferation stops soon after birth. Therefore, it is difficult to cure systolic dysfunction (heart failure), particularly in cases accompanied by myocardial injury, such as myocardial infarction and cardiomyopathy. Drug therapy for heart failure using β-blockers and angiotensin-converting enzyme inhibitors improves the short-term prognosis, but the 5-year survival rate of patients with severe heart failure is less than 50%. To overcome this situation, regenerative therapy has been investigated worldwide wherein cardiovascular cell differentiation of stem cells has been induced and applied to supplement and regenerate the affected regions. As a source of cells for regenerative medicine, somatic stem cells such as bone-marrow-derived cells and vascular endothelial precursor cells, and pluripotent stem cells, such as embryonic stem (ES) cells and induced pluripotent stem (iPS) cells, have been considered. Studies on direct transformation of somatic cells to the target cells, such as cardiomyocytes, have also been progressing.

In parallel with this trend, an important role of microRNA (miRNA or miR) in the maintenance, proliferation, differentiation, and reprogramming of stem cells has recently been revealed. The small RNA present in *C. elegans* was the miRNA initially discovered and reported in *Cell* in 1993 by Ambros and Ruvkun,[1,2] in which the phenomenon currently accepted widely had already been pointed out: this RNA has an antisense sequence complementary to a specific gene sequence and performs the posttranscriptional regulation of gene expression. But the fact that this small RNA is present in plants through to humans and is universally involved in biological activity, that is, the name "microRNA" and its concept, was not established until Fire and Mello *et al.* reported RNA interference in 1998, which led to their being later awarded the Nobel Prize.[3] After the establishment of the concept of microRNAs, the importance of their functions in mammals has been revealed in various fields, and their involvement in cardiogenesis and development of cardiac hypertrophy, heart failure, and arrhythmia has been reported. In mice, impairment of the synthesis of all miRNAs in the heart leads to cardiac hypofunction and subsequently death. Therefore, miRNAs are essential for the heart, and changes in their expressions may also influence pathophysiology in humans. The association between miRNA and myocardial differentiation.

A. Self-replication and Pluripotency of ES Cells and miRNA

ES cells derived from the inner cell mass of fertilized ova before implantation (blastocysts) have the potential to differentiate into all three germ layers (pluripotency) and infinitely self-replicate while maintaining the undifferentiated condition (self-renewal).[4] Accordingly, if appropriate induction of ES cell differentiation in a mass culture becomes possible, the cells can be used as a cell source for regenerative medicine in various fields.

To investigate the function of miRNA in development and differentiation, mice and ES cells in which miRNA synthesis was abolished by knocking out (KO) RNase III gene Dicer1 and double-stranded RNA-binding protein gene Dgcr8 have been prepared. Dicer1 KO mice undergo embryonic death by E7.5, and pluripotency factor Oct4 is not expressed.[5] Similarly, transgenic mice in which heart-muscle-specific KO of Dicer1 starts on E8.5 die because of cardiac-hypoplasia-associated heart failure by E12.5.[6] In Dicer1 KO ES cells, Oct4 is expressed, but their division potential is markedly defective; no endo- and ectodermal markers are expressed even after embryoid body (EB) formation-induced differentiation, and proliferation stops on day 8.[7,8] These findings reveal that miRNA is essential for proliferation and differentiation of cells including cardiogenesis. The phenotype of Dgcr8 KO ES cells is slightly different from that of Dicer1 KO, and proliferation and differentiation continue after 16 days of EB formation. Oct4 and Nanog are expressed in Dgcr8 KO ES cells, suggesting that several miRNAs promote ES cell differentiation by repressing pluripotency factors.

Retinoic acid used to induce myocardial differentiation enhances miR-134, miR-296, and miR-470 expressions. These miRNAs promote ES cell differentiation by targeting Sox2, Nanog, and Oct4.[9] miR-145 directly targets Oct4, Sox2, and Klf4 and inhibits self-replication of ES cells, promoting differentiation.[8] miR-145 is inhibited by Oct4 in undifferentiated cells, but its expression is enhanced with progression of differentiation.

Inversely, the expression levels of many miRNAs specifically expressed in undifferentiated ES cells decrease with EB-formation-induced differentiation.[10] For example, the ES-cell-specific miR-290 cluster indirectly regulates DNA methylation by targeting Rbl-2, but the levels of this miRNA decrease as differentiation progresses.[11,12]

To efficiently induce stem cell differentiation into target cells, it may be necessary to cancel miRNA-dependent controlled self-replication and multipotency (Fig. 1).

B. Development and Formation of the Heart and miRNA

Many transcription factors, such as Nkx2.5, GATA4, Tbx5, myocyte enhancer factor-2 (MEF2), and serum response factor (SRF), have been identified as cardiomyocyte precursor cells or essential genes for morphogenesis of the heart in previous studies on cardiogenesis.[13] They are also known to play important roles in myocardial differentiation of ES cells, but no master gene, such as MyoD

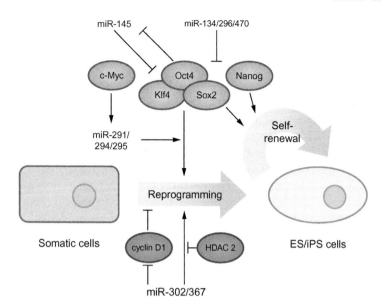

FIG. 1. Self-replication and reprogramming to iPS cells of ES cells and miRNA. miRNA is related to ES cell proliferation, maintenance of multipotency, and reprogramming of somatic cells to iPS cells and target genes. miRNA generally targets several genes, and a single gene may be regulated by several miRNAs. The transcription control network in stem cells may be more complex. (For color version of this figure, the reader is referred to the online version of this chapter.)

in skeletal muscle cells, has been discovered. It has been suggested that the development/differentiation program is operated by gene expression control by networks of several transcription and regulatory factors in the heart and heart muscle. In addition, involvement of miRNA in cardiogenesis and myocardial differentiation has recently been clarified (Fig. 2).

miRNAs are distinguished on the basis of their sequence. The presence of several hundred species has been reported in mammals so far, but, basically, they are synthesized through a common pathway. They are transcribed from genomic DNA, go through two precursors, primary- and precursor-miRNAs (pri- and pre-miRNAs, respectively), and finally become mature miRNAs. When the function of a gene is analyzed, experimental systems of acquisition or inhibition of function, or both, are constructed, but to evaluate the overall function of miRNAs, it is practically impossible to simultaneously knock out/ inhibit or overexpress all miRNAs, and the inhibition of all miRNAs by knocking out molecules essential for the synthesis process may be the most practical method.

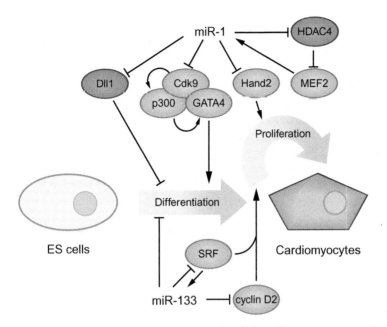

FIG. 2. Cardiogenesis, myocardial differentiation of ES cells, and miRNA. miR-1/133 plays an important role in cardiogenesis and myocardial differentiation of ES cells and target genes. miR-1/133 is controlled by feedback from the target genes. (For color version of this figure, the reader is referred to the online version of this chapter.)

Dicer is a ribonuclease (RNase) essential for the processing of pre-miRNA to mature miRNA, although there is an exception. Dicer-knockout mice have been prepared and the phenotypes have been analyzed. Fetal death occurs in homozygous whole-body knockout mice,[7] while heterozygous mice are normal; hence, the function in the heart is analyzed in homozygous heart-muscle-specific knockout mice employing the Cre-loxP system. It is possible that Dicer has a function other than miRNA processing and its loss has an influence, but, at present, the phenotypes are considered to be due to the inhibition of all miRNA functions. The phenotypes in the heart are outlined in the following sections.

C. Nkx2.5 Cre: Dicer^flox/flox Mice

NKx2.5 is a heart-specific transcription factor essential for cardiogenesis. As it is expressed in cardiomyocyte precursors at 8.5 days of embryonic age, Dicer can be deleted from the heart in the early embryonic period by expressing Cre recombinase under the promoter of NKx2.5. Fetal death from heart failure occurs at 12.5 days of embryonic age, and edema and hypoplasia of the ventricular muscle are observed.[14]

D. α-MHC Cre: Dicer^flox/flox Mice

The α-myosin heavy chain (MHC) is a heart-muscle-specific constrictive protein forming two isoforms with β-MHC. Heart-muscle-specific Dicer deletion can be induced in the mid-embryonic period by expressing Cre recombinase under the α-MHC promoter. In the fetal heart at 14.5 days of embryonic age, Dicer is knocked out and the total miRNA expression level is reduced, but fetal death does not occur. A dilated cardiomyopathy-like condition occurs within 4 days after birth, and the animals die from heart failure.[15]

E. α-MHC-MCM Cre: Dicer^flox/flox Mice

Tamoxifen administration induces Cre recombinase expression under the α-MHC promoter, similar to that in the above mice, which enables heart-muscle-specific Dicer deletion at a specific timing. When Dicer deletion is induced in young mice at 3 weeks of age, mild ventricular remodeling and marked atrial dilation develop, and animals start to suddenly die after 1 week. When it is induced in adult mice at 8 weeks of age, marked bilateral ventricular dilatation accompanied by cardiomyocyte enlargement, muscle fiber disarray, and ventricular fibrosis develop and the systolic function decreases.[16]

The most important factor for cardiogenesis is the miR-1/133 family specifically expressed at a high level in heart and skeletal muscle cells.

1. MIR-1

miR-1 is encoded by two genes (miR-1-1 and miR-1-2), and each gene is simultaneously transcribed with one of two miR-133a genes (miR-133a-1 and miR-133a-2). Their expressions are limited to the heart and skeletal muscle, and their transcriptions are regulated by transcription factors SRF and MEF2/MyoD, respectively.[17] In cardiogenesis, miR-1 is considered to control the balance between the differentiation and proliferation of cardiomyocyte precursors through Hand2.[17] When miR-1 is overexpressed under the β-MHC promoter in fetal cardiomyocytes, the ventricle is thinned at 13.5 days of embryonic age and cardiogenesis stops. Half of homozygous miR-1-2-knockout mice die between the late embryonic period and birth because of a ventricular septal defect.[6] In most of the other half, heart failure develops after birth and sudden cardiac death occurs as a result of arrhythmia without apparent morphological abnormality in the heart.[6]

2. miR-133a

There are three genes of miR-133 (miR-133a-1/miR-133a-2/133b), and these are bicistronically transcribed with miR-1-2, miR-1, and miR-206, respectively. Expression of the two miR-1/133a clusters is regulated by MEF2 and SRF and limited to heart and skeletal muscles. Expression of the miR-206/133b cluster is limited to skeletal muscle.[18]

There are two polycistronically transcribed miR-1/133a clusters: miR-1-1/133a-2 is abundant in the precursor atria and miR-1-2/133a-1 is specifically present in the ventricle. miR-1/133a cluster expression in cardiomyocytes is directly regulated by MEF2 and SRF. There is also the miR-206/133b cluster transcribed from other genetic loci, but its function is skeletal-muscle-specific.

miR-1 targets Hand2 (dHand) to negatively control cardiogenesis, in addition to forming a feedback loop through targeting histone deacetylase (HDAC) 4 to inhibit downstream MEF2. In addition to feedback through targeting SRF, miR-133 inhibits myocardial proliferation by targeting cyclin D2.[16,17] Both miR-1-2-defective mice and miR-133a-1/2 double KO mice have ventricular septal defects and die in the late embryonic or neonatal period.[6,19] In mice overexpressing miR-1 or miR-133a under the β-MHC promoter, myocardial proliferation is inhibited, resulting in embryonic death.[19,20] These findings show that miR-1/133 is essential for division/proliferation of cardiomyocytes and for heart formation in the developmental process.

II. Myocardial Differentiation of ES Cells and miRNA

The embryologic as well as pathophysiologic aspects of myocardial differentiation of ES cells have been investigated. When cardiomyocytes pathologically enlarge because of hemodynamic load stimulation, the gene expression pattern changes from the adult to the fetal type, by which GATA4 acetylation by histone acetyl transferase p300 is promoted and expression of fetal-type myocardial genes, such as atrial natriuretic factor (ANF), is enhanced.[21,22] p300/GATA4 binds to cyclin-dependent kinase (CDK) Cdk9 and transcriptional activity is enhanced by phosphorylation of Cdk9.[23] It has recently been clarified that the p300/GATA4/Cdk9 complex also functions in myocardial differentiation of ES cells.[24]

Myocardial enlargement reaction is regulated by miR-1/133,[25,26] but the question is whether miR-1/133 also functions in determining the destiny of undifferentiated cells, similar to transcription factors.

miR-1/133 is not expressed in undifferentiated ES cells, but its expression level increases after differentiation induction in two-dimensional plate culture. In ES cells induced to express miR-1 or 133, spontaneous myocardial differentiation is inhibited.[27] On day 8 of differentiation induction, the expression levels of Nkx2.5 and β-MHC were significantly decreased in miR-1-expressing ES cells compared to those in the control group, and the ANF expression level was also slightly decreased. Similarly, the expression level of Nkx2.5 significantly decreased and the levels of β-MHC and ANF were slightly decreased in miR-133-transfected ES cells.

It has been reported that miR-1 targets HDAC4 and miR-133 targets the 3′ untranslated region (UTR) of SRF in C2C12 skeletal myoblasts,[28] but no reduction of these proteins was confirmed by reporter analysis of ES cells. It is possible that different systems act in ES and C2C12 cells with regard to RNA degradation or inhibition of translation. It has been confirmed that miR-1 targets the 3′ UTR of Cdk9 in ES cells, and the intranuclear Cdk9 protein level decreases as a result of miR-1 overexpression. Therefore, the myocardial differentiation-inhibitory effect of miR-1 on ES cells may be due to p300/GATA4/Cdk9-dependent reduction of fetal-type heart muscle gene expression through inhibition of Cdk9 translation.

When ES cells were treated with an HDAC inhibitor trichostatin A (TSA), acetylation of histones and GATA4 was enhanced, probably through p300, and the heart muscle gene expression levels and the number of self-pulsating cardiomyocyte colonies were increased.[29,30] TSA-induced reduction of the miR-1/133 expression level was also noted, but this TSA-dependent inhibition did not act on miR-24 expressed in several tissues, suggesting the presence of selectivity/specificity. Although its molecular mechanism has not been elucidated, it may become a new drug candidate when miRNA control through a low molecular weight compound becomes possible.

Regarding differentiation induction by EB formation, the findings on miR-1/133 action on myocardial differentiation in two-dimensional plate culture have been reported to be different from those on spontaneous differentiation.[31] Induced miR-133 expression inhibited myocardial differentiation, which was common to both conditions, but miR-1 promoted myocardial differentiation in EB by targeting Notch ligand Delta-like 1 (Dll-1).

One cause for the inconsistency between the plate culture and EB formation may be differences in intercellular adherence and interaction. ES cell masses do not necessarily accurately reproduce intercellular interactions in embryonic development. *Drosophila* miR-1 also targets Dll-1 homologues, but mesoderma-specific miR-1 overexpression reduces the number of cardiomyocyte precursor cells and the abnormality occurring in cardiac morphogenesis.[32] As Dll-1 functions in intercellular signal transmission through Notch, the influence of miR-1 on cells may change depending on the conditions of the cells.

Similarly, time-course gene expression differs between the plate culture and EB formation. For example, the miR-1 expression level continued to rise for 8 days after differentiation induction in plate culture, but the level reached a peak on day 6 after EB formation. In addition, the expression pattern of GATA4—thebinding factor of Cdk9, the target of miR-1—differs between the two conditions. miRNA does not mechanically target a specific gene, but it functions depending on the influence of the competition between signal feedback and microenvironment of cells. To control stem cell differentiation using or targeting miRNA, it is necessary to investigate and optimize the culture and differentiation induction methods.

III. Somatic Cell Reprogramming and miRNA

With the marked advancement in stem cell research, the strategy of cardiovascular regenerative medicine has also been rapidly progressing, particularly with regard to knowledge and techniques concerning somatic cell reprogramming. iPS cells prepared by introducing multiple transcription factors (Oct4/Sox2/Klf4/c-Myc) into somatic cells, such as fibroblasts, are pluripotent stem cells with self-replication potential and with multipotency similar to ES cells.[33–35] Various iPS cell preparation methods have been proposed, but there are problems to be overcome before clinical application can become a reality, such as extrapolation of transcription factors including cancer genes and insufficient establishment efficiency. To overcome these problems, approaches using miRNA are being attempted (Fig. 1).

When somatic cells were transfected with miR-291-3p/294/295, which is an undifferentiated ES-cell-specific miR-290 cluster subset, the efficiency of Oct4/Sox2/Klf4-induced reprogramming to iPS cells was increased.[36] This effect was not obtained in the presence of c-Myc; c-Myc binds to the promoter region of the miR-290 cluster, suggesting that the contribution of c-Myc in establishing iPS cells is partially through the enhancement of miRNA expression.

It has also been reported that somatic cells could be reprogrammed to iPS cells within a short time with high efficiency by the induced expression of the miR-302/367 cluster.[37] miR-302 controlled by Oct4/Sox2 is expressed in ES cells at a high level and promotes the cell cycle from the G1 to the S phase by targeting cyclin D1.[38] Similar to reprogramming through transcription factors, miR-302/367-induced initialization of somatic cells is also markedly enhanced by inhibition of HDAC2 by valproic acid. The involvement of miRNA in chromatin remodeling in stem cell maintenance and differentiation has been briefly described above. Further studies are necessary to elucidate the details.

Basically, established iPS cells can be induced to differentiate into cardiovascular cells by employing the method used for ES cells.[39,40] However, the differentiation efficiency may vary among iPS cell lines prepared employing different methods and from different somatic cells. For example, the myocardial differentiation efficiency of some iPS cell lines is lower than that of other

ES and iPS cell lines, and these cells also show partial resistance to TSA-induced promotion of myocardial differentiation.[41] One method of overcoming this problem may be direct reprogramming of somatic cells to the target cells. It has already been reported that heart and skin fibroblasts could be transformed to self-pulsating cardiomyocytes by introducing GATA4/MEF2c/Tbx5[42] (Fig. 3).

Considering that somatic cells are reprogrammed to iPS cells by miRNA, it is easy to imagine that miRNA plays an important role in the direct transformation of somatic cells into different types of somatic cells. Further clarification of the molecular mechanism of miRNA may lead to the establishment of a highly efficient, simple, and rapid method of differentiation or reprogramming to cardiovascular cells, paving the way to regenerative medicine for cardiovascular diseases.

IV. Heart Disease and MicroRNAs

The molecular mechanism of cardiac hypertrophy had been actively investigated even before the discovery of miRNA, and various molecules and pathways have been identified. The role of miRNA in cardiac hypertrophy is explained as the targeting of one or a few of many identified molecules, but considering that a single miRNA simultaneously targets several hundreds of genes, this explanation may be insufficient. Moreover, secondary influences of many other target genes should be evaluated, but this evaluation has not yet

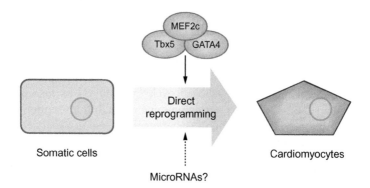

FIG. 3. Direct transformation of somatic cells to cardiomyocytes. Somatic cells can be directly transformed to self-pulsating cardiomyocytes by introducing GATA4/MEF2c/Tbx5 into fibroblasts. miRNA is involved in reprogramming from somatic to iPS cells, suggesting that miRNA also functions in direct reprogramming to cardiomyocytes. (For color version of this figure, the reader is referred to the online version of this chapter.)

been performed in most genes, showing a limitation of the current miRNA studies. However, it is true that a specific miRNA induces cardiac hypertrophy, and studies should be continued to clarify the whole picture of the function. Altered miRNA expression in cardiac hypertrophy has been shown by microarray analysis of mainly a mouse aortic coarctation model. Studies based on this analysis using cultured cardiomyocytes and genetically modified mice have clarified that changes in specific miRNA expression induce cardiac hypertrophy. At present, miRNAs are considered to contribute to the induction of cardiac hypertrophy by controlling gene expression involved in cardiac hypertrophy.

The miRNA expression profiles have been investigated mainly employing microarrays in transverse aortic constriction-induced cardiac enlargement in mice, transplanted human hearts, and cultured cardiomyocytes after hypertrophic stimulation with a reagent. Although not all miRNAs showing changes in expression were confirmed by reproducing cardiac hypertrophy in an experimental system and verifying associated events, the functions of individual miRNAs in cardiac hypertrophy have been investigated by performing an overexpression experiment involving the transfection of cultured cardiomyocytes with a viral vector or double-stranded oligonuclease, a function-inhibition experiment using antagomiR, preparation of knockout mice, cardiac overexpression mice, and antagomiR administration to mice based on the above miRNA expression profiles.

miR-1 plays an important role in cardiogenesis, but at present, it is unclear whether miR-1 contributes to remodeling in human heart failure because miR-1 expression was shown to be enhanced in heart failure in some reports[43] but decreased in others.[44] In a previous report, miR-1 expression was shown to be reduced in an established cardiac hypertrophy model, namely, the calcineurin transgenic mice, and miR-1 overexpression using an adenovirus vector remitted enlargement through reducing calmodulin and MEF2a expressions.[45]

Because the partial knockout of miR-133 with specific antagomiR induces cardiac hypertrophy,[25] there is a possibility of inhibiting cardiac hypertrophy by enhancing miR-133 expression using drugs. On the other hand, the phenotype is normal in mice with knockout of either miR-133a-1 or miR-133a-2. Although the expression level is reduced to half, cardiac hypertrophy develops as a result of the pressure load on the left ventricle.[19] When both miR-133a genes are knocked out, abnormal cardiomyocyte proliferation, apoptosis, and ectopic expression of smooth muscle genes occur in the late embryonic or neonatal period, and animals die of ventricular septal defect.[14] About one-fourth of the animals survive until adulthood, but myocardial fibrosis develops widely without cardiomyocyte enlargement, leading to heart failure or sudden death. The heart-muscle-specific overexpression of miR-133a under the β-MHC promoter inhibits cardiomyocyte proliferation and causes fetal death.[19]

V. Conclusion

miRNAs exhibit few adverse effects because they are composed of RNA present in the body. In addition, several genes can be simultaneously controlled, suggesting their applicability as an effective treatment mode. Actually, studies aiming at clinical application in the future are being actively carried out. However, the modulation of several genes may have an unexpected influence due to differences in the target gene among animal species; more information is necessary for miRNA to be considered safe for treatment. Various problems have to be overcome, which may facilitate application not only in the cardiovascular field but also in many other fields. Attention should be paid to achievements in various fields because a clue to overcome problems may be unexpectedly found.

REFERENCES

1. Lee RC, Feinbaum RL, Ambros V. The *C. elegans* heterochronic gene lin-4 encodes small RNAs with antisense complementarity to lin-14. *Cell* 1993;**75**:843–54.
2. Wightman B, Ha I, Ruvkun G. Posttranscriptional regulation of the heterochronic gene lin-14 by lin-4 mediates temporal pattern formation in *C. elegans*. *Cell* 1993;**75**:855–62.
3. Fire A, Xu S, Montgomery MK, et al. Potent and specific genetic interference by double-stranded RNA in *Caenorhabditis elegans*. *Nature* 1998;**391**:806–11.
4. Evans MJ, Kaufman MH. Establishment in culture of pluripotential cells from mouse embryos. *Nature* 1981;**292**:154–6.
5. Bernstein E, Kim SY, Carmell MA, et al. Dicer is essential for mouse development. *Nat Genet* 2003;**35**:215–7.
6. Zhao Y, Ransom JF, Li A, et al. Dysregulation of cardiogenesis, cardiac conduction, and cell cycle in mice lacking miRNA-1-2. *Cell* 2007;**129**:303–17.
7. Kanellopoulou C, Muljo SA, Kung AL, et al. Dicer-deficient mouse embryonic stem cells are defective in differentiation and centromeric silencing. *Genes Dev* 2005;**19**:489–501.
8. Murchison EP, Partridge JF, Tam OH, et al. Characterization of Dicer-deficient murine embryonic stem cells. *Proc Natl Acad Sci U S A* 2005;**102**:12135–40.
9. Tay Y, Zhang J, Thomson AM, et al. MicroRNAs to *Nanog*, *Oct4* and *Sox2* coding regions modulate embryonic stem cell differentiation. *Nature* 2008;**455**:1124–8.
10. Suh MR, Lee Y, Kim JY, et al. Human embryonic stem cells express a unique set of microRNAs. *Dev Biol* 2004;**270**:488–98.
11. Sinkkonen L, Hugenschmidt T, Berninger P, et al. MicroRNAs control *de novo* DNA methylation through regulation of transcriptional repressors in mouse embryonic stem cells. *Nat Struct Mol Biol* 2008;**15**:259–67.
12. Benetti R, Gonzalo S, Jaco I, et al. A mammalian microRNA cluster controls DNA methylation and telomere recombination via Rbl2-dependent regulation of DNA methyltransferases. *Nat Struct Mol Biol* 2008;**15**:268–79.
13. Srivastava D. Making or breaking the heart: from lineage determination to morphogenesis. *Cell* 2006;**126**:1037–48.
14. Chen JF, Murchison EP, Tang R, et al. Targeted deletion of Dicer in the heart leads to dilated cardiomyopathy and heart failure. *Proc Natl Acad Sci U S A* 2008;**105**:2111–6.

15. da Costa Martins PA, Bourajjaj M, Gladka M, et al. Conditional dicer gene deletion in the postnatal myocardium provokes spontaneous cardiac remodeling. *Circulation* 2008;**118**:1567–76.

16. Cordes KR, Srivastava D. MicroRNA regulation of cardiovascular development. *Circ Res* 2009;**104**:724–32.

17. Liu N, Olson EN. MicroRNA regulatory networks in cardiovascular development. *Dev Cell* 2010;**18**:510–25.

18. Liu N, Williams AH, Kim Y, et al. An intragenic MEF2-dependent enhancer directs muscle-specific expression of microRNAs 1 and 133. *Proc Natl Acad Sci U S A* 2007;**104**:20844–9.

19. Liu N, Bezprozvannaya S, Williams AH. MicroRNA-133a regulates cardiomyocyte proliferation and suppresses smooth muscle gene expression in the heart. *Genes Dev* 2008;**22**:3242–54.

20. Zhao Y, Samal E, Srivastava D. Serum response factor regulates a muscle-specific microRNA that targets Hand2 during cardiogenesis. *Nature* 2005;**436**:214–20.

21. Yanazume T, Hasegawa K, Morimoto T, et al. Cardiac p300 is involved in myocyte growth with decompensated heart failure. *Mol Cell Biol* 2003;**23**:3593–606.

22. Takaya T, Kawamura T, Morimoto T, et al. Identification of p300-targeted acetylated residues in GATA4 during hypertrophic responses in cardiac myocytes. *J Biol Chem* 2008;**283**:9828–35.

23. Sunagawa Y, Morimoto T, Takaya T, et al. Cyclin-dependent kinase-9 is a component of the p300/GATA4 complex required for phenylephrine-induced hypertrophy in cardiomyocytes. *J Biol Chem* 2010;**285**:9556–68.

24. Kaichi S, Takaya T, Morimoto T, et al. Cyclin-dependent kinase 9 forms a complex with GATA4 and is involved in the differentiation of mouse ES cells into cardiomyocytes. *J Cell Physiol* 2011;**226**:248–54.

25. Care A, Catalucci D, Felicetti F, et al. MicroRNA-133 controls cardiac hypertrophy. *Nat Med* 2007;**13**:613–8.

26. Sayed D, Hong C, Chen IY, et al. MicroRNAs play an essential role in the development of cardiac hypertrophy. *Circ Res* 2007;**100**:416–24.

27. Takaya T, Ono K, Kawamura T, et al. MicroRNA-1 and microRNA-133 in spontaneous myocardial differentiation of mouse embryonic stem cells. *Circ J* 2009;**73**:1492–7.

28. Chen JF, Mandel EM, Thomson JM, et al. The role of microRNA-1 and microRNA-133 in skeletal muscle proliferation and differentiation. *Nat Genet* 2006;**38**:228–33.

29. Kawamura T, Ono K, Morimoto T, et al. Acetylation of GATA-4 is involved in the differentiation of embryonic stem cells into cardiac myocytes. *J Biol Chem* 2005;**280**:19682–8.

30. Hosseinkhani M, Hasegawa K, Ono K, et al. Trichostatin A induces myocardial differentiation of monkey ES cells. *Biochem Biophys Res Commun* 2007;**356**:386–91.

31. Ivey KN, Muth A, Arnold J, et al. MicroRNA regulation of cell lineages in mouse and human embryonic stem cells. *Cell Stem Cell* 2008;**2**:219–29.

32. Kwon C, Han Z, Olson EN, Srivastava D. MicroRNA-1 influences cardiac differentiation in *Drosophila* and regulates Notch signaling. *Proc Natl Acad Sci U S A* 2005;**102**:18986–91.

33. Takahashi K, Yamanaka S. Induction of pluripotent stem cells from mouse embryonic and adult fibroblast cultures by defined factors. *Cell* 2006;**126**:663–76.

34. Takahashi K, Tanabe K, Ohnuki M, et al. Induction of pluripotent stem cells from adult human fibroblasts by defined factors. *Cell* 2007;**131**:861–72.

35. Yu J, Vodyanik MA, Smuga-Otto K, et al. Induced pluripotent stem cell lines derived from human somatic cells. *Science* 2007;**318**:1917–20.

36. Judson RL, Babiarz J, Venere M, Blelloch R. Embryonic stem cell specific microRNAs promote induced pluripotency. *Nat Biotechnol* 2009;**27**:459–61.

37. Anokye-Danso F, Trivedi CM, Juhr D, et al. Highly efficient miRNA-mediated reprogramming of mouse and human somatic cells to pluripotency. *Cell Stem Cell* 2011;**8**:376–88.

38. Card DAG, Hebbar PB, Li L, et al. Oct4/Sox2-regulated miR-302 targets cyclin D1 in human embryonic stem cells. *Mol Cell Biol* 2008;**28**:6426–38.

39. Narazaki G, Uosaki H, Teranishi M, et al. Directed and systematic differentiation of cardio-vascular cells from mouse induced pluripotent stem cells. *Circulation* 2008;**118**:498–506.
40. Zwi L, Capsi O, Arbel G, et al. Cardiomyocyte differentiation of human induced pluripotent stem cells. *Circulation* 2009;**120**:1513–23.
41. Kaichi S, Hasegawa K, Takaya T, et al. Cell line-dependent differentiation of induced plurip-otent stem cells into cardiomyocytes in mice. *Cardiovasc Res* 2010;**88**:314–23.
42. Ieda M, Fu JD, Delgado-Olguin P, et al. Direct reprogramming of fibroblasts into functional cardiomyocytes by defined factors. *Cell* 2010;**142**:375–86.
43. Thum T, Galuppo P, Wolf C, et al. MicroRNAs in the human heart: a clue to fetal gene reprogramming in heart failure. *Circulation* 2007;**116**:258–67.
44. Matkovich SJ, Van Booven DJ, Youker KA, et al. Reciprocal regulation of myocardial micro-RNAs and messenger RNA in human cardiomyopathy and reversal of the microRNA signature by biomechanical support. *Circulation* 2009;**119**:1263–71.
45. Ikeda S, Kong SW, Lu J, et al. Altered microRNA expression in human heart disease. *Physiol Genomics* 2007;**31**:367–73.

Wnt Signaling and Cardiac Differentiation

Michael P. Flaherty,[*]
Timothy J. Kamerzell,[†] and
Buddhadeb Dawn[†]

[*]Division of Cardiovascular Medicine,
University of Louisville School of Medicine,
Louisville, Kentucky, USA

[†]Division of Cardiovascular Diseases,
University of Kansas Medical Center,
Kansas City, Kansas, USA

The Wnt family of secreted glycoproteins participates in a wide array of biological processes, including cellular differentiation, proliferation, survival, apoptosis, adhesion, angiogenesis, hypertrophy, and aging. The canonical Wnt signaling primarily utilizes β-catenin-mediated activation of transcription, while the noncanonical mechanisms involve a calcium-dependent protein kinase C-mediated Wnt/Ca^{2+} pathway and a dishevelled-dependent c-Jun N-terminal kinase-mediated planar cell polarity pathway. Although both canonical and noncanonical Wnts have been implicated in cardiac specification, morphogenesis, and differentiation; the molecular events remain unclear and often depend on the cell type and biological context. In this regard, growing evidence indicates that Wnt11 is able to induce cardiogenesis not only during embryonic development but also in adult cells. The cardiogenic properties of Wnt11 may prove useful for preprogramming adult stem cells before myocardial transplantation. Further, elucidation of the molecular steps in Wnt11-induced cardiac differentiation will be necessary to enhance the outcomes of cardiac cell therapy.

I. Introduction

An evolving paradigm supports the notion that transplantation of stem/progenitor cells can repair the damaged myocardium and restore cardiac function. Although the mechanisms underlying the observed benefits continue to unfold, considerable research efforts have been focused on the induction of cardiac differentiation in various cell types, including cells from adult tissues. Recent evidence also indicates that differentiation fates of stem cells are influenced by signaling within their microenvironment.[1] This notion is perhaps best illustrated by studies carried out in cultured embryonic stem cells (ESCs), in which various growth factors and pharmacologic agents can successfully induce cardiomyocytic differentiation.[2–4] However, similar efforts in adult cells have met with variable outcomes.[5–9]

A growing body of evidence suggests that members of the Wnt (derived from *wingless* [*wg*] in *Drosophila* and *Int-1* in mouse[10]) family of secreted glycoproteins play critical roles in cardiac development.[11,12] Expressed in a temporo-spatial manner during embryonic growth, Wnt proteins regulate cardiogenesis in flies, fishes, birds, amphibians, and mammals by binding to specific frizzled (Fz) receptors on target cells and directing signaling via β-catenin-dependent as well as -independent pathways[11–17]. *In vitro*, when cultured murine ESCs or embryonic tissues are exposed to various agonists or antagonists of Wnt signaling *in vitro*, activation of a cardiogenic program is observed.[18–20] While canonical signaling leads to Wnt/β-catenin-dependent activation of Wnt-responsive genes,[21,22] Wnt11-mediated noncanonical signaling utilizes a pathway that involves Wnt/protein kinase C (PKC)–Ca^{2+}/calmodulin-dependent protein kinase II (CamKII) and Wnt/c-Jun N-terminal kinase (JNK) for the induction of cardiogenic gene programs.[11,19,23]

Although the roles of specific molecules differ among species *in vivo* and among cells and models *in vitro*, the potential of Wnt molecules to promote cardiomyogenic differentiation in both embryonic and adult cells appears quite promising. Yet, very little is known about the cardiomyogenic potential of Wnts in adult tissue-specific stem and progenitor cells; or in bone-marrow-derived cells (BMCs) for that matter. Following brief discussions of Wnt signaling pathways, this chapter will focus on the effects of Wnt signaling during cardiogenesis and induction of cardiac differentiation in adult stem and progenitor cells, with particular reference to cell-type-specific differences in outcomes and mechanisms.

II. Wnt Signaling

Our understanding of the Wnt molecules and their participation in various developmental and cellular processes has grown exponentially over recent years. Initially described in the context of viral carcinogenesis in the mouse,[24] the

Wnt proteins have been implicated in cellular patterning and migration, differentiation, proliferation, apoptosis, and other diverse biological events.[21,25–28] Conventionally, Wnt signaling is categorized into "canonical" pathways that involve β-catenin[29,30] and "noncanonical" pathways that utilize the Wnt/Ca^{2+}[31–34] and the planar cell polarity (PCP)[35–38] pathways. Although a number of downstream effector molecules have been identified for each, in many instances, tissue-specific facets of Wnt signaling networks remain incompletely understood with frequent reports describing novel signaling modules.

A. Wnt Family of Proteins

The Wnt family of secreted glycoproteins consists of 19 known members in vertebrates. The expression of *Wnt* genes is observed in the early stages of embryonic development before any mesodermal subdivision.[39,40] During organogenesis, the time- and context-dependent expression of specific Wnts results in a precisely controlled orchestration of several highly complex processes, including segmentation and differentiation. The diverse array of biological activities influenced by Wnt signaling is quite broad, and includes (i) determination of cellular polarity, proliferation, and migration; (ii) craniofacial and neural development; (iii) cardiac development; (iv) tumorigenesis and metastasis; and (v) stem cell differentiation and renewal.[11,21,25–28] Via extensive networks, Wnt signaling is also known to influence homeostasis in numerous organ systems, including neural,[41] renal,[42] bone and marrow,[43,44] gastrointestinal,[45] dermal,[46] retinal,[47] and cardiovascular systems.[11,12,23]

Wnt molecules activate intracellular signaling by binding to the seven-span transmembrane Fz family of cell surface receptors. Based on their ability to induce a secondary body axis in *Xenopus* embryos and on the subcellular signaling elements recruited, the Wnt ligands are classified into two distinct groups. The first group (referred to as the Wnt1 class) includes Wnt1, Wnt2, Wnt3a, and Wnt8, ligands that predominantly use the canonical pathway for intracellular signaling (Fig. 1A).[23] The second group (referred to as the Wnt5a class) includes Wnt4, Wnt5a, Wnt5b, and Wnt11, ligands that use very different intracellular signaling networks, the noncanonical pathways (Fig. 1B).[23]

B. Canonical Wnt Signaling

The canonical Wnt signaling begins with the interaction of Wnt protein with the Fz family of transmembrane receptors[48] in the presence of the low-density lipoprotein receptor-related protein (LRP, isoforms 5 or 6) coreceptor,[49] resulting in a tripartite complex. In the absence of canonical Wnt proteins, β-catenin is bound to Axin and adenomatous polyposis coli tumor suppressor protein and phosphorylated by glycogen synthase kinase (GSK)-3β at the NH2 terminal followed by polyubiqitination and degradation by the proteosome (Fig. 1A).[50,51] Upon Fz binding by canonical Wnts, Dishevelled (Dsh) is recruited to the receptor complex and phosphorylated by casein kinase

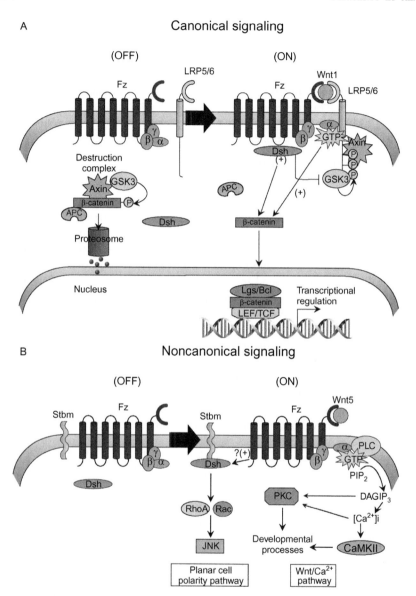

FIG. 1. Schematic diagrams depicting Wnt signal transduction pathways. (A) Canonical signaling: (OFF) In the absence of Wnt, β-catenin remains in the destruction complex comprising adenomatosis polyposis coli and Axin, and is phosphorylated by GSK-3β followed by ubiquitination and proteosomal degradation. (ON) Interaction of Wnt ligand with the Fz–LRP5/6 receptor complex results in the activation of Dishevelled (Dsh), which results in the inactivation of GSK-3β, and cytoplasmic stabilization and nuclear accumulation of β-catenin where it complexes with T-cell factor/lymphoid enhancer factor to modulate transcription. (B) Noncanonical signaling

Iε and forms a complex with Frat1 and inhibits GSK-3β.[52,53] This prevents phosphorylation of β-catenin, which becomes free to translocate to the nucleus and form complexes with the T-cell factor/leukemia enhancer factor (Tcf/Lef) family members (Fig. 1A).[54,55] Along with the Legless family of docking proteins, this complex activates or represses various Wnt target genes, including cyclin D1, c Myc, Dkk-1, and others.

C. Noncanonical Wnt Signaling

The noncanonical signaling is utilized largely by the Wnt5a class members and consists of two primary signal transduction mechanisms: a calcium-dependent PKC-mediated Wnt/Ca^{2+} pathway[31,32]; and a Dsh-dependent c-Jun N-terminal kinase (JNK)-mediated PCP pathway (Fig. 1B).[23,35–37] These second messenger systems responsible for mediating Wnt actions were first described in zebrafish embryos, in which *Xenopus*-derived XWnt5a enhanced intracellular Ca^{2+} transients likely arising from phosphatidylinositol activity that was β-catenin-independent.[56] Subsequently, it was found that Wnts also modulate PKC localization and stimulate PKC activity via a heterotrimeric G-protein-linked, Ca^{2+}-dependent mechanism[57] that involves CaM-KII (Fig. 1B). Although the details remain unclear, the noncanonical Wnts can also activate cytoplasmic Dsh possibly via cooperative interaction between Wnt/Fz and the transmembrane receptor strabismus (Stbm)[37,58] The downstream signaling via CaMKII/calcineurin regulates dorsal–ventral patterning, cytoskeletal processes, specification of embryonic myoblasts, and other developmental events (Wnt/Ca^{2+} pathway), while recruitment of the Rho family of small GTPases (RhoA, Rac, and Cdc42) results in activation of JNK in the PCP pathway (Fig. 1B)[11,36,37,59]

Although the above discussion implies the existence of discrete signaling pathways in the canonical and noncanonical domains, in reality, substantial overlaps exist not only among the noncanonical modules but also between noncanonical and the Wnt/β-catenin signaling.[23] Indeed, the cytoplasmic protein Dsh is involved in both noncanonical as well as Wnt/β-catenin pathways. However, unlike the canonical transduction, noncanonical activation of Dsh results in its recruitment to the cell membrane where it participates in both Wnt/Ca^{2+} and Rho/Rac signaling by itself, in cooperation with Stbm, and as a component of the G-protein-linked Fz receptor pathway.[37,58] In Fz-mediated pathways, there is evidence that Dsh is responsible for the activation of

(represented as the Wnt/Ca^{2+} and PCP pathways): Ca^{2+}/PKC signaling is initiated following Wnt activation of Fz receptors causing heterotrimeric G-protein-dependent Ca^{2+} release, thereby activating PKC and CaMKII; the Rho/JNK signaling occurs via the cooperative interaction between Stbm and Fz receptors, which activates Rho/Rac-mediated activation of JNK through Dsh. (Reproduced from Ref. 28 with permission from Elsevier). (For color version of this figure, the reader is referred to the online version of this chapter.)

Ca^{2+}-dependent PKC and CaMKII signaling by promoting intracellular Ca^{2+} influx, likely via its direct interaction with the $\beta\gamma$ G-protein subunits and/or by interacting directly with PKC forming a Dsh/PKC protein complex that drives PKC-dependent JNK signaling.[37,58,60] Together, these considerations identify Dsh as the central player involved in all key aspects of Wnt signaling. Nevertheless, the molecular details of Wnt signaling remain far from being complete, especially in the context of cardiac development and lineage commitment.

III. Wnt Proteins and Cardiogenesis

Cardiac development is an exquisitely complex process that involves simultaneous as well as sequential participation of numerous known and unknown factors that are produced in specific areas during embryonic growth.[61–63] Utilizing complex and mechanistically diverse intracellular signaling cascades, members of the Wnt family have been shown to play a critical role in cardiac specification and morphogenesis in flies, birds, fishes, amphibians, and mammals.[11,12,23,27] In addition, these cardiogenic properties have also been consistently recapitulated in embryonic as well as adult progenitor cells *in vitro*.[8,19,20,64–67] Nonetheless, the available evidence suggests that the relative importance of signaling via the canonical versus noncanonical Wnt pathways in cardiac specification varies depending on the species and the specific model systems.

A. Role of Canonical Pathways in Cardiogenesis

A critical role of *wg* in cardiac development was suggested by observations in *Drosophila*[68] with subsequent identification of Dsh and the β-catenin homolog Armadillo as mediators of cardiogenic effects of *wg*.[14] In contrast, microinjection of mRNA encoding the canonical Wnt inhibitors Dkk-1 and Crescent resulted in cardiac-specific gene expression and beating in explants from noncardiogenic ventral mesoderm in *Xenopus*,[16] suggesting that inhibition of canonical signaling is necessary for cardiogenesis in amphibians. Further, injection of canonical *Wnt3a* or *Wnt8* into the cardiogenic dorsal marginal zone explants inhibited cardiac induction as evidenced by downregulation of *Nkx2.5* expression.[16] Similar findings were noted in studies with chick embryonic explants where *Dkk-1* or *Crescent* induced cardiogenesis in noncardiogenic posterior mesoderm, and ectopic expression of Wnt3a inhibited cardiac gene expression in the precardiac anterior mesoderm.[17] In another study,[69] canonical Wnt inhibitors (Dkk-1 and Fz8/Fc chimera) promoted, while Wnt3a inhibited, cardiomyocytic differentiation of Flk1+ mesodermal cells *in vitro*. More detailed and stage-specific information was generated in a subsequent study,[70] in which chick embryos were exposed to Wnt3a, lithium chloride

(LiCl, an inhibitor of GSK3β and canonical Wnt activator), and SB415286 (a GSK3β inhibitor) at various stages of cardiac development. The exposure to Wnt3a-conditioned medium, LiCl, or SB415286 at early Hamburger Hamilton (HH) stage 3 resulted mostly in no identifiable cardiac tissue (lack of MF20 expression) in embryos. When embryos were exposed to canonical Wnt activators at mid to late HH stage 3 and at HH stage 4, the inhibition of cardiogenesis was less. Additional evidence along this line was produced in a study with insulin-like growth-factor-binding protein (IGFBP)-4,[71] which binds to Fz8/LRP6 and inhibits Wnt3a activation of β-catenin. IGFBP-4 enhanced cardiomyocyte differentiation of P19CL6 cells *in vitro*, and inhibition of IGFBP-4 reduced cardiomyogenesis both *in vitro* and *in vivo*.[71] Collectively, the above observations[16,17,69–71] suggested that inhibition of canonical Wnt signaling is required for heart development *in vivo*.

Several studies of cardiac differentiation in ESC lines *in vitro*, however, support a cardiogenic role of canonical Wnts. Indeed, Dimethyl sulfoxide (DMSO)-induced cardiac differentiation in the P19CL6 mouse embryonal carcinoma cell line requires the activation of the canonical Wnt pathway,[20] and a role of phosphatidylinositol 3-kinase (PI3-K)–Akt pathway in the maintenance of canonical Wnt signaling has also been reported.[72] Furthermore, β-catenin has been shown to directly regulate the expression of *Islet1* in cardiac progenitors.[73] and be required for the development and proliferation of cardiac progenitors.[74] In contrast, inhibition of β-catenin by Chibby, a nuclear protein that competes with Tcf/Lef for binding to β-catenin, has been shown to promote cardiac differentiation in murine ESCs.[75] Consistent with this, early exposure to BMP-4 followed by a small molecule inhibitor of Wnt/β-catenin signaling has been shown to markedly increase the generation of cardiomyocytes from human ESCs and induced pluripotent stem cells.[76] In another study, a small molecule inhibitor alone was sufficient to generate cardiomyocytes from human ESC-derived mesoderm cells.[77] Although these divergent effects of canonical signaling on cardiac fate determination can be explained in part by cell-type-specific differences, a substantial amount of evidence indicates that canonical Wnt/β-catenin signaling acts biphasically during cardiogenesis.[66,67,69] In the study by Ueno *et al.*,[67] activation of canonical Wnt/β-catenin signaling at the pregastrula stages (up to 5 h postfertilization [hpf]) increased *Nkx2.5* expression, while such activation during gastrulation (6–9 hpf) inhibited the same. Recent reports indicate that, although Wnt/β-catenin signaling promotes the expansion of secondary heart field derivatives,[78,79] it inhibits heart tube formation.[79] Moreover, it is also important to note that, despite this documented efficacy of canonical signaling in cardiac differentiation in embryonic tissue/cells, attempts to induce cardiomyogenic differentiation in adult progenitor cells with canonical Wnt3a have not been successful.[8,64]

B. Role of Noncanonical Pathways in Cardiogenesis

During vertebrate embryonic development, the expression of multiple Wnt genes has been observed in a spatio-temporally regulated manner during gastrulation within or in close proximity to the precardiac mesoderm, suggesting a preeminent role of Wnt proteins in cardiac specification.[15,18,19,80,81] In addition to canonical Wnt1, Wnt3a, and Wnt8, the early expression of Wnt5a and Wnt11 in Hensen's node and in cells close to the primitive streak indicate the importance of noncanonical Wnt signaling in cardiogenesis.[15,81,82]

The first evidence implicating Wnt11 as an important molecular signal during vertebrate cardiogenesis came from a study by Eisenberg and Eisenberg, wherein soluble Wnt11 was sufficient to promote cardiac tissue formation in posterior noncardiac mesoderm of quail embryonic explants.[18] Subsequent studies identified the noncanonical Wnt signaling pathway as the transducing apparatus for Wnt11.[83,84] Using elegant loss- and gain-of-function experiments, Pandur *et al.*[19] subsequently demonstrated that Wnt11 is not only required for cardiogenesis in *Xenopus* embryos but also sufficient to induce a contractile phenotype. This study also showed that in *Xenopus*, noncanonical Wnt11 signal transduction occurs in a JNK-dependent manner, and identified PKC as an upstream kinase.[19] Together, these results indicate that Wnt11, acting via the noncanonical pathway, is indispensable for vertebrate cardiogenesis.[18,23] Similar observations have been reported in *Xenopus* with Wnt-11R, a molecule closely related to mammalian Wnt11.[85] Further, recent data have identified a role of caspase 3-mediated suppression of canonical Wnt activities in Wnt11-induced enhancement of cardiomyocyte differentiation.[86] Consistent with these findings from developmental studies, several lines of evidence have since demonstrated that noncanonical signaling via Wnt11 is also sufficient to induce cardiomyocytic commitment in ESC populations *in vitro*.[19,67,86,87]

C. Induction of Cardiac Differentiation in Adult Cells

The premise that adult stem and progenitor cells from a variety of tissues, including the bone marrow, possess the capacity for cardiomyocytic lineage commitment has led to the emergence of a new and rapidly evolving field of cardiac regenerative medicine and to the development of innovative cell-based therapeutic modalities focused on rejuvenating the damaged human heart. Indeed, the use of autologous adult bone marrow-derived cells (BMCs) for cell-based myocardial repair has already proved to be clinically safe and modestly efficacious.[88–91] Although the mechanisms underlying the cardiac functional and structural benefits derived from cell therapy remain controversial, it is logical to assume that enhanced differentiation of transplanted cells into cardiac lineages can only improve this reparative process. Indeed, if exogenous cells are able to modify the injured myocardial substrate, cardiac reconstitution by these cells is likely to be greatly enhanced by augmentation of their cardiomyogenic potential before myocardial delivery *in vivo*. Accordingly, intense research efforts have been focused on the identification of molecular triggers that may promote the acquisition of a cardiomyocytic phenotype in adult cells from even unrelated tissues.[92,93]

Although several families of signaling molecules, including BMP, FGF, transforming growth factor (TGF)β, and Notch, have been implicated in cardiac development,[61–63] identification of a molecule singularly capable of triggering cardiomyocyte differentiation has been challenging. The interest in Wnt family members for this role was based upon the observations that several Wnt proteins, both canonical and noncanonical, are expressed very early in morphogenesis, and especially in mesodermal regions, giving rise to embryonic heart fields.[15,80–82] Based on this evidence from developmental studies, and the efficacy of Wnt molecules in recapitulating their cardiogenic properties in embryonic cells lines and tissues, induction of cardiac differentiation in adult cells has been attempted using modulators of Wnt pathways.

Although noncanonical Wnt signaling has been shown to play a key role in embryonic heart development, a similar role in adult cells has been investigated in only a small number of studies. Using circulating progenitor cells (CPCs) from adult human blood cocultured with neonatal rat cardiomyocytes, Koyanagi and colleagues examined the ability of Wnt3a- and Wnt11-conditioned media (CM) to promote cardiomyogenic differentiation.[64] While supplementation with Wnt11-CM significantly enhanced the number of cardiac marker-positive cells, Wnt3a-CM failed to do so. Interestingly, the cardiogenic effects were also absent when Wnt11-CM alone was used without neonatal cardiomyocytes.[64] In a similar experiment using murine bone-marrow-derived multipotent adult stem cells (mBM-MASCs) cultured on Wnt-secreting amniotic feeder cell lines, Wnt4, Wnt7a, Wnt7b, and Wnt11 were all able to induce cardiac marker genes (Nkx.25, GATA-4, dHAND, TEF1, and tropomyosine), while only Wnt11 could induce expression of β-myosin heavy chain (MHC) and brain natriuretic peptide (BNP).[65] However, reproducible expression of other typical cardiac marker genes (α-MHC, atrial natriuretic peptide [ANP]) were not observed, and organized contractile apparatus could not be discerned.[65]

The ability of Wnt3a and Wnt11 to induce cardiomyogenic differentiation in unfractionated density-gradient-separated bone marrow mononuclear cells (BMMNCs) was examined in the study by Flaherty *et al.*[8] The Wnt11-producing 293 cells were cultured on an insert, thereby eliminating the impact of cellular contact.[8] The exposure to Wnt11 alone resulted in a marked increase in the expression of cardiac markers at mRNA (Fig. 2)

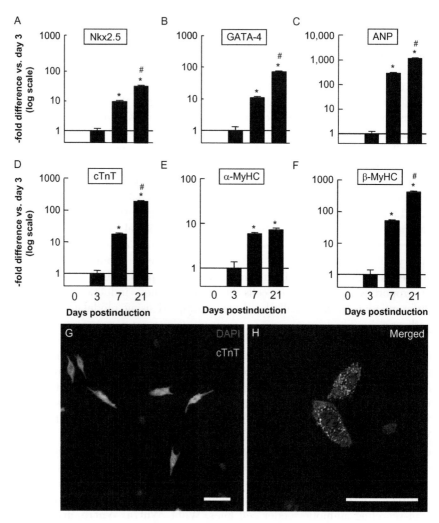

FIG. 2. Induction of the cardiac gene program and cardiac markers in unfractionated bone marrow mononuclear cells (BMMNCs) by Wnt11. BMMNCs were cultured either in the presence of recombinant Wnt3a protein or in coculture with Wnt11-secreting HEK-293 cells or control HEK-293 cells. Quantitative assessment of expression of cardiac transcription factors (GATA-4 and Nkx2.5), atrial natriuretic peptide (ANP), and contractile proteins (cTnT, α- and β-MHC isoforms) was performed by quantitative real-time reverse transcription polymerase chain reaction (qPT-PCR) at days 0, 3, 7, and 21 and by immunohistochemistry at day 21. (A–F) The mRNA levels in unfractionated BMMNCs cultured in the presence of Wnt11 are normalized to 18S rRNA (internal control) and expressed as -fold differences, relative to first detectable mRNA expression after induction (day 3). (G–H) Expression of cardiac troponin T (cTnT, panel G, green), cTnT and cardiac myosin heavy chain (panel H, red), and connexin43 (panel H, green) in Wnt11-treated BMMNCs, indicating the acquisition of a cardiomyocytic phenotype. Data are mean ± SEM; *$P < 0.001$ versus respective levels on day 3; #$P < 0.001$ versus respective levels on day 7. Scale bar = 25 μm. (Reproduced from Ref. 8 with permission from Wolters Kluwer Health/Lippincott Williams & Wilkins). (For interpretation of the references to color in this figure legend, the reader is referred to the online version of this chapter.)

and protein levels, while Wnt3a promoted pluripotency and hematopoietic potential (CD45 expression). The temporal pattern of cardiac-specific contractile proteins and ANP expression appeared to recapitulate the cardiac fetal gene program. This phenomenon has been described by others in cultured embryoid bodies and MSCs[5,94,95]: (i) upregulation of ANP and (ii) a differential expression of the β-isoform of MIIC, which predominates in fetal life, over the adult α-isoform.

In contrast to the above observations with noncanonical Wnt proteins, and despite the documented role of canonical β-catenin-dependent cardiogenic signaling in embryonic cardiac development and ESCs *in vitro*,[14,20,68,72] the attempts to induce a cardiac phenotype in adult cells using canonical signaling modulators have been largely unsuccessful.[8,64]

D. Mechanisms of Wnt11-Induced Cardiac Differentiation

The so-called canonical and noncanonical pathways utilize disparate subcellular signaling machineries to execute Wnt functions, although emerging evidence indicates that the Wnt/β-catenin and the Wnt/Ca^{2+} and Wnt/PCP modules do share common molecules and feedback loops.[34,96] For instance, a biphasic role for Wnt/β-catenin signaling in cardiac specification in ESCs[66,67] and zebrafish has been demonstrated, whereby, early canonical signaling is repressed by Wnt11 in favor of noncanonical signaling later during embryonic development.[67] Indeed, several mechanisms for Wnt11-mediated repression of the canonical Wnt signaling pathways, and vice versa, have also been reported.[67,97]

Even though not all of the molecular details have been definitively established the available evidence favors the notion that, for cardiac specification to occur during embryonic development, canonical Wnt signaling needs to be suppressed.[11,16,17,23] This is accomplished by antagonists of β-catenin pathway components (such as Dkk-1, Crescent, and Wnt11), or by expression/activation of other molecules that exert negative feedback inhibition of canonical signaling and/or promote cardiogenesis by themselves (such as BMPs and Smads), or by a combination of these possibilities. In this regard, ectopic overexpression of Wnt11 attenuates expression of both Wnt3a and β-catenin in a manner similar to that described in P19 embryonal carcinoma cells (Fig. 3A and B).[8,16,98] Although the "canonical inhibition" paradigm for cardiogenesis may portray a facilitative role of Dkk-1, Crescent, or Wnt11, newer evidence supports a more direct (or inductive) role of noncanonical signaling in this process.[8,86] Also, Dkk-1 and other canonical Wnt inhibitors have been shown to induce a homeodomain transcription factor Hex, which induces the expression of

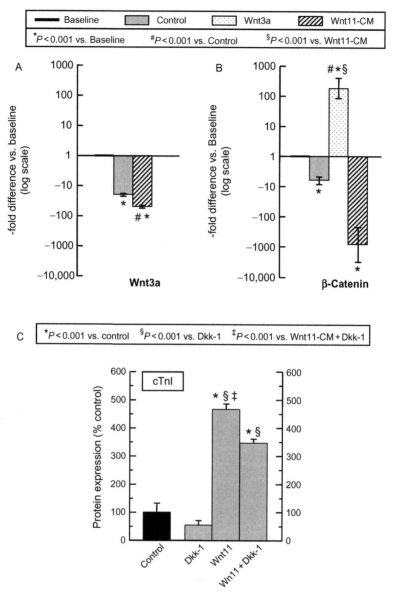

FIG. 3. Wnt and β-catenin expression in unfractionated bone marrow mononuclear cells (BMMNCs). Quantitative assessment of Wnt3a (A) and β-catenin (B) expression was performed by qRT-PCR in freshly isolated BMMNCs (baseline), and in BMMNCs cultured with control HEK-293 cells (control), or recombinant Wnt3a protein (Wnt3a), or in presence of Wnt11-secreting HEK-293 cells (Wnt11-CM). mRNA levels from unfractionated BMMNCs following culture are normalized to 18S rRNA (internal control) and expressed as -fold difference relative to

cardiac markers in *Xenopus* embryos in a non-cell-autonomous fashion.[99] However, further details of the pathway and downstream effector(s) remains to be identified.

Several lines of evidence emerging from both developmental studies and differentiation experiments with ESC or adult cells indicate that activation of PKC and JNK is critical for the induction of cardiac differentiation by Wnt11.[8,19,64,65] Following the activation of specific Fz receptors, the Wnt5a members are able to trigger intracellular Ca^{2+} release,[56] which then activates calcium-sensitive kinases PKC and CaMKII.[57,83] Subsequent to these events, or concurrently, JNK can be activated by Dsh itself, by PKC, or via the Rho kinases.[60,100] Although JNK inhibition did not block Wnt11-induced enhancement of cardiac differentiation in CPCs,[64] in *Xenopus* animal caps[19] and adult murine BMMNCs[8] inhibition of PKC or JNK could effectively block cardiac commitment. Further, activation of JNK was attenuated by PKC inhibitors, indicating that JNK is downstream of PKC in the context of noncanonical Wnt signaling that induces cardiac differentiation in these cells.[8,19] However, JNK inhibition did not reduce the cardiogenic effects of Wnt11 in CPCs,[64] while in BMMNCs, the expression of cardiac genes was reduced only partially by JNK inhibition,[8] suggesting the existence of additional signaling elements downstream of PKC and perhaps in parallel with JNK.

The other pathways and molecules that have been implicated as effectors of noncanonical signaling via PKC include the activating transcription factor/cAMP responsive element binding protein (ATF/CREB) family-dependent recruitment of TGFβ2 signaling,[101] and the GATA family of transcription factors.[102] Consistent with this notion, GATA-4/6 have been indicated as the integrators of Wnt signaling in cardiogenesis.[103] In this study, GATA-4/6 expression and cardiogenesis in *Xenopus* animal caps were inhibited by β-catenin, whereas Wnt11 expression was required for GATA-4/6-induced cardiogenesis. However, in the study by Belama Bedada *et al.*[65] PKC inhibition slightly increased *GATA-4* expression in adult mBM-MASCs, yet abrogated the expression of other cardiac genes. Lastly, in the study by Flaherty *et al.*[8] canonical inhibitor Dkk-1 could partially inhibit the induction of cardiac marker troponin I by Wnt11 in BMMNCs (Fig. 3C), suggesting the existence of additional intersections between the Wnt/β-catenin and noncanonical pathways in cardiogenic

baseline (calibrator). No expression of Wnt11 was observed in unfractionated BMMNCs or in empty vector (control) cells (therefore data are not shown). (C) Quantitative assessment of cardiac troponin I (cTnI) protein levels was performed using enzyme-linked immunosorbent assay (ELISA) in BMMNCs cultured with no addition (control), with the canonical Wnt inhibitor Dickkopf-1 (Dkk-1), or in presence of Wnt11-conditioned medium (CM) without (Wnt11-CM) or with Dkk-1 (Wnt11-CM + Dkk-1). Protein levels are expressed as percent of control. Data are mean ± SEM. (Reproduced from Ref. 8 with permission from Wolters Kluwer Health/Lippincott Williams & Wilkins).

differentiation of adult progenitors. Collectively, these apparently discordant findings underscore the cell-type-specific differences in the intracellular signaling cascades triggered by different Wnt ligands and Fz receptors.

In addition to the uncertainties regarding molecular pathways, several unresolved issues persist regarding the role of Wnt11 as a true "inducer" of cardiogenesis.[27] The evidence from studies *in vitro* with adult cells is helpful in this regard, as the issue of "induction" could be addressed more directly in these studies by eliminating the complex microenvironment inherent in embryonic tissues, and by using Wnt proteins alone. The candidacy of Wnt11 as an inducer of cardiac differentiation is supported by its ability to direct the expression of cardiac markers in adult human CPCs,[64] mBM-MASCs,[65] and unfractionated BMMNCs,[8] However, it is important to point out that in all of these instances, cardiac differentiation was confirmed largely by expression of cardiac markers alone. Moreover, in mBM-MASCs,[65] Wnt11 did not induce the expression of α-MHC, an important marker of the adult cardiac phenotype, and in Wnt11-treated BMMNCs,[8] contraction and sarcomeric organization were not documented despite the expression of other markers of advanced cardiac differentiation, including cardiac MHC, troponin I, and troponin T. Finally, in adult human CPCs, Wnt11 provoked cardiac differentiation only in coculture with neonatal rat cardiomyocytes, and not by itself.[64] Therefore, it is logical to infer that, although Wnt11 alone is sufficient to activate a cardiomyogenic gene program in some cell types, synergistic and perhaps sequential signaling initiated by other cardiogenic factors is critical in rendering the mature cardiomyocytic phenotype.

E. Cell-Specific Differences in Cardiogenic Effects of Noncanonical/Wnt11 Signaling

Although Wnt/β-catenin-directed cardiomyogenesis has been validated *in vitro* with the use of ESCs and the P19CL6 cells,[20,66,67] preservation of ESC pluripotency and self-renewal via Wnt3a signal transduction has also been reported.[104] However, these cardiogenic properties of canonical Wnt signaling appear to be restricted to pluripotent model systems and have yet to be applied to adult progenitors,[8,64] indicating variations in canonical Wnt actions depending on cell types. The observations with the noncanonical Wnt proteins also suggest cell-type-specific differences in biological actions of noncanonical Wnt signaling. Figure 4 illustrates the differential impact of Wnt11 on cell fate determination in embryonic, cord blood-, bone-marrow-, and adult blood-derived progenitor cell populations.[23] Consistent with their roles during embryonic development, the initiators of noncanonical signaling were also able to induce a cardiomyogenic program in pluripotent ESCs.[19,86,87] In contrast, in hematopoietic progenitor cell-enriched avian fetal bone marrow cells and quail

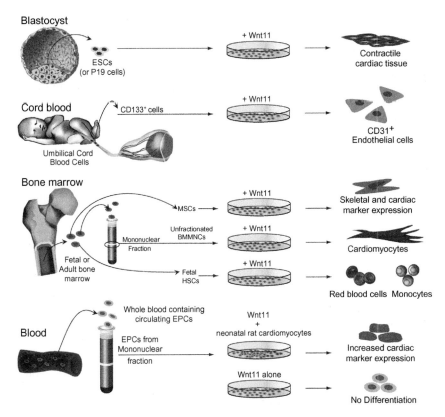

FIG. 4. Differential impact of Wnt11 signaling on cell fate *in vitro*. Pluripotent ESCs give rise to cardiac tissue with a contractile phenotype with Wnt11 treatment. However, Wnt11-treated cord-blood-derived CD133+ progenitors undergo endothelial lineage commitment; and Wnt11-treated bone-marrow-derived cells show considerable divergence with regard to lineage commitment and acquisition of phenotypic characteristics. In fetal HSCs, Wnt11 promotes red blood cell commitment and favors monocyte precursor formation. Wnt11-treated bone-marrow-derived multipotent adult stem cells express skeletal and cardiac markers, and Wnt11-treated CPCs adopt a cardiac phenotype when cocultured with neonatal rat cardiomyocytes. BMMNC, bone marrow mononuclear cell; EPC, endothelial progenitor cell; ESC, embryonic stem cell; HSC, hematopoietic stem cell; MSC, mesenchymal stem cell. (Reproduced from Ref. 28 with permission from Elsevier). (See Color Insert.)

mesodermal cell line QCE6, Wnt11 treatment promoted the expansion of red blood cells and monocytes at the expense of macrophages.[105] In another study that examined the effects of Wnts on cord blood-derived CD133 + cells, Wnt11 promoted the development of an endothelial-like phenotype, evidenced by increased number of CD31 + cells, at the expense of blast cells.[106]

The effects of Wnt11 in cells derived from adult tissues have also been variable. Although culture on Wnt11-producing amniotic feeder layer induced the expression of cardiac transcription factors, β-MHC, troponin T, and BNP in mBM-MASCs, reproducible induction of α-MHC, a common marker of adult cardiomyocyte phenotype, and ANP, another well-established marker of cardiac commitment, was not observed.[65] Moreover, in the study by Koyanagi *et al.*, Wnt11 could enhance cardiomyogenic differentiation in adult human CPCs only with coculture with neonatal rat cardiomyocytes, but not by itself.[64] However, Wnt11 alone (produced by HEK-293 cells in the absence of cellular contact) was able to induce the expression of cardiac markers in a temporal pattern reminiscent of the fetal gene program in adult BMMNCs.[8] These differences likely underscore the differential use of available intracellular signaling pathways by Wnt11 in a cell-type-specific manner.

IV. Concluding Remarks

The Wnt proteins participate in a wide range of biological processes that include key events during embryonic development as well as cellular differentiation, proliferation, adhesion, survival and apoptosis, angiogenesis, hypertrophy, and aging. With regard to cardiac differentiation, both canonical and noncanonical Wnt pathways play important roles, and the available evidence indicates that canonical Wnt signaling promotes heart formation in flies, while its inhibition is critical for cardiogenesis in vertebrates. Although canonical Wnt molecules can induce cardiac differentiation in select ESCs, a similar effect could not be produced in adult cells. Noncanonical signaling, however, has been effective in guiding several adult somatic cell populations toward a cardiomyogenic lineage commitment. However, the intracellular signaling recruited by Wnt11 in this context remains poorly understood, and its influence on cellular fate varies from one cell type to another. Nonetheless, the cardiomyogenic effects of Wnt11/noncanonical signaling in BMMNCs seem to be transduced by PKC, with JNK being a downstream mediator perhaps in combination with other molecules. Finally, while Wnt11 suppresses Wnt3a and β-catenin expression in unfractionated BMMNCs, inhibition of canonical signaling attenuates the cardiomyogenic potential of Wnt11, indicating cooperative and likely synergistic cross-talk between both pathways in the determination of differentiation fate of these cells.

Although the molecular basis remains to be elucidated in greater detail, the efficacy of noncanonical/Wnt11 signaling in cardiac differentiation of adult cells has considerable translational relevance in the light of accumulating data from clinical trials of cardiac repair with BMMNCs that have shown modest success.[88,90,91,107] Indeed, the easy availability of autologous BMMNCs in large numbers and their safety profile support the potential usefulness of BMMNCs for therapeutic cardiac repair.[93] The *in vivo* cardiac reparative capabilities of these cells can potentially be greatly enhanced through the use of Wnt11 to preprogram BMMNCs toward a cardiomyocytic fate prior to their cardiac delivery. A better understanding of the intracellular signaling cascades that follow Wnt11/Fz interaction will be critical to further improve the efficacy of cardiomyogenesis by Wnt11 possibly in combination with other modulators of Wnt pathways.

ACKNOWLEDGMENTS

This work was supported in part by NIH grant R01 HL-89939. We gratefully acknowledge the expert secretarial assistance of Ms. Renee Falsken.

REFERENCES

1. Czyz J, Wobus A. Embryonic stem cell differentiation: the role of extracellular factors. *Differentiation* 2001;**68**:167–74.
2. Wobus AM, Kaomei G, Shan J, Wellner MC, Rohwedel J, Ji G, et al. Retinoic acid accelerates embryonic stem cell-derived cardiac differentiation and enhances development of ventricular cardiomyocytes. *J Mol Cell Cardiol* 1997;**29**:1525–39. doi:S0022282897904338 [pii].
3. Behfar A, Zingman LV, Hodgson DM, Rauzier JM, Kane GC, Terzic A, et al. Stem cell differentiation requires a paracrine pathway in the heart. *FASEB J* 2002;**16**:1558–66.
4. Fukuda K, Yuasa S. Stem cells as a source of regenerative cardiomyocytes. *Circ Res* 2006;**98**:1002–13.
5. Makino S, Fukuda K, Miyoshi S, Konishi F, Kodama H, Pan J, et al. Cardiomyocytes can be generated from marrow stromal cells in vitro. *J Clin Invest* 1999;**103**:697–705.
6. Hattan N, Kawaguchi H, Ando K, Kuwabara E, Fujita J, Murata M, et al. Purified cardiomyocytes from bone marrow mesenchymal stem cells produce stable intracardiac grafts in mice. *Cardiovasc Res* 2005;**65**:334–44.
7. Abdel-Latif A, Zuba-Surma EK, Case J, Tiwari S, Hunt G, Ranjan S, et al. TGF-beta1 enhances cardiomyogenic differentiation of skeletal muscle-derived adult primitive cells. *Basic Res Cardiol* 2008;**103**:514–24.
8. Flaherty MP, Abdel-Latif A, Li Q, Hunt G, Ranjan S, Ou Q, et al. Noncanonical Wnt11 signaling is sufficient to induce cardiomyogenic differentiation in unfractionated bone marrow mononuclear cells. *Circulation* 2008;**117**:2241–52.
9. Rose RA, Jiang H, Wang X, Helke S, Tsoporis JN, Gong N, et al. Bone marrow-derived mesenchymal stromal cells express cardiac-specific markers, retain the stromal phenotype, and do not become functional cardiomyocytes in vitro. *Stem Cells* 2008;**26**:2884–92. doi:10.1634/stemcells.2008-0329 2008-0329 [pii].
10. Nusse R, Brown A, Papkoff J, Scambler P, Shackleford G, McMahon A, et al. A new nomenclature for int-1 and related genes: the Wnt gene family. *Cell* 1991;**64**:231. doi:0092-8674(91)90633-A [pii].
11. Eisenberg LM, Eisenberg CA. Wnt signal transduction and the formation of the myocardium. *Dev Biol* 2006;**293**:305–15.

12. Gessert S, Kuhl M. The multiple phases and faces of wnt signaling during cardiac differentiation and development. *Circ Res* 2010;**107**:186–99. doi: 10.1161/CIRCRESAHA.110.221531 107/2/186 [pii].

13. Augustine K, Liu ET, Sadler TW. Antisense attenuation of Wnt-1 and Wnt-3a expression in whole embryo culture reveals roles for these genes in craniofacial, spinal cord, and cardiac morphogenesis. *Dev Genet* 1993;**14**:500–20. doi:10.1002/dvg.1020140611.

14. Park M, Wu X, Golden K, Axelrod JD, Bodmer R. The wingless signaling pathway is directly involved in Drosophila heart development. *Dev Biol* 1996;**177**:104–16. doi:10.1006/dbio.1996.0149 S0012-1606(96)90149-9 [pii].

15. Eisenberg CA, Gourdie RG, Eisenberg LM. Wnt-11 is expressed in early avian mesoderm and required for the differentiation of the quail mesoderm cell line QCE-6. *Development* 1997;**124**:525–36.

16. Schneider VA, Mercola M. Wnt antagonism initiates cardiogenesis in *Xenopus laevis*. *Genes Dev* 2001;**15**:304–15.

17. Marvin MJ, Di Rocco G, Gardiner A, Bush SM, Lassar AB. Inhibition of Wnt activity induces heart formation from posterior mesoderm. *Genes Dev* 2001;**15**:316–27.

18. Eisenberg CA, Eisenberg LM. WNT11 promotes cardiac tissue formation of early mesoderm. *Dev Dyn* 1999;**216**:45–58.

19. Pandur P, Lasche M, Eisenberg LM, Kuhl M. Wnt-11 activation of a non-canonical Wnt signalling pathway is required for cardiogenesis. *Nature* 2002;**418**:636–41.

20. Nakamura T, Sano M, Songyang Z, Schneider MD. A Wnt- and beta-catenin-dependent pathway for mammalian cardiac myogenesis. *Proc Natl Acad Sci USA* 2003;**100**:5834–9.

21. Cadigan KM, Nusse R. Wnt signaling: a common theme in animal development. *Genes Dev* 1997;**11**:3286–305.

22. Rao TP, Kuhl M. An updated overview on Wnt signaling pathways: a prelude for more. *Circ Res* 2010;**106**:1798–806, doi: 10.1161/CIRCRESAHA.110.219840 106/12/1798 [pii].

23. Flaherty MP, Dawn B. Noncanonical Wnt11 signaling and cardiomyogenic differentiation. *Trends Cardiovasc Med* 2008;**18**:260–8. doi:10.1016/j.tcm.2008.12.001 S1050-1738(08) 00132-1 [pii].

24. Nusse R, Varmus HE. Many tumors induced by the mouse mammary tumor virus contain a provirus integrated in the same region of the host genome. *Cell* 1982;**31**:99–109. doi:0092-8674(82)90409-3 [pii].

25. Moon RT, Kohn AD, De Ferrari GV, Kaykas A. WNT and beta-catenin signalling: diseases and therapies. *Nat Rev Genet* 2004;**5**:691–701. doi:10.1038/nrg1427 nrg1427 [pii].

26. Logan CY, Nusse R. The Wnt signaling pathway in development and disease. *Annu Rev Cell Dev Biol* 2004;**20**:781–810.

27. Eisenberg LM, Eisenberg CA. Evaluating the role of Wnt signal transduction in promoting the development of the heart. *ScientificWorldJournal* 2007;**7**:161–76.

28. Klaus A, Birchmeier W. Wnt signalling and its impact on development and cancer. *Nat Rev Cancer* 2008;**8**:387–98. doi:nrc2389 [pii] 10.1038/nrc2389.

29. Barker N. The canonical Wnt/beta-catenin signalling pathway. *Methods Mol Biol* 2008;**468**:5–15. doi:10.1007/978-1-59745-249-6_1.

30. MacDonald BT, Tamai K, He X. Wnt/beta-catenin signaling: components, mechanisms, and diseases. *Dev Cell* 2009;**17**:9–26. doi:10.1016/j.devcel.2009.06.016 S1534-5807(09) 00257-3 [pii].

31. Slusarski DC, Corces VG, Moon RT. Interaction of Wnt and a Frizzled homologue triggers G-protein-linked phosphatidylinositol signalling. *Nature* 1997;**390**:410–3. doi:10.1038/37138.

32. Kuhl M, Sheldahl LC, Park M, Miller JR, Moon RT. The Wnt/Ca2+ pathway: a new vertebrate Wnt signaling pathway takes shape. *Trends Genet* 2000;**16**:279–83. doi:S0168-9525(00)02028-X [pii].

33. Kuhl M. The WNT/calcium pathway: biochemical mediators, tools and future requirements. *Front Biosci* 2004;**9**:967–74.

34. Kohn AD, Moon RT. Wnt and calcium signaling: beta-catenin-independent pathways. *Cell Calcium* 2005;**38**:439–46. doi:10.1016/j.ceca.2005.06.022 S0143-4160(05)00109-0 [pii].

35. Boutros M, Paricio N, Strutt DI, Mlodzik M. Dishevelled activates JNK and discriminates between JNK pathways in planar polarity and wingless signaling. *Cell* 1998;**94**:109–18. doi: S0092-8674(00)81226-X [pii].

36. Tada M, Smith JC. Xwnt11 is a target of Xenopus Brachyury: regulation of gastrulation movements via Dishevelled, but not through the canonical Wnt pathway. *Development* 2000;**127**:2227–38.

37. Veeman MT, Axelrod JD, Moon RT. A second canon. Functions and mechanisms of beta-catenin-independent Wnt signaling. *Dev Cell* 2003;**5**:367–77.

38. Fanto M, McNeill H. Planar polarity from flies to vertebrates. *J Cell Sci* 2004;**117**:527–33. doi:10.1242/jcs.00973117/4/527 [pii].

39. Hume CR, Dodd J. Cwnt-8C: a novel Wnt gene with a potential role in primitive streak formation and hindbrain organization. *Development* 1993;**119**:1147–60.

40. Bouillet P, Oulad-Abdelghani M, Ward SJ, Bronner S, Chambon P, Dolle P. A new mouse member of the Wnt gene family, mWnt-8, is expressed during early embryogenesis and is ectopically induced by retinoic acid. *Mech Dev* 1996;**58**:141–52.

41. Malaterre J, Ramsay RG, Mantamadiotis T. Wnt-Frizzled signalling and the many paths to neural development and adult brain homeostasis. *Front Biosci* 2007;**12**:492–506. doi:2077 [pii].

42. Nelson PJ, von Toerne C, Grone HJ. Wnt-signaling pathways in progressive renal fibrosis. *Expert Opin Ther Targets* 2011;**15**:1073–83. doi:10.1517/14728222.2011.588210.

43. Macsai CE, Foster BK, Xian CJ. Roles of Wnt signalling in bone growth, remodelling, skeletal disorders and fracture repair. *J Cell Physiol* 2008;**215**:578–87. doi:10.1002/jcp. 21342.

44. Gregory CA, Gunn WG, Reyes E, Smolarz AJ, Munoz J, Spees JL, et al. How Wnt signaling affects bone repair by mesenchymal stem cells from the bone marrow. *Ann N Y Acad Sci* 2005;**1049**:97–106. doi:10.1196/annals.1334.010 1049/1/97 [pii].

45. Doucas H, Garcea G, Neal CP, Manson MM, Berry DP. Changes in the Wnt signalling pathway in gastrointestinal cancers and their prognostic significance. *Eur J Cancer* 2005;**41**:365–79. doi:10.1016/j.ejca.2004.11.005 S0959-8049(04)00921-9 [pii].

46. Bergmann C, Akhmetshina A, Dees C, Palumbo K, Zerr P, Beyer C, et al. Inhibition of glycogen synthase kinase 3beta induces dermal fibrosis by activation of the canonical Wnt pathway. *Ann Rheum Dis* 2011;**70**:2191–8. doi:10.1136/ard.2010.147140 ard.2010.147140 [pii].

47. Lad EM, Cheshier SH, Kalani MY. Wnt-signaling in retinal development and disease. *Stem Cells Dev* 2009;**18**:7–16. doi:10.1089/scd.2008.0169.

48. Vinson CR, Conover S, Adler PN. A Drosophila tissue polarity locus encodes a protein containing seven potential transmembrane domains. *Nature* 1989;**338**:263–4. doi:10.1038/338263a0.

49. Pinson KI, Brennan J, Monkley S, Avery BJ, Skarnes WC. An LDL-receptor-related protein mediates Wnt signalling in mice. *Nature* 2000;**407**:535–8. doi:10.1038/35035124.

50. Aberle H, Bauer A, Stappert J, Kispert A, Kemler R. beta-catenin is a target for the ubiquitin-proteasome pathway. *EMBO J* 1997;**16**:3797–804. doi:10.1093/emboj/16.13.3797.

51. Ikeda S, Kishida S, Yamamoto H, Murai H, Koyama S, Kikuchi A. Axin, a negative regulator of the Wnt signaling pathway, forms a complex with GSK-3beta and beta-catenin and promotes GSK-3beta-dependent phosphorylation of beta-catenin. *EMBO J* 1998;**17**:1371–84. doi:10.1093/emboj/17.5.1371.

52. Lee JS, Ishimoto A, Yanagawa S. Characterization of mouse dishevelled (Dvl) proteins in Wnt/Wingless signaling pathway. *J Biol Chem* 1999;**274**:21464–70.

53. Kishida M, Hino S, Michiue T, Yamamoto H, Kishida S, Fukui A, et al. Synergistic activation of the Wnt signaling pathway by Dvl and casein kinase Iepsilon. *J Biol Chem* 2001;**276**:33147–55. doi:10.1074/jbc.M103555200M103555200 [pii].

54. Behrens J, von Kries JP, Kuhl M, Bruhn L, Wedlich D, Grosschedl R, et al. Functional interaction of beta-catenin with the transcription factor LEF-1. *Nature* 1996;**382**:638–42 doi:10.1038/382638a0.

55. Ishitani T, Ninomiya-Tsuji J, Matsumoto K. Regulation of lymphoid enhancer factor 1/T-cell factor by mitogen-activated protein kinase-related Nemo-like kinase-dependent phosphorylation in Wnt/beta-catenin signaling. *Mol Cell Biol* 2003;**23**:1379–89.

56. Slusarski DC, Yang-Snyder J, Busa WB, Moon RT. Modulation of embryonic intracellular Ca^{2+} signaling by Wnt-5A. *Dev Biol* 1997;**182**:114–20.

57. Sheldahl LC, Park M, Malbon CC, Moon RT. Protein kinase C is differentially stimulated by Wnt and Frizzled homologs in a G-protein-dependent manner. *Curr Biol* 1999;**9**:695–8.

58. Sheldahl LC, Slusarski DC, Pandur P, Miller JR, Kuhl M, Moon RT. Dishevelled activates Ca2+ flux, PKC, and CamKII in vertebrate embryos. *J Cell Biol* 2003;**161**:769–77.

59. Miller JR, Hocking AM, Brown JD, Moon RT. Mechanism and function of signal transduction by the Wnt/beta-catenin and Wnt/Ca2+ pathways. *Oncogene* 1999;**18**:7860–72.

60. Kinoshita N, Iioka H, Miyakoshi A, Ueno N. PKC delta is essential for Dishevelled function in a noncanonical Wnt pathway that regulates Xenopus convergent extension movements. *Genes Dev* 2003;**17**:1663–76.

61. Zaffran S, Frasch M. Early signals in cardiac development. *Circ Res* 2002;**91**:457–69.

62. Brand T. Heart development: molecular insights into cardiac specification and early morphogenesis. *Dev Biol* 2003;**258**:1–19.

63. Olson EN, Schneider MD. Sizing up the heart: development redux in disease. *Genes Dev* 2003;**17**:1937–56.

64. Koyanagi M, Haendeler J, Badorff C, Brandes RP, Hoffmann J, Pandur P, et al. Non-canonical Wnt signaling enhances differentiation of human circulating progenitor cells to cardiomyogenic cells. *J Biol Chem* 2005;**280**:16838–42.

65. Belema Bedada F, Technau A, Ebelt H, Schulze M, Braun T. Activation of myogenic differentiation pathways in adult bone marrow-derived stem cells. *Mol Cell Biol* 2005;**25**:9509–19.

66. Naito AT, Shiojima I, Akazawa H, Hidaka K, Morisaki T, Kikuchi A, et al. Developmental stage-specific biphasic roles of Wnt/beta-catenin signaling in cardiomyogenesis and hematopoiesis. *Proc Natl Acad Sci USA* 2006;**103**:19812–7.

67. Ueno S, Weidinger G, Osugi T, Kohn AD, Golob JL, Pabon L, et al. Biphasic role for Wnt/beta-catenin signaling in cardiac specification in zebrafish and embryonic stem cells. *Proc Natl Acad Sci USA* 2007;**104**:9685–90.

68. Wu X, Golden K, Bodmer R. Heart development in Drosophila requires the segment polarity gene wingless. *Dev Biol* 1995;**169**:619–28.

69. Yamashita JK, Takano M, Hiraoka-Kanie M, Shimazu C, Peishi Y, Yanagi K, et al. Prospective identification of cardiac progenitors by a novel single cell-based cardiomyocyte induction. *FASEB J* 2005;**19**:1534–6.

70. Manisastry SM, Han M, Linask KK. Early temporal-specific responses and differential sensitivity to lithium and Wnt-3A exposure during heart development. *Dev Dyn* 2006;**235**:2160–74.

71. Zhu W, Shiojima I, Ito Y, Li Z, Ikeda H, Yoshida M, et al. IGFBP-4 is an inhibitor of canonical Wnt signalling required for cardiogenesis. *Nature* 2008;**454**:345–9 doi:10.1038/nature07027 nature07027 [pii].

72. Naito AT, Akazawa H, Takano H, Minamino T, Nagai T, Aburatani H, et al. Phosphatidylinositol 3-kinase-Akt pathway plays a critical role in early cardiomyogenesis by regulating canonical Wnt signaling. *Circ Res* 2005;**97**:144–51.

73. Lin L, Cui L, Zhou W, Dufort D, Zhang X, Cai CL, et al. Beta-catenin directly regulates Islet1 expression in cardiovascular progenitors and is required for multiple aspects of cardiogenesis. *Proc Natl Acad Sci USA* 2007;**104**:9313–8. doi:10.1073/pnas.0700923104 0700923104 [pii].

74. Kwon C, Arnold J, Hsiao EC, Taketo MM, Conklin BR, Srivastava D. Canonical Wnt signaling is a positive regulator of mammalian cardiac progenitors. *Proc Natl Acad Sci USA* 2007;**104**:10894–9. doi:10.1073/pnas.0704044104 0704044104 [pii].

75. Singh AM, Li FQ, Hamazaki T, Kasahara H, Takemaru K, Terada N. Chibby, an antagonist of the Wnt/beta-catenin pathway, facilitates cardiomyocyte differentiation of murine embryonic stem cells. *Circulation* 2007;**115**:617–26.

76. Ren Y, Lee MY, Schliffke S, Paavola J, Amos PJ, Ge X, et al. Small molecule Wnt inhibitors enhance the efficiency of BMP-4-directed cardiac differentiation of human pluripotent stem cells. *J Mol Cell Cardiol* 2011;**51**:280–7 doi:10.1016/j.yjmcc.2011.04.012 S0022-2828(11) 00170-2 [pii].

77. Willems E, Spiering S, Davidovics H, Lanier M, Xia Z, Dawson M, et al. Small-molecule inhibitors of the Wnt pathway potently promote cardiomyocytes from human embryonic stem cell-derived mesoderm. *Circ Res* 2011;**109**:360–4. doi:10.1161/CIRCRESAHA.111.249540 CIRCRESAHA.111.249540 [pii].

78. Ai D, Fu X, Wang J, Lu MF, Chen L, Baldini A, et al. Canonical Wnt signaling functions in second heart field to promote right ventricular growth. *Proc Natl Acad Sci USA* 2007;**104**:9319–24.

79. Klaus A, Saga Y, Taketo MM, Tzahor E, Birchmeier W. Distinct roles of Wnt/beta-catenin and Bmp signaling during early cardiogenesis. *Proc Natl Acad Sci USA* 2007;**104**:18531–6.

80. Kispert A, Vainio S, Shen L, Rowitch DH, McMahon AP. Proteoglycans are required for maintenance of Wnt-11 expression in the ureter tips. *Development* 1996;**122**:3627–37.

81. Baranski M, Berdougo E, Sandler JS, Darnell DK, Burrus LW. The dynamic expression pattern of frzb-1 suggests multiple roles in chick development. *Dev Biol* 2000;**217**:25–41. doi:10.1006/dbio.1999.9516 S0012-1606(99)99516-7 [pii].

82. Chapman SC, Brown R, Lees L, Schoenwolf GC, Lumsden A. Expression analysis of chick Wnt and frizzled genes and selected inhibitors in early chick patterning. *Dev Dyn* 2004;**229**:668–76. doi:10.1002/dvdy.10491.

83. Kuhl M, Sheldahl LC, Malbon CC, Moon RT. Ca(2+)/calmodulin-dependent protein kinase II is stimulated by Wnt and Frizzled homologs and promotes ventral cell fates in Xenopus. *J Biol Chem* 2000;**275**:12701–11.

84. Heisenberg CP, Tada M, Rauch GJ, Saude L, Concha ML, Geisler R, et al. Silberblick/Wnt11 mediates convergent extension movements during zebrafish gastrulation. *Nature* 2000;**405**:76–81.

85. Garriock RJ, D'Agostino SL, Pilcher KC, Krieg PA. Wnt11-R, a protein closely related to mammalian Wnt11, is required for heart morphogenesis in Xenopus. *Dev Biol* 2005;**279**:179–92.

86. Abdul-Ghani M, Dufort D, Stiles R, De Repentigny Y, Kothary R, Megeney LA. Wnt11 promotes cardiomyocyte development by caspase-mediated suppression of canonical Wnt signals. *Mol Cell Biol* 2011;**31**:163–78. doi:10.1128/MCB.01539–09 MCB.01539-09 [pii].

87. Terami H, Hidaka K, Katsumata T, Iio A, Morisaki T. Wnt11 facilitates embryonic stem cell differentiation to Nkx2.5-positive cardiomyocytes. *Biochem Biophys Res Commun* 2004;**325**: 968–75.

88. Strauer BE, Brehm M, Zeus T, Kostering M, Hernandez A, Sorg RV, et al. Repair of infarcted myocardium by autologous intracoronary mononuclear bone marrow cell transplantation in humans. *Circulation* 2002;**106**:1913–8.

89. Assmus B, Honold J, Schachinger V, Britten MB, Fischer-Rasokat U, Lehmann R, et al. Transcoronary transplantation of progenitor cells after myocardial infarction. *N Engl J Med* 2006;**355**:1222–32.

90. Schachinger V, Erbs S, Elsasser A, Haberbosch W, Hambrecht R, Holschermann H, et al. Intracoronary bone marrow-derived progenitor cells in acute myocardial infarction. *N Engl J Med* 2006;**355**:1210–21.

91. Abdel-Latif A, Bolli R, Tleyjeh IM, Montori VM, Perin EC, Hornung CA, et al. Adult bone marrow-derived cells for cardiac repair: a systematic review and meta-analysis. *Arch Intern Med* 2007;**167**:989–97.

92. Chugh AR, Zuba-Surma EK, Dawn B. Bone marrow-derived mesenchymal stems cells and cardiac repair. *Minerva Cardioangiol* 2009;**57**:185–202.

93. Dawn B, Abdel-Latif A, Sanganalmath SK, Flaherty MP, Zuba-Surma EK. Cardiac repair with adult bone marrow-derived cells: the clinical evidence. *Antioxid Redox Signal* 2009;**11**: 1865–82. doi:10.1089/ARS.2009.2462 10.1089/ARS.2009.2462 [pii].

94. Auda-Boucher G, Bernard B, Fontaine-Perus J, Rouaud T, Mericksay M, Gardahaut MF. Staging of the commitment of murine cardiac cell progenitors. *Dev Biol* 2000;**225**:214–25.

95. Xu C, Police S, Hassanipour M, Gold JD. Cardiac bodies: a novel culture method for enrichment of cardiomyocytes derived from human embryonic stem cells. *Stem Cells Dev* 2006;**15**:631–9.

96. Kuhl M, Geis K, Sheldahl LC, Pukrop T, Moon RT, Wedlich D. Antagonistic regulation of convergent extension movements in Xenopus by Wnt/beta-catenin and Wnt/Ca^{2+} signaling. *Mech Dev* 2001;**106**:61–76. doi:S0925477301004166 [pii].

97. Maye P, Zheng J, Li L, Wu D. Multiple mechanisms for Wnt11-mediated repression of the canonical Wnt signaling pathway. *J Biol Chem* 2004;**279**:24659–65.

98. Lev S, Kehat I, Gepstein L. Differentiation pathways in human embryonic stem cell-derived cardiomyocytes. *Ann N Y Acad Sci* 2005;**1047**:50–65.

99. Foley AC, Mercola M. Heart induction by Wnt antagonists depends on the homeodomain transcription factor Hex. *Genes Dev* 2005;**19**:387–96.

100. Li L, Yuan H, Xie W, Mao J, Caruso AM, McMahon A, et al. Dishevelled proteins lead to two signaling pathways. Regulation of LEF-1 and c-Jun N-terminal kinase in mammalian cells. *J Biol Chem* 1999;**274**:129–34.

101. Zhou W, Lin L, Majumdar A, Li X, Zhang X, Liu W, et al. Modulation of morphogenesis by noncanonical Wnt signaling requires ATF/CREB family-mediated transcriptional activation of TGFbeta2. *Nat Genet* 2007;**39**:1225–34.

102. Ventura C, Zinellu E, Maninchedda E, Fadda M, Maioli M. Protein kinase C signaling transduces endorphin-primed cardiogenesis in GTR1 embryonic stem cells. *Circ Res* 2003;**92**:617–22. doi:10.1161/01.RES.0000065168.31147.5B [pii].

103. Afouda BA, Martin J, Liu F, Ciau-Uitz A, Patient R, Hoppler S. GATA transcription factors integrate Wnt signalling during heart development. *Development* 2008;**135**:3185–90.

104. Singla DK, Schneider DJ, LeWinter MM, Sobel BE. wnt3a but not wnt11 supports self-renewal of embryonic stem cells. *Biochem Biophys Res Commun* 2006;**345**:789–95.

105. Brandon C, Eisenberg LM, Eisenberg CA. WNT signaling modulates the diversification of hematopoietic cells. *Blood* 2000;**96**:4132–41.

106. Nikolova T, Wu M, Brumbarov K, Alt R, Opitz H, Boheler KR, et al. WNT-conditioned media differentially affect the proliferation and differentiation of cord blood-derived CD133+ cells in vitro. *Differentiation* 2007;**75**:100–11.

107. Strauer BE, Brehm M, Zeus T, Bartsch T, Schannwell C, Antke C, et al. Regeneration of human infarcted heart muscle by intracoronary autologous bone marrow cell transplantation in chronic coronary artery disease: the IACT Study. *J Am Coll Cardiol* 2005;**46**:1651–8.

Cross Talk Between the Notch Signaling and Noncoding RNA on the Fate of Stem Cells

Yaoliang Tang, Yingjie Wang,
Lijuan Chen,
Neal Weintraub, and
Yaohua Pan

Division of Cardiovascular Disease,
Internal Medicine, University of Cincinnati,
Cincinnati, Ohio, USA

The Notch signal pathway directly controls stem cell survival, proliferation, and differentiation. Noncoding RNAs, including small microRNAs (miRNAs) and long noncoding RNAs (LncRNAs), are responsible for fine regulation of gene expression. It is also becoming increasingly evident that miRNAs have a profound impact on the Notch signaling pathway; conversely, Notch signaling can regulate a handful of miRNAs that are involved in the stem cell function. In this chapter, we summarize our current knowledge of Notch-mediated stem cell differentiation and the cross talk between miRNAs and the Notch signal pathway on determining the fate of stem cells.

Progress in Molecular Biology
and Translational Science, Vol. 111
http://dx.doi.org/10.1016/B978-0-12-398459-3.00008-3

I. Introduction

A. Notch Signal Pathway

Notch signaling is a highly conserved signaling pathway that allows for communication between two adjacent cells and can elicit many downstream responses, including cell-fate specification, and progenitor cell differentiation, maintenance, and proliferation.[1] In canonical Notch signaling, a Notch trans-membrane receptor interacts extracellularly with a canonical Notch transmembrane ligand on a contacting cell, initiating proteolytic cleavage of the receptor and the subsequent release of the Notch intracellular domain (N-ICD) of the receptor. N-ICD then translocates to the nucleus where it interacts with an RBPJ (recombination signal binding protein for immunoglobulin kappa J region) and initiates the transcription of Notch target genes (Fig. 1). The deletion of RBPJ results in embryonic lethality due to a collapsed endocardium and the lack of mesenchymal cushion cells during heart development.[2] In mammals, four different Notch receptors have been identified (Notch 1–4). Notch receptors are activated by five type I transmembrane ligands: two Serrate/Jagged (Jag1 and Jag2) and three Delta-like (Dll1, Dll3, and Dll4). RBPJ is a key transcription factor in the signaling pathway of all four mammalian Notch receptors.

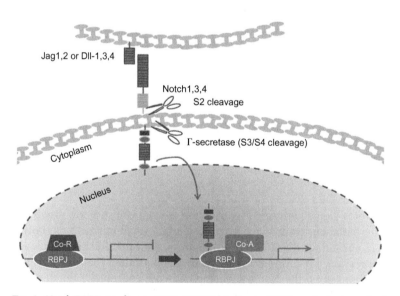

Fɪɢ. 1. Notch/RBPJ signaling in CPC. N-ICD binding to RBPJ results in the activation of Notch-dependent gene expression. (For color version of this figure, the reader is referred to the online version of this chapter.)

B. Stem Cells

1. Embryonic Stem Cells and Alternative induced Pluripotent Stem Cells

Pluripotent stem cell lines can be derived from the inner cell mass of the 5- to 7-day-old blastocyst. However, human embryonic stem cell (hESC) research is ethically and politically controversial because it involves the destruction of human embryos.[3]

The hottest area in stem cell biology is the study of induced pluripotent stem (iPS) cells. Right now, iPS cells are considered a type of pluripotent stem cell artificially derived from adult somatic cells by "forced" expression of reprogramming factors, for example, Oct4, Sox2, Klf4, c-Myc, Nanog, Lin 28, and the miR302–367 cluster.[4–6] iPS cells are similar to natural embryonic stem (ES) cells in many aspects, including gene expression, microRNA (miRNA) expression, chromatin methylation pattern, teratoma formation containing all three embryonic germ-layer-derived tissues upon *in vivo* grafting into immunocompromised mice, and differentiability. However, iPS cells avoid the ethical problems specific to ES cells regarding the destruction of human embryos. The important advance of iPS cells is the possibility that they can be developed for autologous cell therapy application. Generation of iPS cells can use the viral[7] or, the more recent, nonviral approach[8] to deliver reprogramming genes, mRNA, or proteins[9] into cells. However, this field is still facing several significant obstacles, including incomplete reprogramming and the safety of iPS cells, which are most likely to trigger cancer.

2. Adult Tissue-Derived Multipotent Progenitors

While many ethical and immunogenic obstacles hinder the use of hESCs in tissue regeneration, the adult stem cells are an alternative autologous source to overcome those issues. Most adult progenitors are multipotent cells with limited potential to differentiate into specific lineages that exist in different tissues. Adult stem cells are found in almost all tissues, including the bone marrow, heart, brain, liver, pancreas, and skin. Bone-marrow-derived c-kit positive hematopoietic and mesenchymal cells have been implicated in myocardial regeneration with limited functional improvement in infarcted hearts by a paracrine mechanism that is dependent on the secretion of soluble factors and not on transdifferentiation.[10–12]

Resident cardiac stem cells are an alternative source of cells for cardiac repair. CPCs expressing c-kit,[13] Sca-1,[14] Isl1+,[15] cardiosphere-derived cells (CDCs),[16–19] or cardiac side population[20] cells have been identified in adult mouse myocardium and have been implicated in heart regeneration after myocardial infarction. These cardiac-derived progenitors can reconstitute well-differentiated myocardium and, therefore, their real potential as tissue-resident progenitors for specific tissue regeneration is higher than bone-marrow-derived cells.

C. Noncoding RNA

1. MicroRNA

miRNAs are endogenous 21–22-nucleotide (nt) noncoding small RNAs.[21] The precursor miRNAs (pre-miRs) are processed by Dicer into mature miRNAs after a series of processing steps.[22] Mature miRNAs, together with Dicer and Argonaute proteins, form the miRNA-induced silencing complex. Guided by the miRNA through base-pairing, the miRNA complex binds to the 3′-untranslated region (3′UTR) of target genes for cleavage or translational repression.[23] Previous studies have suggested that miRNAs may play important roles in cardiovascular and neural development,[24,25] stem cell differentiation,[26–28] apoptosis,[29] and tumor.[30] miRNA offers completely new possibilities for improving stem cell therapy.

2. Long Noncoding RNA

Although considerable attention has been devoted to miRNA in recent years, the transcriptome also encodes a significant number of potential regulatory RNAs, for example, the long noncoding RNAs (LncRNAs), which are emerging as regulators of many basic cellular pathways during stem cell or progenitor cell development. For example, Vax2os1 is a retina-specific LncRNA and is involved in the control of cell cycle progression of photoreceptor progenitor cells in the ventral retina.[31] In another example, the LncRNA Mistral (Mira) activates transcription of the homeotic genes Hoxa6 and Hoxa7 in mouse embryonic stem cells (mESCs) by recruiting MLL1 to chromatin.[32] Some functional LncRNAs are directly controlled by key mESC transcription factors, including Oct4 and Nonog.[33] Recent studies have revealed that LncRNAs are subject to epigenetic regulation in a similar manner in ES cells and lineage-restricted neuronal progenitor cells and that knockdown of the H3K27me3 methyltransferase Ezh2 results in the upregulation of the expression of these LncRNAs in ES cells.[34]

D. Epigenetics

The acetylation and deacetylation status of histones modifies the chromatin structure, which regulates gene expression as well as stem cell differentiation and survival. Histone deacetylases (HDACs) are a class of enzymes that remove the acetyl group from the N-terminal tails of histones and alter their interaction with DNA, thus serving as epigenetic regulators of gene expression.[35] Based on the structural characteristics of HDACs, the HDAC family of proteins are grouped into classes I, II, and IV.[36] Class I HDACs, including HDAC1, HDAC2, HDAC3, and HDAC8, have ubiquitous tissue distribution and present within the cell nucleus. The class I selective histone deacetylase inhibitor (HDACi) MGCD0103 has significant activity against colon cancer initiating cells (CCICs) and also significantly inhibits non-CCIC CRC cell xenograft formation.[37] HDAC4, HDAC5, HDAC7, and HDAC9 belong to Class IIA HDACs, which exhibit restricted tissue expression and are involved in stem cell differentiation. Karamboulas et al.[38] have reported that Class II HDAC regulates the specification of mesoderm cells into cardiomyoblasts by inhibiting the expression of GATA4 and Nkx2-5 in a P19 stem cell model system. Chen et al.[39] have shown that HDACis trichostatin A (TSA,50 nmol/L) and sodium butyrate (NaB, 200 µmol/L) can cause a pronounced reduction in HDAC4 activity and an upregulation of the expression of cardiac markers including GATA4, MEF2C, Nkx2.5, cardiac actin, and alpha-SMA. Class IIB HDACs, including HDAC6 and HDAC10, are cytoplasmic enzymes. Class IV HDAC, HDAC11, found in both nucleus and cytoplasm, may influence immune activation via IL-10 regulation.[40]

II. Direct Role of Notch Signaling in Stem Cell Maintenance and Differentiation

A. Notch Signal for Cardiovascular Lineage Differentiation of Stem Cells

Notch signaling plays an essential role in cardiovascular development[1,41] Most severe congenital heart malformations have inherited gene variants in Notch1.[42] Notch signal promotes epithelial-to-mesenchymal transition (EMT) via transcriptional induction of the Snail repressor, a potent and evolutionarily conserved mediator of EMT in many tissues. It also promotes transforming growth factor-beta-mediated EMT, which leads to cellularization of developing cardiac valvular primordial.[2] In addition, Notch signaling converts polarized epithelial cells into motile, invasive cells.[43] Koyanagi et al.[44] have reported that Notch is activated upon coculture of endothelial progenitor cells (EPCs) with neonatal rat cardiac myocytes. Gamma-secretase-dependent Notch activation is required for cardiac gene expression in human cells and induces the expression of noncanonical Wnt proteins, which may act in a paracrine manner to further amplify cardiac differentiation.

The mammalian heart has long been considered a postmitotic organ, with little regenerative capacity following heart injury, thereby leading to scar formation. In contrast, zebrafish can fully regenerate their hearts after ventricular resection, with little scarring.[45] The cellular mechanisms underlying regeneration in zebrafish appear to involve activation of endogenous CPC and expansion of cardiomyocytes.[45–48] Recent evidence indicates that the adult mammalian heart (including the human heart) possesses a pool of intermediate cardiac progenitors that can differentiate into cells that phenotypically resemble cardiomyocytes, endothelial cells (ECs), and vascular smooth muscle cells, which are the major components of myocardial tissue.[49,50] Unfortunately, resident mammalian CPC cell populations become acutely depleted following myocardial infarction,[51] which limits their endogenous regenerative capacity. A most recent clinical trial from the Cedars-Sinai Heart Institute and Johns Hopkins Hospital demonstrated that cardiosphere-derived progenitor cells grown from patients' own cardiac tissue can heal their damaged hearts by an increase in muscle mass and a shrinkage of the scar size.[52] However, CPC-treated hearts do not show ejection fraction improvement; therefore, elucidating the mechanisms that are beneficial for CPC cardiac differentiation and proliferation in ischemic myocardium could markedly improve the efficacy of stem-cell-mediated cardiac protection and repair. Recent studies have shown that zebrafish possess a unique yet poorly understood capacity for cardiac regeneration, which involves the execution of a specific genetic program with a marked upregulation of the Delta/Notch signal pathway, indicating a role for the Notch pathway in the activation of the regenerative response.[53,54] Notch-1 gene modification of immature cardiac myocytes (ICMs) has been reported by Collesi et al.[55] Their work demonstrated that the inhibition of Notch signaling in ICMs blocked their proliferation and induced apoptosis; in contrast, its activation by Jagged1 or by the constitutive expression of its activated form using an adeno-associated virus markedly stimulated proliferative signaling and promoted ICM expansion. In mammals, Notch signal is a critical determinant of CPC growth and differentiation.[56] Recent studies have shown that the nuclear translocation of N1-ICD upregulates Nkx2.5 in CPC and promotes the formation of cycling myocytes in vitro. N1-ICD and RBPJ form a protein complex, which in turn binds to the Nkx2.5 promoter, initiating transcription and myocyte differentiation.[54] Cardiosphere-derived-Cells (CDCs) including CPCs, which play critical roles in physiological and pathological turnover of heart tissue, are multipotent progenitors present in the human heart. Our group[57] examined the role of Notch-1/RBPJ signaling in CDC differentiation. We isolated CDCs from mouse cardiospheres and analyzed the differentiation of transduced cells expressing the N1-ICD, the active form of Notch-1, using a terminal differentiation marker PCR array. We found that Notch-1 primarily supported differentiation of CDCs into smooth muscle cells (SMCs), as demonstrated by the upregulation of key SMC proteins, including smooth muscle myosin heavy chain (Myh11) and SM22α (Tagln), in N1-ICD expressing CDC. Conversely, genetic ablation of RBPJ in CDC diminished the expression of SMC differentiation markers, confirming that SMC differentiation of CDC is dependent on RBPJ. These findings underscore the role of Notch signaling for cardiovascular lineage differentiation of stem cells.

B. Notch Signaling for Stem-Cell-Mediated Angiogenesis

Angiogenesis is a complex process of formation of new blood vessels by EC activation, migration, proliferation, and sprouting of the existing vasculature, which are critical for a number of physiologic and pathologic conditions, including healing of damaged tissues. The Notch signaling pathway is an evolutionarily conserved intercellular signaling pathway. Recent studies have demonstrated that Notch signaling plays a central role in angiogenesis by regulating EC specification and vascular smooth muscle cell differentiation and by controlling blood vessel sprouting and branching. Vascular formation includes two rate-limiting steps: vasculogenesis (the *de novo* formation of vessels) and angiogenesis (the growth of new vessels from preexisting vessels by sprouting). Angiogenesis is a complex process mediated by many signaling molecules and their interaction with specific receptors in coordination with multiple cell types. Notch signaling and VEGF signaling are central to activating and modulating angiogenesis. During VEGF-mediated angiogenesis, only a fraction of the ECs in the vasculature acquire the tip cell phenotype; the others form the stalk of the vascular sprout or stay behind to maintain the integrity and perfusion of the growing vascular bed.[58] Endothelial tip cells are characterized by their position at the very tip of angiogenic sprouts and by their extensive filopodia protrusions directed toward attractive angiogenic cues.[59] Notch-mediated angiogenesis is mainly modulated by Notch ligand Delta-like (Dll)4 signaling from tip ECs to the neighboring stalk ECs to limit the extra amount of sprouting by VEGFR2 (VEGF receptor 2) suppression.[60,61] Both Jag1 and Dll4 are critical for angiogenesis, but play opposite roles. Tip cells contain relatively high levels of Delta-like 4 (Dll4), whereas stalk cells preferentially express Jag1, which is a proangiogenic regulator that positively controls the number of sprouts and tips by antagonizing Dll4–Notch signaling. Jagged1 and Dll4 not only maintain differential Notch signaling among tip and stalk cells but also control the levels of VEGF receptor expression and thereby EC growth and proliferation throughout the angiogenic vasculature. Activation of Notch signaling by Dll4 in stalk ECs leads to the downregulation of VEGFR2 expression in these cells.[58,62] Consequently, insufficient DLL4 expression increases the number of tip-like cells and EC proliferation in the sprouting intersegmental vessels in zebrafish. In addition, disruption of Dll4–Notch interactions causes extensive endothelial sprouting.[63] In response to VEGF, ECs adopt stalk cell behavior when exposed to Dll4-expressing cells in their neighbor or have limited access to VEGF. Contrarily, ECs adopt tip-cell fate with low Notch activity and stronger VEGF receptor expression.[58] The behavior of ECs as tip or stalk cells is very transient and determined by the balance of Jag1, VEGF-mediated signaling promoting angiogenesis, and Dll4-mediated signaling suppressing EC proliferation and sprouting.

Notch signal also controls arterial–venous specification of EPCs. Diez *et al.*[64] have reported a hypoxia-mediated Dll4–Notch–Hey2 signaling, which plays an important role in the developmental regulation of arterial cell fate decision of EPCs under hypoxia. They found that EPCs contain high amounts of nuclear receptor subfamily 2, group F member 2 (NR2F2), a regulator of vein identity, while levels of the arterial regulators Dll4 and Hey2 are low in normoxia condition; however, NR2F2 can be replaced by Dll4 and Hey2 in EPC under hypoxia, leading to arterial cell fate of EPCs. Yamamizu *et al.*[65] reported similar findings in Flk1(+) vascular progenitor cells, but with a different mechanism: their work shows that Notch signaling is activated downstream of cAMP through phosphatidylinositol-3 kinase, and forced activation of Notch with VEGF completely reconstitutes cAMP-elicited arterial EC induction and synergistically enhances target gene promoter activity *in vitro* and arterial gene expression during *in vivo* angiogenesis. Another study by Caiado *et al.*[66] proved that Notch signal has a similar effect on bone-marrow-derived vascular precursor cells (BM-PCs). They found that the Notch pathway modulates the adhesion of BM-PC to the extracellular matrix *in vitro* via regulation of integrin α3β1, and that activation of the Notch pathway on BM-PC improved wound-healing *in vivo* through angiogenesis. Interestingly, RBPJ disruption in ECs of adult mice induces a more intensive angiogenic response to injury by modulating BM-PC proliferation and accumulative vessel outgrowth, indicating that in adults, RBPJ-mediated Notch signaling may play an essential role in the maintenance of vascular homeostasis by repressing EC proliferation.[67]

C. Notch and Neural Stem Cell Self-Renewal and Differentiation

Stroke is a leading cause of long-term disability worldwide, with few therapeutic options available for improving behavioral recovery. That endogenous neural stem and progenitor cells are capable of promoting reparative responses following brain injury and stroke makes these cells attractive therapeutic targets for stimulating cell replacement and neuronal plasticity.[68]

In neural stem cells (NSCs), Notch signaling prevents terminal differentiation and maintains self-renewal of brain-derived NSCs.[69,70] Mizutani et al. found that knockdown of RBPJ promotes the conversion of NSCs to intermediate neural progenitors (INPs) and that activation of RBPJ is insufficient to convert INPs back to NSCs.[71] Similarly, Breunig et al.[72] found that activated Notch-1 overexpression induces proliferation, whereas abrogation of Notch signaling in vivo or in vitro leads to a transition from neural stem or precursor cells to transit-amplifying cells or neurons. Ables et al.[73] reported similar findings using tamoxifen-inducible Notch-1 knockout mice to eliminate Notch-1 and concomitantly express yellow fluorescent protein in nestin-expressing Type-1 NSCs in the adult hippocampal subgranular zone, suggesting that Notch-1 signaling is required to maintain a reservoir of undifferentiated cells and ensure continuity of adult hippocampal neurogenesis. Neurons can be continuously generated from stem cells lining the lateral ventricles in the adult mammalian brain; however, stem cells were depleted by stroke and failed to self-renew sufficiently to maintain their own population. Carlen et al.[74] found that inhibition of the Notch signaling pathway in uninjured animals allowed ependymal stem cells to enter the cell cycle and produce olfactory bulb neurons, whereas forced Notch signaling was sufficient to block the ependymal stem cell response to stroke. Recently, Haupt et al.[75] reported that protein transduction to directly deliver recombinant N-ICD into mouse and human NSCs can accomplish transgene-free Notch activation and keep cells in a proliferative state.

NSCs reside in specialized microenvironments or "niches" that regulate their function.[76] The hypoxia niche is important for maintaining the undifferentiated state of stem cells. Gustafsson et al.[77] found that hypoxia blocks neuronal and myogenic differentiation in a Notch-dependent manner. Hypoxia activates Notch-responsive promoters and increases expression of Notch direct downstream genes. Qiang et al.[78] reported similar findings and observed that hypoxia-inducible factor-1alpha-induced activation of the Notch pathway is essential for hypoxia-mediated maintenance of glioblastoma stem cells (GSCs), and that inhibition of the phosphatidylinositol 3-kinase–Akt and ERK1/2 pathways partly reduces the hypoxia-induced activation of the Notch pathway and subsequent GSC maintenance.

III. Cross Talk Between MicroRNA and Notch Signal on Stem Cell Fate (Fig. 2)

There is a dearth of information regarding the cross talk between microRNA expression and Notch signaling in stem cell differentiation. Recent studies have demonstrated that Notch signal pathways also regulate the

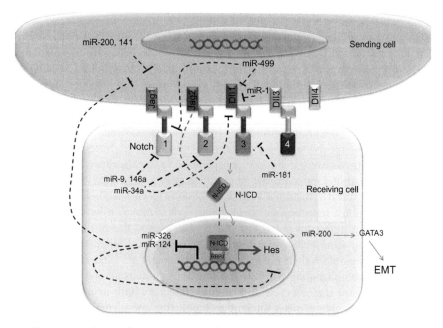

Fɪɢ. 2. A mechanism for the integration of microRNA and Notch signaling by direct interaction between individual microRNA and Notch receptors or Notch ligands. (See Color Insert.)

expression of various miRNAs in hematopoiesis.[79] In this section, we review the importance of miRNA-mediated stem cell differentiation via Notch signaling regulation in the control of many fundamental Notch activation processes.

A. MicroRNAs Control Hematopoietic Progenitor Differentiation via Notch Regulation

miRNAs play an important role in regulating hematopoiesis. miR-181a and miR-181b belong to the miR-181 family, which are clustered and located on chromosomes 1 and 9. A very recent study has shown that miR-181 expression levels have a profound impact on the development of human NK cells from CD34(+) hematopoietic progenitor cells (HPCs). Enforced expression of the miR-181a/b in HPC resulted in an increase in CD56+ NK cells at day 28 compared with scrambled controls. Conversely, inhibition of miR-181a or miR-181b led to a decrease in the percentage of mature CD56+ NK cells. Furthermore, miR-181a/b transcript levels increase during NK cell development from stage 1 NK cell progenitors (CD34+ CD117− CD94− CD56−) to stage 2 progenitors (CD34+ CD117+ CD94− CD56−) to stage 3 progenitors

(CD34− CD117+ CD94− CD56lo/−) to final lineage-committed CD56+ NK cells, supporting the important role of the expression of miR-181a and miR-181b for the differentiation of precursors through the NK cell lineage. Indeed, overexpression of the individual miR-181 members promotes NK cell development by directly targeting and downregulating the NLK protein level, a negative regulator of Notch signaling, via miR 181 binding sites located within their 3′UTRs, resulting in increased NK cell development.[80] miR-150 may control T-cell development via Notch 3 regulation. Ghisi et al.[81] have shown that forced expression of miR-150 reduces NOTCH3 levels in T-cell lines and has adverse effects on their proliferation and survival.

B. MicroRNAs Control Neuronal Progenitor Proliferation and Differentiation via Notch Regulation

The cross talk between miRNAs and the Notch pathway plays a key role in neuronal progenitor cell differentiation and proliferation. Kefas et al.[82] found a miR-326/Notch axis, in which Notch-1 knockdown upregulates miR-326 expression; conversely, miR-326 also inhibits Notch proteins and activity, indicating a feedback loop. Interestingly, transfection of miR-326 into stem-cell-like glioma lines was cytotoxic, and rescue was obtained with Notch restoration. Human miR-124 is located on chromosome 8. miR-124 is highly expressed in the nervous system of humans,[83] mice,[84] Drosophila,[85] and C. elegans.[86] Previous work has shown that delivering miR-124 causes the expression profile to shift toward that of the brain, the organ in which miR-124 is preferentially expressed,[87] and when introducing miR-124 into HeLa cells, it decreases the levels of hundreds of nonneuronal transcripts and promotes a neuronal-like mRNA profile.[88] A recent in vitro study showed that miR-124 regulates adult neurogenesis in the subventricular zone (SVZ) stem cell niche, which is the largest neurogenic niche in the adult mammalian brain.[89] Cheng et al. have used FACS-purified stem cell astrocytes from the neurogenic SVZ of the mouse brain and found that miR-124 knockdown by penetratin-conjugated 2′-O-methyl antisense miR-124 increased the number of dividing neural precursors and formed neurospheres, but reduced the total number of postmitotic neurons. Conversely, miR-124 overexpression caused a substantial decrease in the overall proportion of dividing cells and an increase in the number of postmitotic neurons. Chen et al.[90] searched through predicted targets for known conserved Notch pathway genes and found miR-124 canonical seed sites in the 3′UTRs of all three Ciona Hairy/Enhancer-of-Split genes (Hes1, HesB, HesL) and in Neuralized, a ubiquitin ligase that degrades the Notch ligand Delta. They verified these mRNA targets of miR-124 and discovered that there is a feedback interaction between miR-124 and Notch signaling;

Notch signaling silences miR-124 in epidermal midline cells, whereas in PNS midline cells, miR-124 silences Notch, Neuralized, and all three Ciona Hairy/Enhancer-of-Split genes. In addition, ectopic expression of miR-124 is sufficient to convert epidermal midline cells into PNS neurons, consistent with a role for miR-124 in modulating Notch signaling. All these findings suggest that miR-124 might interact with Notch signaling to regulate PNS specification.[90] It was also found that the only Notch pathway gene in humans containing an miR-124 seed sequence is JAG1, a Notch ligand. These results suggest a critical role for miR-124 in driving neuronal differentiation across species. miRNAs may facilitate the differentiation of bone-marrow-derived stromal cells into neurons via Notch regulation. Jing *et al.*[91] reported that miR-9 facilitates the differentiation of bone-marrow-derived mesenchymal stem cells (MSCs) into neurons via Notch-1 repression, implying the possibility of developing a promising approach using an miRNA-based cocktail to induce stem cell differentiation into neurons. Furthermore, miR-34a has been reported to downregulate Notch-1and Notch-2 protein expression in glioma stem cells via targeting DLL1, and Notch-1 and Notch-2 3′UTR.[92,93] Expression of miR-34 can inhibit cell invasion. Guessous *et al.*[94] found that miR-34a expression induces glioma stem cell differentiation. Mei *et al.*[95] documented that Notch-1 is a direct target of miR-146a, which modulates NSC proliferation and differentiation and reduces the formation and migration of glioma stem-like cells.

C. miR-449 and Respiratory Ciliate Cell Progenitor Differentiation

Human miR-449a, 449b, and 449c belong to the miR-449 family, which are clustered and located on chromosome 5. Recent studies have reported that the miR-499 family are the most strongly induced miRNAs during epithelium differentiation in human airway mucociliary epithelial cells (HAECs) grown at an air/liquid interface and the mucociliary epidermis of *Xenopus* embryos.[96] Enforced overexpression of the miR-449 in epidermis promotes centriole multiplication and multiciliogenesis by directly repressing the expression of Delta/Notch pathway. Conversely, invalidation of miR-449 leads to an increase in the expression of Dll1 and Notch 1. Furthermore, overexpression of Dll1 mRNA lacking miR-449 target sites repressed multiciliogenesis, whereas both Dll1 and Notch1 knockdown rescued multiciliogenesis in miR-449-deficient cells, supporting an important role and conserved mechanism whereby Notch signaling must undergo miR-449-mediated inhibition to permit differentiation of ciliated cell progenitors.[97] Indeed, miR-449-binding sites located within 3′UTRs of Notch-1 and its ligand Dll1 were identified and validated using a target protection assay in HAECs.

D. miR-1, miR-133, and Cardiac Progenitor Differentiation

Specific miRNAs help refine and limit gene expression and can manipulate ES cell differentiating into cardiomyocytes. miR-1 and miR-133 are cardiac-muscle-specific bicistronic miRNAs. Recent studies have shown that lentiviral-mediated miR-1 and miR-133 overexpression can promote the mesoderm differentiation of mESCs but have a negative impact on differentiation into the ectodermal or endodermal lineages.[98] Expression of miR-1 can increase the expression of Nkx2.5, an early cardiac-specific transcription factor from day 6 to day 10; however, expression of miR-133 does not have this effect. Enforced expression of the miR-1 in mES results in repression of Dll1. Consistently, Dll1 knockdown in mouse ES cells facilitates cardiac mesoderm while inhibiting endoderm and ectoderm differentiation, similar to ES cells with miR-1 over-expression, supporting the important role of miR-1 and miR-133 in cardiac mesoderm differentiation of ES cells.

E. Notch Signal Regulates Epithelial-to-Mesenchymal Transition via MicroRNA

The precise mechanisms of Notch-mediated EMT have not been fully defined. Yang et al.[99] reported a novel Jagged/miR-200-dependent pathway that mediated EMT. Mechanistically, Jagged2 was found to promote EMT by increasing the expression of GATA3, which suppressed expression of miRNA-200, which targets the transcriptional repressors that drive EMT, and thereby induced EMT. Reciprocally, miR-200 inhibited expression of Gata3, which reversed EMT, suggesting that Gata3 and miR-200 are mutually inhibitory and have opposing effects on EMT. In humans, both miR-200c and miR-141 directly inhibited Jagged1, impeding proliferation of human cells.[100]

IV. Epigenetic Regulation of Stem Cell Fate via Notch Signaling

HDACs are enzymes that modulate gene expression and progenitor fate by deacetylating histones and nonhistone proteins. Yamaguchi et al.[101] found that Hdac1 is required for the switch from proliferation to differentiation in the zebrafish retina via inhibition of Hes, a major Notch target. Zhou et al.[102] studied the effects of HDACis (SAHA and sodium butyrate) on cell cycle progression and proliferation of NSCs; they found that HDACis directly regulate cdk inhibitor P27 and P21 transcription with downregulated Notch effector Hes genes. Most recently, Singh et al.[103] studied the role of HDAC3 in cardiac neural-crest development, and found that Hdac3 plays a critical and specific regulatory role in the neural-crest-derived smooth muscle lineage and that its effect is associated with the downregulation of the Notch ligand Jagged1, a key driver of smooth muscle differentiation of progenitors.

V. Conclusions

It is clear that the Notch signal pathway is essential for stem cell differentiation and progenitor cell maintenance, yet this is clearly still the beginning of our understanding. We still do not know how to control stem-cell-targeted differentiation via precisely regulating Notch activation in a temporally and spatially controlled manner. In addition, knowledge is lacking on the cross talk between LncRNA and Notch signaling. For sure, it is worth the effort to use the novel inducible vector system for Notch-mediated stem cell differentiation studies.

ACKNOWLEDGEMENTS

This work was supported by American Heart Association Beginning Grant-in-Aid 0765094Y (to Y. T.) and NIH grant HL086555 (to Y. T.).

REFERENCES

1. High FA, Epstein JA. The multifaceted role of Notch in cardiac development and disease. *Nat Rev Genet* 2008;**9**:49–61.
2. Timmerman LA, Grego-Bessa J, Raya A, Bertran E, Perez-Pomares JM, Diez J, et al. Notch promotes epithelial-mesenchymal transition during cardiac development and oncogenic transformation. *Genes Dev* 2004;**18**:99–115.
3. Lo B, Parham L. Ethical issues in stem cell research. *Endocr Rev* 2009;**30**:204–13.
4. Okita K, Matsumura Y, Sato Y, Okada A, Morizane A, Okamoto S, et al. A more efficient method to generate integration-free human iPS cells. *Nat Methods* 2011;**8**:409–12.
5. Miyoshi N, Ishii H, Nagano H, Haraguchi N, Dewi DL, Kano Y, et al. Reprogramming of mouse and human cells to pluripotency using mature microRNAs. *Cell Stem Cell* 2011;**8**:633–8.
6. Yu J, Hu K, Smuga-Otto K, Tian S, Stewart R, Slukvin II, et al. Human induced pluripotent stem cells free of vector and transgene sequences. *Science* 2009;**324**:797–801.
7. Okita K, Ichisaka T, Yamanaka S. Generation of germline-competent induced pluripotent stem cells. *Nature* 2007;**448**:313–7.
8. Okita K, Nakagawa M, Hyenjong H, Ichisaka T, Yamanaka S. Generation of mouse induced pluripotent stem cells without viral vectors. *Science* 2008;**322**:949–53.
9. Zhou H, Wu S, Joo JY, Zhu S, Han DW, Lin T, et al. Generation of induced pluripotent stem cells using recombinant proteins. *Cell Stem Cell* 2009;**4**:381–4.
10. Tang YL, Zhao Q, Zhang YC, Cheng L, Liu M, Shi J, et al. Autologous mesenchymal stem cell transplantation induce VEGF and neovascularization in ischemic myocardium. *Regul Pept* 2004;**117**:3–10.

11. Tang YL, Zhao Q, Qin X, Shen L, Cheng L, Ge J, et al. Paracrine action enhances the effects of autologous mesenchymal stem cell transplantation on vascular regeneration in rat model of myocardial infarction. *Ann Thorac Surg* 2005;**80**:229–36 [Discussion 236–227].

12. Gnecchi M, He H, Noiseux N, Liang OD, Zhang L, Morello F, et al. Evidence supporting paracrine hypothesis for Akt-modified mesenchymal stem cell-mediated cardiac protection and functional improvement. *FASEB J* 2006;**20**:661–9.

13. Beltrami AP, Barlucchi L, Torella D, Baker M, Limana F, Chimenti S, et al. Adult cardiac stem cells are multipotent and support myocardial regeneration. *Cell* 2003;**114**:763–76.

14. Oh H, Chi X, Bradfute SB, Mishina Y, Pocius J, Michael LH, et al. Cardiac muscle plasticity in adult and embryo by heart-derived progenitor cells. *Ann N Y Acad Sci* 2004;**1015**:182–9.

15. Laugwitz KL, Moretti A, Lam J, Gruber P, Chen Y, Woodard S, et al. Postnatal isl1+ cardioblasts enter fully differentiated cardiomyocyte lineages. *Nature* 2005;**433**:647–53.

16. Tang YL, Shen L, Qian K, Phillips MI. A novel two-step procedure to expand cardiac Sca-1+ cells clonally. *Biochem Biophys Res Commun* 2007;**359**:877–83.

17. Tang YL, Zhu W, Cheng M, Chen L, Zhang J, Sun T, et al. Hypoxic preconditioning enhances the benefit of cardiac progenitor cell therapy for treatment of myocardial infarction by inducing CXCR4 expression. *Circ Res* 2009;**104**:1209–16.

18. Cheng K, Li TS, Malliaras K, Davis DR, Zhang Y, Marban E. Magnetic targeting enhances engraftment and functional benefit of iron-labeled cardiosphere-derived cells in myocardial infarction. *Circ Res* 2010;**106**:1570–81.

19. Messina E, De Angelis L, Frati G, Morrone S, Chimenti S, Fiordaliso F, et al. Isolation and expansion of adult cardiac stem cells from human and murine heart. *Circ Res* 2004;**95**:911–21.

20. Pfister O, Oikonomopoulos A, Sereti KI, Sohn RL, Cullen D, Fine GC, et al. Role of the ATP-binding cassette transporter Abcg2 in the phenotype and function of cardiac side population cells. *Circ Res* 2008;**103**:825–35.

21. Bartel DP. MicroRNAs: genomics, biogenesis, mechanism, and function. *Cell* 2004;**116**:281–97.

22. Zeng Y. Principles of micro-RNA production and maturation. *Oncogene* 2006;**25**:6156–62.

23. Engels BM, Hutvagner G. Principles and effects of microRNA-mediated post-transcriptional gene regulation. *Oncogene* 2006;**25**:6163–9.

24. Liu N, Olson EN. MicroRNA regulatory networks in cardiovascular development. *Dev Cell* 2010;**18**:510–25.

25. Vo NK, Cambronne XA, Goodman RH. MicroRNA pathways in neural development and plasticity. *Curr Opin Neurobiol* 2010;**20**:457–65.

26. Takaya T, Ono K, Kawamura T, Takanabe R, Kaichi S, Morimoto T, et al. MicroRNA-1 and MicroRNA-133 in spontaneous myocardial differentiation of mouse embryonic stem cells. *Circ J* 2009;**73**:1492–7.

27. Ren J, Jin P, Wang E, Marincola FM, Stroncek DF. MicroRNA and gene expression patterns in the differentiation of human embryonic stem cells. *J Transl Med* 2009;**7**:20.

28. Tzur G, Levy A, Meiri E, Barad O, Spector Y, Bentwich Z, et al. MicroRNA expression patterns and function in endodermal differentiation of human embryonic stem cells. *PLoS One* 2008;**3**:e3726.

29. Shi L, Zhang S, Feng K, Wu F, Wan Y, Wang Z, et al. MicroRNA-125b-2 confers human glioblastoma stem cells resistance to. *Int J Oncol* 2012;**40**:119–29.

30. Wu N, Zhao X, Liu M, Liu H, Yao W, Zhang Y, et al. Role of microRNA-26b in glioma development and its mediated regulation on EphA2. *PLoS One* 2011;**6**:e16264.

31. Meola N, Pizzo M, Alfano G, Surace EM, Banfi S. The long noncoding RNA Vax2os1 controls the cell cycle progression of photoreceptor progenitors in the mouse retina. *RNA* 2012;**18**:111–23.

32. Bertani S, Sauer S, Bolotin E, Sauer F. The noncoding RNA Mistral activates Hoxa6 and Hoxa7 expression and stem cell differentiation by recruiting MLL1 to chromatin. *Mol Cell* 2011;**43**:1040–6.

33. Sheik Mohamed J, Gaughwin PM, Lim B, Robson P, Lipovich L. Conserved long noncoding RNAs transcriptionally regulated by Oct4 and Nanog modulate pluripotency in mouse embryonic stem cells. *RNA* 2010;**16**:324–37.

34. Wu SC, Kallin EM, Zhang Y. Role of H3K27 methylation in the regulation of lncRNA expression. *Cell Res* 2010;**20**:1109–16.

35. Yoo DY, Kim W, Nam SM, Kim DW, Chung JY, Choi SY, et al. Synergistic effects of sodium butyrate, a histone deacetylase inhibitor, on increase of neurogenesis induced by pyridoxine and increase of neural proliferation in the mouse dentate gyrus. *Neurochem Res* 2011;**36**:1850–7.

36. Chatterjee TK, Idelman G, Blanco V, Blomkalns AL, Piegore Jr. MG, Weintraub DS, et al. Histone deacetylase 9 is a negative regulator of adipogenic differentiation. *J Biol Chem* 2011;**286**:27836–47.

37. Sikandar S, Dizon D, Shen X, Li Z, Besterman J, Lipkin SM. The class I HDAC inhibitor MGCD0103 induces cell cycle arrest and apoptosis in colon cancer initiating cells by upregulating Dickkopf-1 and non-canonical Wnt signaling. *Oncotarget* 2010;**1**:596–605.

38. Karamboulas C, Swedani A, Ward C, Al-Madhoun AS, Wilton S, Boisvenue S, et al. HDAC activity regulates entry of mesoderm cells into the cardiac muscle lineage. *J Cell Sci* 2006;**119**:4305–14.

39. Chen HP, Denicola M, Qin X, Zhao Y, Zhang L, Long XL, et al. HDAC inhibition promotes cardiogenesis and the survival of embryonic stem cells through proteasome-dependent pathway. *J Cell Biochem* 2011;**112**:3246–55.

40. Villagra A, Cheng F, Wang HW, Suarez I, Glozak M, Maurin M, et al. The histone deacetylase HDAC11 regulates the expression of interleukin 10 and immune tolerance. *Nat Immunol* 2009;**10**:92–100.

41. Kokubo H, Miyagawa-Tomita S, Johnson RL. Hesr, a mediator of the Notch signaling, functions in heart and vessel development. *Trends Cardiovasc Med* 2005;**15**:190–4.

42. Iascone M, Ciccone R, Galletti L, Marchetti D, Seddio F, Lincesso A, et al. Identification of de novo mutations and rare variants in hypoplastic left heart syndrome. *Clin Genet* 2011;**81**:542–54.

43. Grego-Bessa J, Diez J, Timmerman L, de la Pompa JL. Notch and epithelial-mesenchyme transition in development and tumor progression: another turn of the screw. *Cell Cycle* 2004;**3**:718–21.

44. Koyanagi M, Bushoven P, Iwasaki M, Urbich C, Zeiher AM, Dimmeler S. Notch signaling contributes to the expression of cardiac markers in human circulating progenitor cells. *Circ Res* 2007;**101**:1139–45.

45. Poss KD, Wilson LG, Keating MT. Heart regeneration in zebrafish. *Science* 2002;**298**:2188–90.

46. Lepilina A, Coon AN, Kikuchi K, Holdway JE, Roberts RW, Burns CG, et al. A dynamic epicardial injury response supports progenitor cell activity during zebrafish heart regeneration. *Cell* 2006;**127**:607–19.

47. Kikuchi K, Holdway JE, Werdich AA, Anderson RM, Fang Y, Egnaczyk GF, et al. Primary contribution to zebrafish heart regeneration by gata4(+) cardiomyocytes. *Nature* 2010;**464**:601–5.

48. Kikuchi K, Gupta V, Wang J, Holdway JE, Wills AA, Fang Y, et al. tcf21+ epicardial cells adopt non-myocardial fates during zebrafish heart development and regeneration. *Development* 2011;**138**:2895–902.

49. Urbanek K, Torella D, Sheikh F, De Angelis A, Nurzynska D, Silvestri F, et al. Myocardial regeneration by activation of multipotent cardiac stem cells in ischemic heart failure. *Proc Natl Acad Sci USA* 2005;**102**:8692–7.

50. Kajstura J, Gurusamy N, Ogorek B, Goichberg P, Clavo-Rondon C, Hosoda T, et al. Myocyte turnover in the aging human heart. *Circ Res* 2010;**107**:1374–86.

51. Mouquet F, Pfister O, Jain M, Oikonomopoulos A, Ngoy S, Summer R, et al. Restoration of cardiac progenitor cells after myocardial infarction by self-proliferation and selective homing of bone marrow-derived stem cells. *Circ Res* 2005;**97**:1090–2.
52. Makkar RR, Smith RR, Cheng K, Malliaras K, Thomson LE, Berman D, et al. Intracoronary cardiosphere-derived cells for heart regeneration after myocardial infarction (CADUCEUS): a prospective, randomised phase 1 trial. *Lancet* 2012;**379**:895–904.
53. Raya A, Koth CM, Buscher D, Kawakami Y, Itoh T, Raya RM, et al. Activation of Notch signaling pathway precedes heart regeneration in zebrafish. *Proc Natl Acad Sci USA* 2003;**100** (Suppl 1):11889–95.
54. Boni A, Urbanek K, Nascimbene A, Hosoda T, Zheng H, Delucchi F, et al. Notch1 regulates the fate of cardiac progenitor cells. *Proc Natl Acad Sci USA* 2008;**105**:15529–34.
55. Collesi C, Zentilin L, Sinagra G, Giacca M. Notch1 signaling stimulates proliferation of immature cardiomyocytes. *J Cell Biol* 2008;**183**:117–28.
56. Urbanek K, Cabral-da-Silva MC, Ide-Iwata N, Maestroni S, Delucchi F, Zheng H, et al. Inhibition of notch1-dependent cardiomyogenesis leads to a dilated myopathy in the neonatal heart. *Circ Res* 2010;**107**:429–41.
57. Chen L, Ashraf M, Wang Y, Zhou M, Zhang J, Qin G, et al. The role of Notch1 activation in cardiosphere derived cell differentiation. *Stem Cells Dev.* 2012. PMID:22239539.
58. Eilken HM, Adams RH. Dynamics of endothelial cell behavior in sprouting angiogenesis. *Curr Opin Cell Biol* 2010;**22**:617–25.
59. Gerhardt H, Golding M, Fruttiger M, Ruhrberg C, Lundkvist A, Abramsson A, et al. VEGF guides angiogenic sprouting utilizing endothelial tip cell filopodia. *J Cell Biol* 2003;**161**:1163–77.
60. Leslie JD, Ariza-McNaughton L, Bermange AL, McAdow R, Johnson SL, Lewis J. Endothelial signalling by the Notch ligand Delta-like 4 restricts angiogenesis. *Development* 2007;**134**:839–44.
61. Phng LK, Gerhardt H. Angiogenesis: a team effort coordinated by notch. *Dev Cell* 2009;**16**:196–208.
62. Thurston G, Kitajewski J. VEGF and Delta-Notch: interacting signalling pathways in tumour angiogenesis. *Br J Cancer* 2008;**99**:1204–9.
63. Adams RH, Eichmann A. Axon guidance molecules in vascular patterning. *Cold Spring Harb Perspect Biol* 2010;**2**:a001875.
64. Diez H, Fischer A, Winkler A, Hu CJ, Hatzopoulos AK, Breier G, et al. Hypoxia-mediated activation of Dll4-Notch-Hey2 signaling in endothelial progenitor cells and adoption of arterial cell fate. *Exp Cell Res* 2007;**313**:1–9.
65. Yamamizu K, Matsunaga T, Uosaki H, Fukushima H, Katayama S, Hiraoka-Kanie M, et al. Convergence of Notch and beta-catenin signaling induces arterial fate in vascular progenitors. *J Cell Biol* 2010;**189**:325–38.
66. Caiado F, Real C, Carvalho T, Dias S. Notch pathway modulation on bone marrow-derived vascular precursor cells regulates their angiogenic and wound healing potential. *PLoS One* 2008;**3**:e3752.
67. Dou GR, Wang YC, Hu XB, Hou LH, Wang CM, Xu JF, et al. RBP-J, the transcription factor downstream of Notch receptors, is essential for the maintenance of vascular homeostasis in adult mice. *FASEB J* 2008;**22**:1606–17.
68. Cunningham LA, Candelario K, Li L. Roles for HIF-1alpha in neural stem cell function and the regenerative response to stroke. *Behav Brain Res* 2011;**227**:410–7.
69. Alexson TO, Hitoshi S, Coles BL, Bernstein A, van der Kooy D. Notch signaling is required to maintain all neural stem cell populations—irrespective of spatial or temporal niche. *Dev Neurosci* 2006;**28**:34–48.
70. Wang J, Wang C, Meng Q, Li S, Sun X, Bo Y, et al. siRNA targeting Notch-1 decreases glioma stem cell proliferation and tumor growth. *Mol Biol Rep* 2012;**39**:2497–503.

71. Mizutani K, Yoon K, Dang L, Tokunaga A, Gaiano N. Differential Notch signalling distinguishes neural stem cells from intermediate progenitors. *Nature* 2007;**449**:351-5.

72. Breunig JJ, Silbereis J, Vaccarino FM, Sestan N, Rakic P. Notch regulates cell fate and dendrite morphology of newborn neurons in the postnatal dentate gyrus. *Proc Natl Acad Sci USA* 2007;**104**:20558-63.

73. Ables JL, Decarolis NA, Johnson MA, Rivera PD, Gao Z, Cooper DC, et al. Notch1 is required for maintenance of the reservoir of adult hippocampal stem cells. *J Neurosci* 2010;**30**:10484-92.

74. Carlen M, Meletis K, Goritz C, Darsalia V, Evergren E, Tanigaki K, et al. Forebrain ependymal cells are Notch-dependent and generate neuroblasts and astrocytes after stroke. *Nat Neurosci* 2009;**12**:259-67.

75. Haupt S, Borghese L, Brustle O, Edenhofer F. Non-genetic modulation of notch activity by artificial delivery of notch intracellular domain into neural stem cells. *Stem Cell Rev* 2012 Jan 31 [Epub ahead of print], DOI: 10.1007/s12015-011-9335-6.

76. Mazumdar J, O'Brien WT, Johnson RS, LaManna JC, Chavez JC, Klein PS, et al. O2 regulates stem cells through Wnt/beta-catenin signalling. *Nat Cell Biol* 2010;**12**:1007-13.

77. Gustafsson MV, Zheng X, Pereira T, Gradin K, Jin S, Lundkvist J, et al. Hypoxia requires notch signaling to maintain the undifferentiated cell state. *Dev Cell* 2005;**9**:617-28.

78. Qiang L, Wu T, Zhang HW, Lu N, Hu R, Wang YJ, et al. HIF-1alpha is critical for hypoxia-mediated maintenance of glioblastoma stem cells by activating Notch signaling pathway. *Cell Death Differ* 2012;**19**:284-94.

79. Hamidi H, Gustafason D, Pellegrini M, Gasson J. Identification of novel targets of CSL-dependent Notch signaling in hematopoiesis. *PLoS One* 2011;**6**:e20022.

80. Cichocki F, Felices M, McCullar V, Presnell SR, Al-Attar A, Lutz CT, et al. Cutting edge: microRNA-181 promotes human NK cell development by regulating Notch signaling. *J Immunol* 2011;**187**:6171-5.

81. Ghisi M, Corradin A, Basso K, Frasson C, Serafin V, Mukherjee S, et al. Modulation of microRNA expression in human T-cell development: targeting of NOTCH3 by miR-150. *Blood* 2011;**117**:7053-62.

82. Kefas B, Comeau L, Floyd DH, Seleverstov O, Godlewski J, Schmittgen T, et al. The neuronal microRNA miR-326 acts in a feedback loop with notch and has therapeutic potential against brain tumors. *J Neurosci* 2009;**29**:15161-8.

83. Sempere LF, Freemantle S, Pitha-Rowe I, Moss E, Dmitrovsky E, Ambros V. Expression profiling of mammalian microRNAs uncovers a subset of brain-expressed microRNAs with possible roles in murine and human neuronal differentiation. *Genome Biol* 2004;**5**:R13.

84. Lagos-Quintana M, Rauhut R, Yalcin A, Meyer J, Lendeckel W, Tuschl T. Identification of tissue-specific microRNAs from mouse. *Curr Biol* 2002;**12**:735-9.

85. Aboobaker AA, Tomancak P, Patel N, Rubin GM, Lai EC. Drosophila microRNAs exhibit diverse spatial expression patterns during embryonic development. *Proc Natl Acad Sci USA* 2005;**102**:18017-22.

86. Clark AM, Goldstein LD, Tevlin M, Tavare S, Shaham S, Miska EA. The microRNA miR-124 controls gene expression in the sensory nervous system of Caenorhabditis elegans. *Nucleic Acids Res* 2010;**38**:3780-93.

87. Lim LP, Lau NC, Garrett-Engele P, Grimson A, Schelter JM, Castle J, et al. Microarray analysis shows that some microRNAs downregulate large numbers of target mRNAs. *Nature* 2005;**433**:769-73.

88. Conaco C, Otto S, Han JJ, Mandel G. Reciprocal actions of REST and a microRNA promote neuronal identity. *Proc Natl Acad Sci USA* 2006;**103**:2422-7.

89. Cheng LC, Pastrana E, Tavazoie M, Doetsch F. miR-124 regulates adult neurogenesis in the subventricular zone stem cell niche. *Nat Neurosci* 2009;**12**:399-408.

90. Chen JS, Pedro MS, Zeller RW. miR-124 function during Ciona intestinalis neuronal development includes extensive interaction with the Notch signaling pathway. *Development* 2011;**138**:4943–53.

91. Jing L, Jia Y, Lu J, Han R, Li J, Wang S, et al. MicroRNA-9 promotes differentiation of mouse bone mesenchymal stem cells into neurons by Notch signaling. *Neuroreport* 2011;**22**:206–11.

92. Li Y, Guessous F, Zhang Y, Dipierro C, Kefas B, Johnson E, et al. MicroRNA-34a inhibits glioblastoma growth by targeting multiple oncogenes. *Cancer Res* 2009;**69**:7569–76.

93. Hughes DP. How the NOTCH pathway contributes to the ability of osteosarcoma cells to metastasize. *Cancer Treat Res* 2009;**152**:479–96.

94. Guessous F, Zhang Y, Kofman A, Catania A, Li Y, Schiff D, et al. microRNA-34a is tumor suppressive in brain tumors and glioma stem cells. *Cell Cycle* 2010;**9**:1031–6.

95. Mei J, Bachoo R, Zhang CL. MicroRNA-146a inhibits glioma development by targeting Notch1. *Mol Cell Biol* 2011;**31**:3584–92.

96. Marcet B, Chevalier B, Luxardi G, Coraux C, Zaragosi LE, Cibois M, et al. Control of vertebrate multiciliogenesis by miR-449 through direct repression of the Delta/Notch pathway. *Nat Cell Biol* 2011;**13**:693–9.

97. Marcet B, Chevalier B, Coraux C, Kodjabachian L, Barbry P. MicroRNA-based silencing of Delta/Notch signaling promotes multiple cilia formation. *Cell Cycle* 2011;**10**:2858–64.

98. Ivey KN, Muth A, Arnold J, King FW, Yeh RF, Fish JE, et al. MicroRNA regulation of cell lineages in mouse and human embryonic stem cells. *Cell Stem Cell* 2008;**2**:219–29.

99. Yang Y, Ahn YH, Gibbons DL, Zang Y, Lin W, Thilaganathan N, et al. The Notch ligand Jagged2 promotes lung adenocarcinoma metastasis through a miR-200-dependent pathway in mice. *J Clin Invest* 2011;**121**:1373–85.

100. Vallejo DM, Caparros E, Dominguez M. Targeting Notch signalling by the conserved miR-8/200 microRNA family in development and cancer cells. *EMBO J* 2011;**30**:756–69.

101. Yamaguchi M, Tonou-Fujimori N, Komori A, Maeda R, Nojima Y, Li H, et al. Histone deacetylase 1 regulates retinal neurogenesis in zebrafish by suppressing Wnt and Notch signaling pathways. *Development* 2005;**132**:3027–43.

102. Zhou Q, Dalgard CL, Wynder C, Doughty ML. Histone deacetylase inhibitors SAHA and sodium butyrate block G1-to-S cell cycle progression in neurosphere formation by adult subventricular cells. *BMC Neurosci* 2011;**12**:50.

103. Singh N, Trivedi CM, Lu M, Mullican SE, Lazar MA, Epstein JA. Histone deacetylase 3 regulates smooth muscle differentiation in neural crest cells and development of the cardiac outflow tract. *Circ Res* 2011;**109**:1240–9.

Myocardial Regeneration: The Role of Progenitor Cells Derived from Bone Marrow and Heart

XIAOHONG WANG, ARTHUR H.L. FROM, AND JIANYI ZHANG

Department of Medicine, University of Minnesota Medical School, Minneapolis, Minnesota, USA

In animal models of myocardial infarction (MI), transplantation of various types of progenitor cells has been reported to (i) improve left ventricular (LV) function, (ii) decrease LV remodeling, (iii) limit fibrosis of noninfarcted LV regions, and (iv) in some cases, reduce infarct scar size. Moreover, in some reports these beneficial effects were present despite very low rates of long-term engraftment and transdifferentiation of transplanted cells into cardiomyocytes. In contrast, in other reports, significant numbers of transplanted cells do appear to have transdifferentiated into cardiomyocytes and vascular cells. Paracrine signals emanating from transplanted cells also appear to be very important because they protect injured cardiomyocytes and may activate endogenous cardiac progenitor cells (CPCs) to generate cardiomyocytes and vascular cells. Herein, we review evidence that transplanted bone-marrow- or cardiac-derived CPCs and/or *in situ* CPCs can be stimulated to propagate, differentiate, and partially replace cardiomyocytes damaged during AMI. The possibility that preexisting cardiomyocytes can be induced to reenter the cell cycle and regenerate replacement cardiomyocytes is also discussed.

Progress in Molecular Biology and Translational Science, Vol. 111
http://dx.doi.org/10.1016/B978-0-12-398459-3.00009-5

195

I. Introduction

As is well known, adult zebrafish and amphibians can regenerate lost cardiomyocytes by reactivating the cell cycle in uninjured cardiomyocytes.[1,2] Moreover, it has recently been reported that the neonatal mouse heart retains this capacity for a few days post-parturition.[3] Cardiac regeneration in response to injury may result from (i) the activation of endogenous cardiac progenitor cells (CPCs) that migrate to the site of injury, propagate, and then differentiate into cardiomyocytes and vascular cells (reviewed in Ref. 4)[5–7]; (ii) bone-marrow (BM)-derived progenitor cells transplanted into injured myocardium[8–11]; and (iii) the reentry of adult cardiomyocytes into the cell cycle in response to injury and stimuli.[12–17]

Consistent with the first possibility, convincing data suggest that c-kit[+] CPCs are the source of the replacement of "worn" cardiomyocytes that occurs throughout life; however, there is significant controversy regarding the rate of this process.[4,18–24] Unfortunately, in the absence of administration of exogenous growth factors following MI, the endogenous regenerative process is modest.[5,7] Consequently, following an MI resulting in moderate to severe damage to the left ventricle (LV), a large number of animal and human subjects ultimately develop LV chamber enlargement, hypertrophy of uninjured cardiomyocytes, and contractile dysfunction. These processes often culminate in the development of heart failure. Because animal experimentation and clinical trials have often, but not always, shown that cell therapy significantly improves LV function, the mechanisms underlying this therapeutic response are of great interest. Possible beneficial mechanisms will be considered in relation to the regenerative capabilities of transplanted BM-derived cells and transplanted and *in situ* CPCs. It should be appreciated that much of the data concerning the direct regenerative properties of transplanted BM-derived progenitor cells and CPCs should be considered provisional. We regret that the excellent works of many groups are not cited because of the space constraints inherent in a brief review of a large subject area.

II. Is There Cardiomyocyte Regeneration from Endogenous CPCs Post-MI?

Even in the absence of therapeutic interventions, it is thought that cardiomyocyte regeneration from endogenous CPCs occurs post-MI and that this process can (albeit modestly) contribute to the preservation of LV structure and function. The possibility that this reparative mechanism exists is supported by evidence that significant rates of cardiomyocyte replacement occur throughout life in normal animal and human hearts (see Refs. 19,20,25 and references cited therein) and preliminary data from our group supports this view (see below).

We and others (Ref. 26 and references cited therein) have reported that Sca-1$^+$/CD31$^-$ cells exist in adult mouse heart. The Sca-1 receptor is an established marker of progenitor cells in mice and is often coexpressed with the c-kit receptor. We have reported that Sca-1$^+$/CD31$^-$ cells comprised $10.8 \pm 2.4\%$ of the cardiomyocyte-free cell population extracted from the mouse heart[26] (Fig. 1). *In vitro* culture of isolated Sca-1$^+$/CD31$^-$ cells lead to the formation of significant microtubular structures by day 3. Alkaline phosphatase staining was positive in 50% of these microtubules and caveolin-1 staining was also present. Induction of Sca-1$^+$/CD31$^-$ cells with VEGF for 14 days caused a significant increase of CD31, caveolin-1, and vWF protein, and mRNA expressions in these cells, suggesting that they could differentiate into endothelial cells *in vitro*. Low mRNA expression of these endothelial cell markers was also present after 14 days of culture in the absence of VEGF, indicating that prolonged culture without added growth factors could induce a modest level of differentiation.

The Sca-1$^+$/CD31$^-$ cells also had the ability to differentiate into cardiomyocytes following exposure to DKK-1, DMSO, BMP-2, FGF-4, FGF-8, and 5-azacytidine over a 14-day culture period (Fig. 2). Moreover, coculture of these cells with neonatal rat cardiomyocytes was associated with their further differentiation, as evidenced by the expression of troponin T and phospholamban. The latter finding supports the commonly held notion that the cellular microenvironment of CPCs plays a crucial role in cardiomyocyte differentiation.

In the same report, we showed that, 7 days after MI, the Sca-1$^+$/CD31$^-$ content of the cardiomyocyte-free cell population extracted from the mouse heart increased from $10.8 \pm 2.4\%$ to $21.8 \pm 1.6\%$ ($p < 0.01$) (Fig. 3).[26] Interestingly, the increase of this putative CPC population after MI was transient and the number of these cells returned to baseline values within 14–21 days post-MI. Similarly, c-kit$^+$ CPC numbers have been reported to be increased in hypertrophied and cardiomyopathic human hearts.[7,27,28] In our study, the majority of the additional Sca-1$^+$/CD31$^-$ cells were found in the MI border zone (BZ), reflecting their tendency to home in to an injured area.

In further support of the notion that Sca-1$^+$ CPCs may be supporting a small but functionally significant degree of cardiomyocyte replacement in the post-MI mouse heart, we examined the effect of a heterozygous knockout (KO) of myocardial Sca-1 antigen on the course of post-MI LV remodeling (Ref. 29 and unpublished data). It was observed that the post-MI LV remodeling in the KO animals was significantly more severe than that present in wild-type (WT)

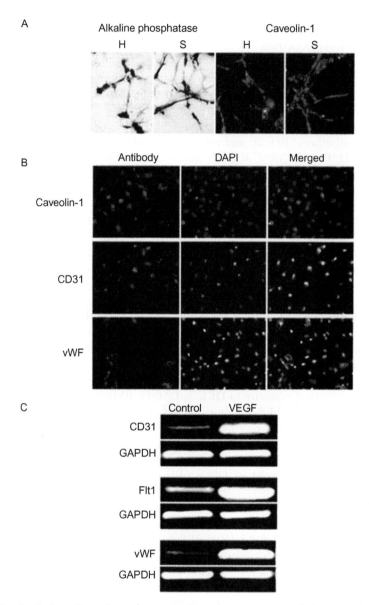

FIG. 1. The heart-derived Sca-1$^+$/CD31$^-$ cell population possesses endothelial cell differentiation capacity. (A) *Left*, Sca-1$^+$/CD31$^-$ cells were cultured in growth factor-reduced Matrigel and stained for alkaline phosphatase (purple). *Right*, H-derived and S-derived Sca-1$^+$/CD31$^-$ cells were cultured in growth factor-reduced Matrigel; immunofluorescence staining for caveolin-1 (red) was positive in both H and S. (B) Immunofluorescence staining showed heart-derived Sca-1$^+$/CD31$^-$ cells expressed endothelial cell markers, including caveolin-1 (red), CD31(red), and vWF (red), after being induced by 10 ng/ml VEGF for 14 days. (C) Expression of CD31, Flt1, and vWF mRNA was induced in heart-derived Sca-1$^+$/CD31$^-$ cells by 14 days of culture with VEGF. *Abbreviations*: DAPI, 4,6-diamidino-2-phenylindol dihydrochloride; GAPDH, glyceraldehyde-3-phosphate dehydrogenase; H, heart; S, skeletal muscle; VEGF, vascular endothelial growth factor; vWF, von Willebrand factor.[26] (For interpretation of the references to color in this figure legend, the reader is referred to the online version of this chapter.)

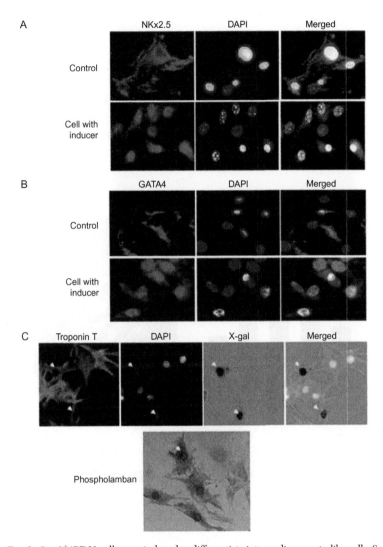

F<small>IG</small>. 2. Sca-1$^+$/CD31 cells were induced to differentiate into cardiomyocyte-like cells. Sca-1$^+$/
CD31$^-$ cells were cultured in the basal medium containing 10 ng/ml Dickkopf-1, 0.75% dimethyl-
sulfoxide, 10 ng/ml BMP2, 100 ng/ml fibroblast growth factor 4 (FGF4), and 10 ng/ml FGF8 in
combination (with the addition of 10 μM 5′-azacytizine for the initial 3 days); total time in culture
was 14 days. (A) Immunofluorescence staining of induced Sca-1$^+$/CD31$^-$ cells showed nuclear
expression of NKx2.5-stained (left, red) and DAPI-stained (middle) nuclei. *Right*, merged view of
NKx2.5-stained and DAPI-stained nuclei. (B) GATA-4-stained (left, red) and DAPI-stained (middle) nuclei. *Right*, merged view of GATA-4-stained and DAPI-stained nuclei. (C) Sca-1$^+$/CD31$^-$
cells cocultured with neonatal rat cardiomyocytes were induced to express cardiac-specific markers.
Top row of photomicrographs shows troponin T (red), DAPI (light blue), X-gal (dark blue), and a
merged view of the preceding photomicrographs. Arrows point to triple-positive cells. The bottom
photomicrograph shows an X-gal-positive cell (blue) costained for phospholamban (brown). *Abbreviations*: BMP, bone morphogenetic protein; DAPI, 4,6-diamidino-2-phenylindol dihydrochloride; DKK1, Wnt antagonist, Dickkopf-1; NKx2.5, Homeobox protein NKx2.5; X-gal, 5-bromo-4-
chloro-3-indolyl-β-D-galactoside.[26] (For interpretation of the references to color in this figure
legend, the reader is referred to the online version of this chapter.)

A

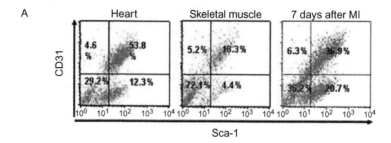

B

C

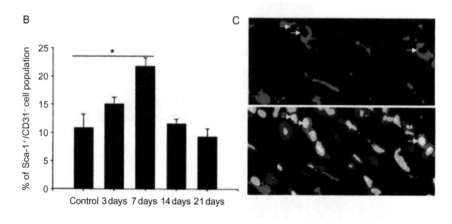

D

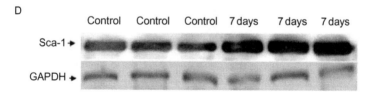

FIG. 3. A significant increase of Sca-1$^+$/CD31$^-$ cell population after myocardial infarction. (A) Representative fluorescence-activated cell sorting analyses of cardiomyocyte-depleted cell suspensions obtained from a normal heart (left), a normal skeletal muscle (middle), and a heart that was infarcted 7 days prior to cell harvest (right). Significantly more endogenous Sca-1$^+$/CD31$^-$ cells (right bottom quadrant) are present in normal myocardium than in skeletal muscle. The number of Sca-1$^+$/CD31$^-$ cells in myocardium increased after myocardial infarction. (B) The time course of the increase of the myocardial Sca-1$^+$/CD31$^-$ cell population following infarction. (C) Numerous endogenous Sca-1$^+$ cells distributed in the peri-infarct region 7 days after MI. Arrows point to Sca-1 (red) positive cells in upper photomicrograph and double-positive cells in merged picture of the same section stained with DAPI (blue). (D) Representative Western blot demonstrating enhanced Sca-1 protein expression in hearts 7 days after MI (top); same blot reprobed with GAPDH antibody as controls for equal loading (bottom). *Abbreviations*: GAPDH, glyceraldehyde-3-phosphate dehydrogenase; MI, myocardial infarction.[26] (For interpretation of the references to color in this figure legend, the reader is referred to the online version of this chapter.)

animals, suggesting that endogenous cardiac Sca-1$^+$/CD31$^-$ progenitor cells do support modest preservation of LV function post-MI. Similarly, in homozygous Sca-1-null mice, there is late spontaneous deterioration of LV function as compared to WT controls and the null mice also have more deterioration of LV function following transverse aortic constriction.[30] Taken together, these data suggest that cardiac endogenous Sca-1$^+$ cells are cardioprotective, but the mechanism of protection remains unclear.

III. Paracrine Effects of Transplanted Cells in the Injured Heart

In many animal studies, it has been reported that only a small percentage of transplanted cells (especially those of BM origin) show long-term engraftment in recipient myocardium,[31–37] and an even smaller fraction of the engrafted cells appear to transdifferentiate into cardiomyocytes or vascular cells suggesting that a paracrine effect contributes significantly to the functional beneficial effects of the cellular therapies.[31–33]

A striking example of this sequence of events has been reported by Dzau's group. These investigators, employing a mouse I–R model, showed that early (within an hour of reperfusion) intramyocardial injection of Akt overexpressing BM-derived mesenchymal stem cells (MSCs) markedly attenuated the size of the resulting infarct.[37] Crucially, injection of only the culture medium conditioned by these MSCs yielded the same improvements.[34,35] Moreover, the major protective effect preceded by several days the increased capillary density that was also induced by injection of MSCs or conditioned medium. Whether significant cardiomyocyte regeneration from endogenous CPCs occurred in this model is unknown. However, in the aforementioned studies, the paracrine effects associated with either MSC or conditioned medium injection could have stimulated *in situ* CPCs to regenerate new cardiomyocytes, as will be discussed below.

IV. IGF + HGF Administration Can Activate *In Situ* CPCs to Generate Cardiomyocytes

In a series of studies using rodent and larger animal models, Anversa's group has presented substantial evidence that the relatively ineffective reparative responses resulting from MI-associated activation of endogenous CPCs can be substantially enhanced by the acute (i.e., immediately after MI) or somewhat delayed (1–4 weeks) injection of growth factors (insulin-related growth factor, IGF and hepatocyte growth factor, HGF) into the MI BZ.[5,7]

These investigators reported that administration of the growth factors caused CPCs to migrate to the injury zone, proliferate, and differentiate into vascular cells and small cardiomyocytes. A striking feature of these studies was that the regeneration of even a relatively thin layer of immature cardiomyocytes within the scar region was accompanied by a striking decrease in the severity of LV structural remodeling and contractile dysfunction. The direct role that the administered growth factors themselves played in limiting the apoptosis of injured cardiomyocytes was not established in these studies. Importantly, this work suggests that the functional benefits of cardiomyocyte regeneration may be disproportionate to the relatively small amounts of myocardium regenerated.

In a conceptually related mouse MI study, we asked whether enhanced mobilization and activation of the cardiac endogenous Sca-1$^+$ CPCs and/or c-kit$^+$ CPCs might be beneficial post-MI. We hypothesized that stromal-cell-derived factor-1-alpha (SDF-1α), a CXCR4 receptor ligand and stem cell attractant, would enhance CPCs numbers in injured myocardium and, thereby, foster cardiomyocyte regeneration. A PEGylated fibrin patch containing bound SDF-1α was placed on the epicardial surface of the ischemic region immediately following LAD ligation in a mouse MI model.[38] Preliminary *in vitro* studies indicated that SDF-1α was released from the patch for at least 10 days. The percentage of c-kit$^+$ cells within the injury zone at 2 weeks post-MI was significantly higher in the SDF-1α-treated MI animals than in the MI controls ($11.20 \pm 1.7\%$ vs. $4.22 \pm 0.96\%$, $p < 0.05$) and increased numbers of these cells persisted for up to 1 month after MI. LV function was also significantly better maintained in the SDF-1α patch-treated hearts than in non-treated hearts and late scar expansion was also decreased in that group. Collectively, these observations indicated that the cardiac endogenous Sca-1$^+$ (putative) CPCs were not responsive to SDF-1α but did not exclude the possibility that c-kit$^+$ CPC-derived cardiomyocyte regeneration may have occurred. These data indicated that c-kit$^+$ CPCs are responsive to activation by SDF-1α.

V. Do Transplanted BM-Derived and/or Transplanted or *In Situ* CPCs Transdifferentiate into Cardiomyocytes and Vascular Cells?

In a number of post-MI studies, favorable structural and functional results have been reported following implantation of either BM-derived MSCs or cardiac-derived CPCs and evidence of cardiac regeneration has been present in some of these studies.

A. MSCs (BM Derived)

A number of groups have reported beneficial responses following intra-myocardial or intracoronary or intravenous BM-derived MSCs in small and large animal models of MI.[10,39–44] Moreover, several clinical trials on the effectiveness of MSC transplantation have also reported that administration of these cells post-MI is associated with decreased LV structural and functional remodeling.[45,46] Additionally, in some animal studies, evidence has been presented that transplanted engrafted MSCs can differentiate into cardiomyocytes and vascular cells.[10,40] We have carried out a number of studies with intramyocardial injection of MSCs in large animal (swine) models of post-MI remodeling. It was consistently found that LV ejection fraction (EF) was significantly increased after cell transplantation.[33,36,47] In one study, we directly injected allogeneic marrow-derived multipotent progenitor cells (MPCs) into the MI BZ at the time of coronary ligation[33] (Fig. 4). In the cell-treated hearts, EF and myocardial energetic characteristics (^{31}P magnetic resonance spectroscopy) measured during the terminal study a few weeks later were improved as reflected by a higher ratio of phosphocreatine to ATP. However, in that study, only 0.35% of the transplanted cells remained engrafted at the time of sacrifice and only ~2% of these cells had cardiomyocyte markers suggestive of differentiation. Histological sections from the infarct zones did show an increased

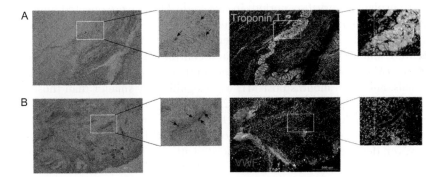

FIG. 4. Engraftment and differentiation of pMultistem *in vivo*. Infarcted swine hearts with LacZ-labeled pMultistem injection were harvested and dissected into 10-Î¼m sections. Dissected samples were fixed in zinc fixative and stained for both LacZ (blue) and troponin T or von Willebrand factor (vWF). Nuclei were stained by DAPI. (A) IZ of a Lac-Z⁺pMultistem-treated heart expressing cardiac myocyte phenotype troponin T. The left two pictures are phase-contrast images with magnifications ×10 and ×40, respectively. The right two pictures are fluorescence images with magnifications ×10 and ×40, respectively. Patches of spared myocytes were observed only in the pMultistem-treated hearts, not in untreated hearts (shown in B). (For interpretation of the references to color in this figure legend, the reader is referred to the online version of this chapter.)

presence of cardiomyocytes within the scars as compared to the scars in the untreated group. These data support the view that the transplanted cells, by means of paracrine signaling, inhibited native cardiomyocyte apoptosis, increased peri-MI region capillary density, and possibly stimulated the differentiation of some endogenous CPCs into new cardiomyocytes within the infarct and BZ regions (Fig. 4).

In a porcine MI model that utilized a 60-min period of no flow ischemia followed by reperfusion, the intracoronary injection of MSC during the subsequent reperfusion period was associated with improved structural, functional, and myocardial energetic characteristics at the time of terminal study 4 weeks after the cell transplantation, despite the fact that only 0.55% of the infused cells were engrafted.[36] Moreover, only a small fraction of the engrafted cells costained positive for cardiogenic markers. Histological sections from the peri-infarct area again revealed increased capillary numbers in the BZ and increased numbers of cardiomyocytes within the scarred region.

Using a swine model of I/R and a novel fibrin patch for enhanced delivery of autologous MSCs, we found a significant cell engraftment rate 4 weeks after the transplantation with evidence of myocyte regeneration from engrafted MSCs (Fig. 5).[48] In a separate study in swine, we also found that MPC injected into the BZ of an acute MI can stimulate endogenous CPCs to regenerate myocardium.[47] A unique feature of this study was the availability of serial MRI measurements of the LV structure and function over a 4-month posttransplant observation period. In this study, MPCs were injected into the BZ following permanent ligation of the distal LAD. Although apparent scar sizes (defined by delayed gadolinium enhancement imaging) at 10 days post-MI were similar in cell-transplanted and control MI groups (approximately 10% of LV wall area), scar sizes subsequently decreased in the cell-treated animals, and, by 4 months post-MI, were significantly smaller than those in the control group ($4.6 \pm 1\%$ vs. $8.6 \pm 2.4\%$ of LV wall area in control group, $p < 0.05$). Histological sections from the BZ and infarct areas of the cell-treated group revealed spared cardiomyocytes within the scarred region. However, despite transplantation-associated attenuation of post-MI structural and functional deterioration, no engraftment of transplanted cells was visualized at the time of sacrifice 4 months post-MI. Thus, the rate of transdifferentiation of transplanted cells into cardiomyocytes was minimal. In this study, if the primary mechanism of scar size reduction induced by cell transplantation was only the sparing of native cardiomyocytes that were otherwise destined to die as a consequence inhibition of apoptosis, it would be expected that reduction of scar size would have been seen at the time of the 10 day post-MI MRI study because enhanced early apoptosis is characteristic of most MI models. However, the possibility also exists that "scar" definition (i.e., between irreversibly damaged vs. severely injured volumes of

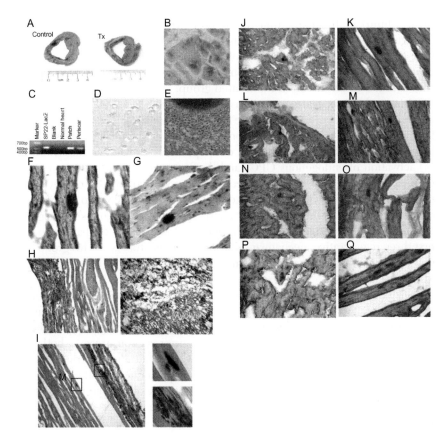

FIG. 5. Control heart (A) compared with a Tx heart. (B) X-gal staining shows that "blue" cells are homing in to the anterior LV wall. (C) 560-bp products were found only in the patch and periscar area of Tx hearts. Primers were designed for specifically identifying the Ad5RSV-*lacZ* vector sequence. (D) In stem cell culture medium, the blue cells migrated out of the specimens taken from the periscar area. (E) Typical specimen maintained in stem cell culture medium on *day* 3 before the second change of the medium. The cells migrated out of the specimen and continued dividing. (F) Typical triple staining of hematoxylin (nuclei), cardiac-specific troponin T, and X-gal (β-galactosidase). Striation of myocytes can be clearly observed by staining of cardiac-specific troponin T. The size of this blue cell is demonstrated by the ratio of the blue cell to the nuclei of native myocytes (hematoxylin and eosin). (G) A heart 18 days after the stem cell transplantation; histological examination showed that blue cells distributed sparsely in the periscar area and most of them accumulated in the scar area. (H) Low-power view of periscar (left) and scar area (right). N, necrosis; M, spared myocardium. (I) In periscar area, a few large rod-shaped blue cells were identified (left). High-power view of respective box zone (right panel). Double staining of the sections from the periscar areas where the rod-shaped blue cells were identified for β-galactosidase, and cardiac-specific proteins showed that β-galactosidase colocalized with α-actin (J, K), desmin (L, M), phospholamban (N, O), and cardiac-specific troponin T (P, Q). J, L, N, and P and K, M, O, and Q are cross-sectional and longitudinal sections, respectively. These data suggest that these few blue rod-shaped cells derived from exogenous MSCs were likely cardiomyocytes.[48] (See Color Insert.)

myocardium) at the early time point post-MI may have been somewhat ambiguous as compared to the specificity of the MRI data obtained 4 months post-MI. In any event, stimulation of endogenous CPCs in response to cell transplantation to form new cardiomyocytes may have played a significant role in scar size reduction in the treated group.

As noted above, Hare's group has extensively studied the effects of MSC transplantation in acute and subacute swine models of MI. These investigators have reported evidence that MSCs contribute to post-MI repair processes by at least four mechanisms: (i) reduction of apoptosis of stressed cardiomyocytes, (ii) activation of endogenous CPCs and the consequent regenerations of cardiomyocytes from that cellular source, (iii) reactivation of the cell cycle in preexisting cardiomyocytes, and (iv) differentiation of MSCs into cardiomyocytes and vascular cells (see Ref. 11 for review). Moreover, paracrine signals emanating from the transplanted cells appeared to strongly contribute to the first three regenerative mechanisms observed in their studies.[11] This group has also reported similar favorable responses to cell therapy in a chronic swine MI model.[42]

Suzuki *et al.*[49] have recently reported that transplantation of autologous MSCs into stable, chronically hibernating swine myocardium resulted in functional improvement in the hibernating region and increased myocardial mass and myocyte counts, but decreased myocyte cross-section areas in both the hibernating and remote regions. Notably, MSC injection did not trigger these responses in normal swine hearts. Evidence was presented that the LV hypertrophy in the treated hibernation group was a result of cardiac regeneration resulting from activation of c-kit$^+$ cells resident in the myocardium. Why similar hypertrophic responses occurred in both the hibernating and remote regions remains to be elucidated.

In summary, in small and large animal model studies, the results of transplantation of BM-derived cells have been mixed. In some cases, the effects appear to be mainly paracrine and a result of decreased native cardiomyocyte apoptosis and increased capillary generation; in others, evidence has been presented that the transplanted MSCs may also differentiate into cardiomyocytes and vascular cells and/or activate endogenous CPCs to regenerate cardiomyocytes. This variation in responses may be related to variations in the nature of the MSCs isolated from BM, their culture conditions and, in some cases, the effects of preconditioning the MSCs by exposure to hypoxia or growth factors prior the implantation.[8,10,50,51]

B. Cardiac Progenitor Cells

As noted, most clinical progenitor cell transplantation studies reported thus far have involved the administration of BM-derived progenitor cells. However, a phase 1 clinical trial of autologous c-Kit$^+$ CPCs transplantation has now been reported.[52] In that study, CPCs were obtained from cardiac biopsies performed during coronary artery bypass surgery in patients with reduced LV function. The derived CPCs were expanded and, approximately 4 months post-surgery, they were infused into the relevant coronary arteries. The main messages presented were (i) the transplantation procedure was safe and there were no complications associated with the procedure during follow-up of up to 1 year post-procedure and (ii) the treated hearts had significant increases of LV EF and significant decreases of infarct volume over the period of observation in contrast to the stability of these variables in the untreated hearts. The mechanistic basis of these effects was not examined in this preliminary study, but in our view, the positive results obtained certainly justify more definitive phase 2 and 3 clinical trials. The following discussion will examine the possible mechanistic basis of these clinical findings by summarizing animal studies relevant to this question.

Anversa's and Marban's groups among others have examined the effects of injecting c-kit$^+$ CPCs into the BZ following MI. In the reports from Anversa's group, CPCs were extracted from cardiac homogenates, expanded, and then used for injection.[4,53–55] Marban and associates derived their CPCs from cultures of adult human (or porcine) cardiac biopsies.[24,56] Over a period of *in vitro* culture, cells emerged from the explants, propagated, and formed cardiospheres, and these cardiospheres and/or cardiosphere-derived cells (CDCs) were then used for cell transplantation. Of note, CDCs were found to be comprised of both noncardiac- and cardiac-committed cells and secreted growth factors and cytokines under both *in vitro* and *in vivo* conditions.

In the reports from Anversa's group, the injection of CPCs in acute or subacute rat MI models was associated with the reduction of scar size, evidence of cardiomyocyte regeneration from the injected CPCs, increased vascularity, decreased LV remodeling, and, substantially, improved LV performance, and these effects of cell transplantation were enhanced by coadministration of growth factors.[4,53–55]

Based on reports from Anversa's group, we tested the hypothesis that the concomitant delivery of IGF + HGF and heart-derived Sca-1$^+$/CD31$^-$ cells at the time of MI would better preserve LV structure and function than would cell transplantation alone.[29] In a mouse model, nuclear LacZ-labeled cardiac-derived Sca-1$^+$/CD31$^-$ cells were injected into the BZ immediately after coronary ligation with or without the addition of recombinant mouse IGF + HGF to the cell suspensions. Sca-1$^+$/CD31$^-$ cell transplantation limited infarct size and attenuated post-MI LV structural and functional deterioration. LV EFs were control, 53.3 ± 3.2; MI, 18.9 ± 3.9; MI + Cells, 29.1 ± 5.1 ($n = 6$, $p < 0.05$ MI vs. all others). Combined growth factor and cell administration yielded even better results as evidenced by the fact that EF was the highest in the cell transplantation plus growth factor group (38.3 ± 3.1, $n = 6$, $p < 0.05$ vs.

cell transplantation alone). Although IGF + HGF supplementation promoted survival of transplanted Sca-1$^+$/CD31$^-$ cells, the frequency of transdifferentiation of engrafted cells into cardiac or vascular cells was quite low in all treatment groups. However, in both groups that received cells (but most prominently in combined treatment group), native cardiomyocyte apoptosis rates were substantially reduced and LV function was improved. Angiogenesis was also enhanced, most prominently in the combined treatment group. Because differentiation of the engrafted cells into cardiomyocytes was minimal, the improved LV function was likely a consequence of direct effects of the growth factors themselves and paracrine effects emanating from the transplanted cells. However, this study did not exclude the possibility that cardiomyocyte regeneration arising from the differentiation of endogenous CPCs or from reactivation of the cell cycle in preexisting cardiomyocytes contributed to the therapeutic response.

Recently, Tang et al.[55] addressed the question of whether delayed cell therapy (i.e., late after MI) would be effective. They reported that intracoronary infusion of syngeneic c-kit$^+$ cardiac-derived progenitor cells in a rat model of chronic (30-day) MI was associated with modest but significant LV structural and functional improvement. They further suggested that this improvement was primarily due to the stimulation of proliferation and differentiation of endogenous CPCs. In contrast, only rarely did the infused progenitor cells appear to differentiate into cardiomyocytes and vascular cells and the latter phenomenon occurred in only a minority of transplanted animals.[55] In an earlier study from the same laboratory in which cell transplantation was performed 20 days post-MI, the number of transplanted cells that differentiated into cardiomyocytes was significantly larger.[54] Taken together, the two studies suggest that there may be an optimal time for cell transplantation during which the unfavorable early post-MI myocardial milieu for engraftment and differentiation has receded. Once that favorable "window" has passed (i.e., perhaps after the scar is further consolidated), the milieu may no longer be as supportive for transplantation-associated myocardial regeneration. Taken together, the data from these (and other) rodent studies suggest that transplanted CPCs and CDCs can differentiate into cardiomyocytes as well as stimulate endogenous CPCs to do so.

Similarly, at the time of coronary ligation, Marban's group injected CDCs into the hearts of immunocompromised mice.[57] They reported that the infarct size was reduced and LV function was improved in the treated animals. Moreover, there was evidence that both injected CDCs and (recruited) endogenous CPCs differentiated into cardiomyocytes and also contributed to the generation of capillaries in the BZ; the latter responses were at least partially attributed to the paracrine effects of the transplanted cells. In an earlier report, this group also showed that transplantation of unfractionated CDCs yielded better results than did injection of purified samples of either noncardiac-committed or cardiac-committed CDCs.[24]

Because the data obtained in large animal models seem to have more predictive value with regard to potential clinical responses than do rodent studies, CDCs experiments in swine MI models have been recently reported by Marban's group. In one study, autologous CDCs were injected into the infarct-associated coronary artery 4 weeks post-MI and the animals were then observed for an additional 8 weeks.[56] Before transplantation, LV structural and functional characteristics including infarct size (delayed gadolinium enhancement MRI) were comparable in placebo and CDC-treated groups. At the terminal study 8 weeks posttransplantation, infarct size relative to LV mass was decreased in the CDCs group although absolute infarct size was unchanged in both groups. The decrease of the relative infarct size was a consequence of significantly increased LV mass in the CDC group. Although there was evidence of differentiation of engrafted CDCs to vascular cells and cardiomyocytes, the quantitative contributions of the engrafted cells to cardiomyocyte regeneration appeared low relative to the observed LV mass increase. Hence, some cardiomyocyte regeneration may have been a consequence of endogenous CPC differentiation and/or stimulation of already differentiated cardiomyocytes to propagate. Notably, there was no evidence of either tumor development or infarction induced by the infused CPCs.

In a later study, the effects of intramyocardial injection of CDCs, cardiospheres, or placebo into the BZ of swine hearts 4 weeks post-MI were compared 8 weeks posttransplantation.[58] Echocardiographically measured end-systolic and end-diastolic LV chamber volumes were stabilized by cardiosphere transplantation but not by the other interventions, and the reduction of end-diastolic volumes was also significant in the group that received cardiospheres. In contrast, although EF was not affected by any intervention in that study, E_{max} measurements obtained from families of pressure–volume loops were significantly increased in the cardiosphere-treated group. The LV functional effects of the cardiosphere injection appeared to be disproportionately more prominent than those of CDCs and no tumors developed in any treated animal. Unfortunately, detailed studies of the fate of transplanted CPCs or cardiospheres were not done. These data indicated that cardiosphere transplantation was more efficacious than that of CDCs, but, as noted, did not address the mechanisms of the beneficial effects in a comprehensive fashion.

In a more mechanistic study, human CPCs (in the form of cardiospheres or CDCs) were shown to be capable of direct cardiac regeneration *in vivo* and CDCs were also shown to secrete vascular endothelial growth factor, HGF, and IGF into the myocardium when transplanted into a SCID mouse acute MI model.[57] This study also reported that CDCs injected into the MI BZ increased the expression of Akt, decreased the apoptotic rate and caspase 3 levels, and increased the capillary numbers. Based on the number of human-specific cells relative to overall increases in capillary density and myocardial viability, the authors suggested that direct differentiation of CDCs quantitatively accounted for 20–50% of the observed effects. Unfortunately, similar detailed data evaluating the differentiation potential of CDCs transplanted into injured large animal hearts are not available.

VI. Do CDCs and/or MSCs Stimulate Endogenous CPCs to Regenerate Cardiomyocytes and Vascular Cells?

As noted, some studies suggest that MSCs, in addition to transdifferentiating into cardiomyocytes and vascular cells, also stimulate endogenous CPC to do the same. In contrast, Lee's laboratory has reported that only c-kit$^+$ cells but not MSCs (both derived from bone marrow) have this capability.[59] This response of CPCs to the transplanted cells is almost certainly due to paracrine, cytokine, chemokine, and growth factor signaling emanating from the transplanted c-kit$^+$ cells. However, as pointed out in a recent review, this controversy is not yet resolved.[60]

VII. Can Differentiated Cardiomyocytes Be Induced to Dedifferentiate and Reenter the Cell Cycle?

As indicated earlier, following injury, myocardial regeneration from preexisting cardiomyocytes occurs in zebrafish and, surprisingly, also in neonatal mice (but only for a few days following birth in the latter).[1,3] Recently, several reports have suggested that, under certain experimental conditions, dedifferentiation and proliferation of already differentiated cardiomyocytes can also occur. In an *in vitro* study, Marban's laboratory reported that atrial and ventricular cardiomyocytes extracted from adult rat hearts and cultured in a mitogen-rich medium for ~11 days dedifferentiated to a degree that some expressed stem-cell-associated antigens and had reduced levels of cell cycle inhibitory molecules.[61] Additionally, a fair number of these myocyte-derived cells subsequently formed spheres and redifferentiated into cardiomyocytes and endothelial cells. It follows that, if this type of dedifferentiation and cell cycle reentry could be activated in the *in vivo* heart, then it might be possible to regenerate a significant number of cardiomyocytes that are lost to injury.

In this context, Bersell *et al.*[13] recently reported that the growth factor neuregulin 1 acting on its tyrosine kinase receptor ErbB4 induced *in vitro* mononuclear cardiomyocytes to undergo both karyokinesis and cytokinesis. In mice, they found that inactivation of ErbB4 reduced cardiomyocyte proliferation and the converse was true in mice in which ErbB4 was activated. Moreover, they showed that the new cardiomyocytes were not derived from endogenous CSCs or BM-derived progenitor cells. Of potentially more clinical relevance, they also showed that in mice that had survived a coronary ligation induced MI, 12 weeks of daily neuregulin 1 injections begun 1 week post-MI resulted in considerable myocardial regeneration. The latter was associated with a significantly reduced infarct size and a marked improvement of LV performance. These and other reports in this new area of investigation raise the possibility that cardiac regeneration from *in situ* cardiomyocytes as well as from CPCs may be possible in the future.

VIII. Conclusions and Future Perspectives

The clinical studies involving injection of (primarily) BM-derived cells either into the myocardium, venous system, or coronary arteries of patients following an MI have yielded positive (albeit modest) results vis-à-vis the recovery of LV function. Unfortunately, detailed clinical studies of the mechanistic basis of the improved LV function are not yet available. In contrast, many of the animal studies that we have discussed strongly support the view that myocardial regeneration can be a consequence of transplantation of BM- or cardiac-derived progenitor cells if the experimental conditions are "right." In this context, "right" includes selection of the proper cell type for transplantation, the details of cell preparation (including the use of ischemic preconditioning and/or growth factor preincubation), and the timing of the transplantation in relation to the myocardial injury. Further, the intramyocardial administration of growth factors at the time of MI has been reported to activate endogenous CPCs to generate new cardiomyocytes in some small and large animal studies. Post-MI cardiomyocyte regeneration is currently proposed to be due to combinations of a number of mechanisms including (i) paracrine effects exerted by the transplanted cells (the benefits of which could persist even after the injected cells were lost to apoptosis, rejection, etc.), (ii) transdifferentiation of implanted cells into cardiomyocytes and vascular cells, (iii) paracrine stimulation of endogenous progenitor cells to differentiate into cardiomyocytes and vascular cells, and (iv) paracrine stimulation of preexisting cardiomyocytes to reenter the cell cycle and propagate. Additional beneficial paracrine effects of transplanted cells include prevention of apoptosis of potentially salvagable cardiomyocytes and stimulation of angiogenesis. Collectively, the available data support the view that induction of cardiac regeneration following myocardial injury may be an attainable goal and one worth a continuing research effort.

Acknowledgments

This work was supported by U.S. Public Health Service Grants HL50470, HL67828, and HL95077, and by National Centers for Research Resources (NCRR), National Institutes of Health Grant P41RR08079. X. W. was supported AHA Greater Midwest Grant-in-Aid.

References

1. Poss KD, Wilson LG, Keating MT. Heart regeneration in zebrafish. *Science* 2002;**298**:2188–90.
2. Singh BN, Koyano-Nakagawa N, Garry JP, Weaver CV. Heart of newt: a recipe for regeneration. *J Cardiovasc Transl Res* 2010;**3**:397–409.
3. Porrello ER, Mahmoud AI, Simpson E, Hill JA, Richardson JA, Olson EN, et al. Transient regenerative potential of the neonatal mouse heart. *Science* 2011;**331**:1078–80.
4. Leri A, Kajstura J, Anversa P, Frishman WH. Myocardial regeneration and stem cell repair. *Curr Probl Cardiol* 2008;**33**:91–153.
5. Linke A, Muller P, Nurzynska D, Casarsa C, Torella D, Nascimbene A, et al. Stem cells in the dog heart are self-renewing, clonogenic, and multipotent and regenerate infarcted myocardium, improving cardiac function. *Proc Natl Acad Sci USA* 2005;**102**:8966–71.
6. Torella D, Rota M, Nurzynska D, Musso E, Monsen A, Shiraishi I, et al. Cardiac stem cell and myocyte aging, heart failure, and insulin-like growth factor-1 overexpression. *Circ Res* 2004;**94**:514–24.
7. Urbanek K, Torella D, Sheikh F, De Angelis A, Nurzynska D, Silvestri F, et al. Myocardial regeneration by activation of multipotent cardiac stem cells in ischemic heart failure. *Proc Natl Acad Sci USA* 2005;**102**:8692–7.
8. Afzal MR, Haider H, Idris NM, Jiang S, Ahmed RP, Ashraf M. Preconditioning promotes survival and angiomyogenic potential of mesenchymal stem cells in the infarcted heart via NF-kappaB signaling. *Antioxid Redox Signal* 2010;**12**:693–702.
9. Kudo M, Wang Y, Wani MA, Xu M, Ayub A, Ashraf M. Implantation of bone marrow stem cells reduces the infarction and fibrosis in ischemic mouse heart. *J Mol Cell Cardiol* 2003;**35**:1113–9.
10. Pasha Z, Wang Y, Sheikh R, Zhang D, Zhao T, Ashraf M. Preconditioning enhances cell survival and differentiation of stem cells during transplantation in infarcted myocardium. *Cardiovasc Res* 2008;**77**:134–42.
11. Williams AR, Hare JM. Mesenchymal stem cells: biology, pathophysiology, translational findings, and therapeutic implications for cardiac disease. *Circ Res* 2011;**109**:923–40.
12. Barazzoni R, Zanetti M, Sturnega M, Stebel M, Semolic A, Pirulli A, et al. Insulin down-regulates SIRT1 and AMPK activation and is associated with changes in liver fat, but not in inflammation and mitochondrial oxidative capacity, in streptozotocin-diabetic rat. *Clin Nutr* 2011;**30**:384–90.
13. Bersell K, Arab S, Haring B, Kuhn B. Neuregulin1/ErbB4 signaling induces cardiomyocyte proliferation and repair of heart injury. *Cell* 2009;**138**:257–70.
14. Braun T, Dimmeler S. Breaking the silence: stimulating proliferation of adult cardiomyocytes. *Dev Cell* 2009;**17**:151–3.
15. Breen DM, Giacca A. Effects of insulin on the vasculature. *Curr Vasc Pharmacol* 2011;**9**:321–32.
16. Campa VM, Gutierrez-Lanza R, Cerignoli F, Diaz-Trelles R, Nelson B, Tsuji T, et al. Notch activates cell cycle reentry and progression in quiescent cardiomyocytes. *J Cell Biol* 2008;**183**:129–41.

17. Di Stefano V, Giacca M, Capogrossi MC, Crescenzi M, Martelli F. Knockdown of cyclin-dependent kinase inhibitors induces cardiomyocyte re-entry in the cell cycle. *J Biol Chem* 2011;**286**:8644–54.

18. Bearzi C, Rota M, Hosoda T, Tillmanns J, Nascimbene A, De Angelis A, et al. Human cardiac stem cells. *Proc Natl Acad Sci USA* 2007;**104**:14068–73.

19. Bergmann O, Bhardwaj RD, Bernard S, Zdunek S, Barnabe-Heider F, Walsh S, et al. Evidence for cardiomyocyte renewal in humans. *Science* 2009;**324**:98–102.

20. Kajstura J, Urbanek K, Perl S, Hosoda T, Zheng H, Ogorek B, et al. Cardiomyogenesis in the adult human heart. *Circ Res* 2010;**107**:305–15.

21. Kajstura J, Urbanek K, Rota M, Bearzi C, Hosoda T, Bolli R, et al. Cardiac stem cells and myocardial disease. *J Mol Cell Cardiol* 2008;**45**:505–13.

22. Laugwitz KL, Moretti A, Lam J, Gruber P, Chen Y, Woodard S, et al. Postnatal isl1+ cardioblasts enter fully differentiated cardiomyocyte lineages. *Nature* 2005;**433**:647–53.

23. Messina E, De Angelis L, Frati G, Morrone S, Chimenti S, Fiordaliso F, et al. Isolation and expansion of adult cardiac stem cells from human and murine heart. *Circ Res* 2004;**95**:911–21.

24. Smith RR, Barile L, Cho HC, Leppo MK, Hare JM, Messina E, et al. Regenerative potential of cardiosphere-derived cells expanded from percutaneous endomyocardial biopsy specimens. *Circulation* 2007;**115**:896–908.

25. Hsieh PC, Segers VF, Davis ME, MacGillivray C, Gannon J, Molkentin JD, et al. Evidence from a genetic fate-mapping study that stem cells refresh adult mammalian cardiomyocytes after injury. *Nat Med* 2007;**13**:970–4.

26. Wang X, Hu Q, Nakamura Y, Lee J, Zhang G, From AH, et al. The role of the sca-1+/CD31- cardiac progenitor cell population in postinfarction left ventricular remodeling. *Stem Cells* 2006;**24**:1779–88.

27. Kubo H, Jaleel N, Kumarapeli A, Berretta RM, Bratinov G, Shan X, et al. Increased cardiac myocyte progenitors in failing human hearts. *Circulation* 2008;**118**:649–57.

28. Urbanek K, Quaini F, Tasca G, Torella D, Castaldo C, Nadal-Ginard B, et al. Intense myocyte formation from cardiac stem cells in human cardiac hypertrophy. *Proc Natl Acad Sci USA* 2003;**100**:10440–5.

29. Wang X, Li Q, Braunlin E, From AH, Zhang J. The beneficial role of endogenous Sca-1+/CD31- cells in post-infarction left ventricular remodeling. *Circulation* 2009;**120**:S790.

30. Rosenblatt-Velin N, Ogay S, Felley A, Stanford WL, Pedrazzini T. Cardiac dysfunction and impaired compensatory response to pressure overload in mice deficient in stem cell antigen-1. *FASEB J* 2012;**26**:229–39.

31. Iso Y, Spees JL, Serrano C, Bakondi B, Pochampally R, Song YH, et al. Multipotent human stromal cells improve cardiac function after myocardial infarction in mice without long-term engraftment. *Biochem Biophys Res Commun* 2007;**354**:700–6.

32. Noiseux N, Gnecchi M, Lopez-Ilasaca M, Zhang L, Solomon SD, Deb A, et al. Mesenchymal stem cells overexpressing Akt dramatically repair infarcted myocardium and improve cardiac function despite infrequent cellular fusion or differentiation. *Mol Ther* 2006;**14**:840–50.

33. Zeng L, Hu Q, Wang X, Mansoor A, Lee J, Feygin J, et al. Bioenergetic and functional consequences of bone marrow-derived multipotent progenitor cell transplantation in hearts with postinfarction left ventricular remodeling. *Circulation* 2007;**115**:1866–75.

34. Gnecchi M, He H, Liang OD, Melo LG, Morello F, Mu H, et al. Paracrine action accounts for marked protection of ischemic heart by Akt-modified mesenchymal stem cells. *Nat Med* 2005;**11**:367–8.

35. Gnecchi M, He H, Noiseux N, Liang OD, Zhang L, Morello F, et al. Evidence supporting paracrine hypothesis for Akt-modified mesenchymal stem cell-mediated cardiac protection and functional improvement. *FASEB J* 2006;**20**:661–9.

36. Wang X, Jameel MN, Li Q, Mansoor A, Qiang X, Swingen C, et al. Stem cells for myocardial repair with use of a transarterial catheter. *Circulation* 2009;**120**:S238–46.
37. Mangi AA, Noiseux N, Kong D, He H, Rezvani M, Ingwall JS, et al. Mesenchymal stem cells modified with Akt prevent remodeling and restore performance of infarcted hearts. *Nat Med* 2003;**9**:1195–201.
38. Zhang G, Nakamura Y, Wang X, Hu Q, Suggs LJ, Zhang J. Controlled release of stromal cell-derived factor-1 alpha in situ increases c-kit + cell homing to the infarcted heart. *Tissue Eng* 2007;**13**:2063–71.
39. Nakamura Y, Wang X, Xu C, Asakura A, Yoshiyama M, From AH, et al. Xenotransplantation of long-term-cultured swine bone marrow-derived mesenchymal stem cells. *Stem Cells* 2007;**25**:612–20.
40. Quevedo HC, Hatzistergos KE, Oskouei BN, Feigenbaum GS, Rodriguez JE, Valdes D, et al. Allogeneic mesenchymal stem cells restore cardiac function in chronic ischemic cardiomyopathy via trilineage differentiating capacity. *Proc Natl Acad Sci USA* 2009;**106**:14022–7.
41. Schuleri KH, Amado LC, Boyle AJ, Centola M, Saliaris AP, Gutman MR, et al. Early improvement in cardiac tissue perfusion due to mesenchymal stem cells. *Am J Physiol Heart Circ Physiol* 2008;**294**:H2002–11.
42. Schuleri KH, Feigenbaum GS, Centola M, Weiss ES, Zimmet JM, Turney J, et al. Autologous mesenchymal stem cells produce reverse remodelling in chronic ischaemic cardiomyopathy. *Eur Heart J* 2009;**30**:2722–32.
43. Shujia J, Haider HK, Idris NM, Lu G, Ashraf M. Stable therapeutic effects of mesenchymal stem cell-based multiple gene delivery for cardiac repair. *Cardiovasc Res* 2008;**77**:525–33.
44. Xu M, Uemura R, Dai Y, Wang Y, Pasha Z, Ashraf M. In vitro and in vivo effects of bone marrow stem cells on cardiac structure and function. *J Mol Cell Cardiol* 2007;**42**:441–8.
45. Hare JM, Traverse JH, Henry TD, Dib N, Strumpf RK, Schulman SP, et al. A randomized, double-blind, placebo-controlled, dose-escalation study of intravenous adult human mesenchymal stem cells (prochymal) after acute myocardial infarction. *J Am Coll Cardiol* 2009;**54**:2277–86.
46. Williams AR, Trachtenberg B, Velazquez DL, McNiece I, Altman P, Rouy D, et al. Intramyocardial stem cell injection in patients with ischemic cardiomyopathy: functional recovery and reverse remodeling. *Circ Res* 2011;**108**:792–6.
47. Jameel MN, Li Q, Mansoor A, Qiang X, Sarver A, Wang X, et al. Long term functional improvement and gene expression changes after bone marrow derived multipotent progenitor cell transplantation in myocardial infarction. *Am J Physiol Heart Circ Physiol* 2010;**298**: H1348–56.
48. Liu J, Hu Q, Wang Z, Xu C, Wang X, Gong G, et al. Autologous stem cell transplantation for myocardial repair. *Am J Physiol Heart Circ Physiol* 2004;**287**:H501–11.
49. Suzuki G, Iyer V, Lee TC, Canty Jr. JM. Autologous mesenchymal stem cells mobilize cKit + and CD133 + bone marrow progenitor cells and improve regional function in hibernating myocardium. *Circ Res* 2011;**109**:1044–54.
50. Niagara MI, Haider H, Jiang S, Ashraf M. Pharmacologically preconditioned skeletal myoblasts are resistant to oxidative stress and promote angiomyogenesis via release of paracrine factors in the infarcted heart. *Circ Res* 2007;**100**:545–55.
51. Suzuki Y, Kim HW, Ashraf M, Haider H. Diazoxide potentiates mesenchymal stem cell survival via NF-kappaB-dependent miR-146a expression by targeting Fas. *Am J Physiol Heart Circ Physiol* 2010;**299**:H1077–82.
52. Bolli R, Chugh AR, D'Amario D, Loughran JH, Stoddard MF, Ikram S, et al. Cardiac stem cells in patients with ischaemic cardiomyopathy (SCIPIO): initial results of a randomised phase 1 trial. *Lancet* 2011;**378**:1847–57.

53. Padin-Iruegas ME, Misao Y, Davis ME, Segers VF, Esposito G, Tokunou T, et al. Cardiac progenitor cells and biotinylated insulin-like growth factor-1 nanofibers improve endogenous and exogenous myocardial regeneration after infarction. *Circulation* 2009;**120**:876–87.

54. Rota M, Padin-Iruegas ME, Misao Y, De Angelis A, Maestroni S, Ferreira-Martins J, et al. Local activation or implantation of cardiac progenitor cells rescues scarred infarcted myocardium improving cardiac function. *Circ Res* 2008;**103**:107–16.

55. Tang XL, Rokosh G, Sanganalmath SK, Yuan F, Sato H, Mu J, et al. Intracoronary administration of cardiac progenitor cells alleviates left ventricular dysfunction in rats with a 30-day-old infarction. *Circulation* 2010;**121**:293–305.

56. Johnston PV, Sasano T, Mills K, Evers R, Lee ST, Smith RR, et al. Engraftment, differentiation, and functional benefits of autologous cardiosphere-derived cells in porcine ischemic cardiomyopathy. *Circulation* 2009;**120**:1075–83 1077 p following 1083.

57. Chimenti I, Smith RR, Li TS, Gerstenblith G, Messina E, Giacomello A, et al. Relative roles of direct regeneration versus paracrine effects of human cardiosphere-derived cells transplanted into infarcted mice. *Circ Res* 2010;**106**:971–80.

58. Lee ST, White AJ, Matsushita S, Malliaras K, Steenbergen C, Zhang Y, et al. Intramyocardial injection of autologous cardiospheres or cardiosphere-derived cells preserves function and minimizes adverse ventricular remodeling in pigs with heart failure post-myocardial infarction. *J Am Coll Cardiol* 2011;**57**:455–65.

59. Loffredo FS, Steinhauser ML, Gannon J, Lee RT. Bone marrow-derived cell therapy stimulates endogenous cardiomyocyte progenitors and promotes cardiac repair. *Cell Stem Cell* 2011;**8**:389–98.

60. Malliaras K, Marban E. Cardiac cell therapy: where we've been, where we are, and where we should be headed. *Br Med Bull* 2011;**98**:161–85.

61. Zhang Y, Li TS, Lee ST, Wawrowsky KA, Cheng K, Galang G, et al. Dedifferentiation and proliferation of mammalian cardiomyocytes. *PLoS One* 2010;**5**:e12559.

Role of GATA-4 in Differentiation and Survival of Bone Marrow Mesenchymal Stem Cells

Meifeng Xu,* Ronald W. Millard,[†] and Muhammad Ashraf*

*Department of Pathology and Laboratory Medicine, University of Cincinnati College of Medicine, Cincinnati, Ohio, USA

[†]Pharmacology and Cell Biophysics, University of Cincinnati Medical Center, Cincinnati, Ohio, USA

Cell and tissue regeneration is a relatively new research field and it incorporates a novel application of molecular genetics. Combinatorial approaches for stem-cell-based therapies wherein guided differentiation into cardiac lineage cells and cells secreting paracrine factors may be necessary to overcome the limitations and shortcomings of a singular approach. GATA-4, a GATA zinc-finger transcription factor family member, has been shown to regulate differentiation, growth, and survival of a wide range of cell types. In this chapter, we discuss whether overexpression of GATA-4 increases mesenchymal stem cell (MSC) transdifferentiation into cardiac phenotype and enhances the MSC secretome, thereby increasing cell survival and promoting postinfarction cardiac angiogenesis. MSCs engineered with GATA-4 enhance their capacity to differentiate into cardiac cell phenotypes, improve survival of the cardiac progenitor cells and their offspring, and modulate the paracrine activity of stem cells to support their angiomyogenic potential and cardioprotective effects.

Progress in Molecular Biology
and Translational Science, Vol. 111
http://dx.doi.org/10.1016/B978-0-12-398459-3.00010-1

217

I. Introduction

Myocardial ischemia is caused by an acute reduction of coronary blood flow resulting in diminished nutrient supply to the myocardium. Sustained ischemia leads to myocardial infarction (MI) accompanied by necrosis of large numbers of cardiomyocytes (CMs). MI compromises the heart's basal and dynamic performance range and can lead for heart failure. Numerous approaches are available to prevent or reverse coronary artery disease, including lifestyle changes, dietary modification, regular exercise, pharmacotherapy, and percutaneous coronary interventions. These interventions have significantly decreased the prevalence of coronary artery syndromes and mortality after MI. The cellular mechanism underlying the development of heart failure after MI has long been recognized as rooted in a decreased number of viable CMs and an inability of the remaining viable CMs to compensate for this loss of numbers and their associate role in cardiac contraction and relaxation functions. Current medical/surgical treatments such as coronary artery bypass grafting, angioplasty, and stent placement are performed as conduit artery reconstructions in order to address ischemia with myriad forms of coronary artery syndromes so as to partially or fully ameliorate the extent and consequences of acute and chronic infarction. Cardiac transplantation represents the sole method of replacing the failing heart and its blood supply with a healthy, fully functional pump. However, whole-organ transplantation is limited by an inadequate supply of donor hearts. In addition, heart transplant recipients require life-long immunosuppression pharmacotherapy. Consequently, the potential value and utility of alternative solutions arising from efforts to develop novel therapeutic strategies that facilitate protection of CMs against regional myocardial ischemic injury in order to enhance cardiac muscle regeneration in the diseased heart are obvious. In this regard, stem cell strategies[1,2] have emerged as promising approaches to support and enhance the heart's intrinsic repair system by promoting angiogenesis, myogenesis, and tissue matrix reverse remodeling after MI.

II. Cytotherapy in Myocardial Infarction

The pioneering work of Anversa and colleagues identified resident populations of stem cells, namely, cardiac stem cells (CSCs), in heart tissue. These cells are multipotent and can adopt all critical cell phenotypes found in heart tissue.[3,4] However, the number of CSCs is insufficient for timely and full tissue recovery/reconstruction/restoration in cases of a massive MI where as many as a billion CMs can be lost.[5] Researchers worldwide are pursuing a variety of approaches to enhance the intrinsic regenerative processes in the damaged heart. The focus has largely been on the potential salutary effects of stem cells intended to prevent or reverse heart failure after MI. Stem cells maintain the capacity for undifferentiated self-renewal through mitotic cell division, yet they can serve as progenitors for a diverse range of specialized cell types. The main goal of cardiac tissue cytotherapy is to inhibit tissue degeneration in the acute phases of functional loss following MI by regenerating cardiac muscle in the failing heart in order to improve the basal and dynamic range of ventricular pump performance.

A. Representative Stem Cells

A wide variety of cells have been examined for their capacity to regenerate the myocardium after ischemia-induced infarction and improve cardiac function. The main cell types used in ischemic myocardial treatments have included endogenous CSCs,[3,4,6,7] embryonic stem cells (ESCs),[8–10] skeletal myoblasts (SMs),[11,12] inducible pluripotent stem cells (iPSCs),[13,14] and bone-marrow-derived stem cells (BMSCs).[15–19]

1. ENDOGENOUS CARDIAC STEM CELLS

Endogenous populations of CSCs can replicate and have the potential for heart tissue regeneration.[3,4,6,7] CSCs delivered by intramyocardial injection or intracoronary routes after myocardial ischemic injury in mouse and rat models have been shown to ameliorate reductions in left ventricular (LV) function.[6,7] CSCs found in the mammalian myocardium express one or more cell surface biomarkers including c-kit,[3] stem cell antigen-1 (Sca-1),[20] breast cancer resistance protein (Abcg2),[21] and the transcription factor Islet-1.[22]

i. *c-kit⁺ cells*. Beltrami *et al.*[3] reported the existence of a population of Lin⁻ c-kit⁺ cells in the adult rat heart. These cells exhibit all the properties expected of CSCs, including self-renewal, clonogenicity, and multipotentiality, and are able to differentiate into CM, smooth muscle cells, and endothelial cells. A significant increase in cardiac function is observed after these cells are administered in laboratory models with regional LV ischemia.[3]

ii. *Sca-1⁺ cells*. In 2003, Oh and colleagues[20] reported the existence of a subset of CSCs expressing Sca-1 (spinocerebellar ataxia type-1 protein). These cells express cardiac gene markers after *in vitro* treatment with the DNA demethylation agent 5-azacytidine. Transplantation of CSC Sca-1⁺ into a mouse model of ischemia/reperfusion (I/R) injury increased the production of cardiac contractile proteins.[20]

iii. *Side population (SP) cells.* The adult heart contains SP CSCs expressing Abcg-2 (ATP-binding cassette subfamily G member 2 protein) which can proliferate and differentiate and are capable of participating in myocardial tissue repair.[21,23]

iv. *Islet-1⁺ cells.* CSCs expressing islet-1 (insulin gene enhancer protein) can be induced to differentiate into cardiac phenotypes after coculture with neonatal CMs.[22,24] However, to date no cell-based cardiac therapy experiments have been reported using $CSC^{Islet-1+}$.

2. EMBRYONIC STEM CELLS

ESCs, from the inner cell mass of mammalian blastocysts, can differentiate into CMs that exhibit spontaneous contractile activity and express cardiac transcription factors.[8–10,25,26] ESC-derived CM clusters compare favorably with human fetal, neonatal, and adult atrial and ventricular myocytes in expression patterns and levels of cardiac-specific biomarkers.[9] Most ECS-derived cells have electrophysiological characteristics similar to those of human fetal ventricular myocytes.[10,25,26] Action potentials[10,26] and high levels of the gap junction protein N-cadherin[27] are observed when contracting human-sourced ESC-derived CMs are injected into the left ventricle. These studies provide clear evidence that transplantation or injection of ESCs can consistently repopulate and integrate into the infarcted myocardium to improve LV contractile function.[10,28,29] Although the cardiogenic potential of ESCs is unquestioned, some ethical questions and the potential risk for development of teratoma[30] have hampered the broad adoption of ESCs for human post-MI therapeutic application.

3. SKELETAL MYOBLASTS AUTOLOGOUS

SMs of somatic origin are proposed as an ideal cell type for cell-based therapies based on biological safety and availability, not raising the ethical or religious concerns[31,32] associated with human ESCs. SMs can differentiate into multinucleated myotubes without the risk of tumor formation. Myoblasts have been one of the first cell candidates for cardiac regeneration in experimental[11] and clinical[12] studies. SMs form myotubes after transplantation, and cardiac function is subsequently improved in laboratory animal models of MI.[11,33–35] Teams conducting various clinical studies have reported that SM transplantation improves cardiac function.[31,32,36,37] However, failure to express α-MHC, cardiac troponin I, or atrial natriuretic peptide and the lack of intercalated disk proteins N-cadherin and connexin43,[38] as well as the failure to electromechanically couple with the host myocytes (a major arrhythmogenic issue), are recognized as current challenges related to SMs.[39]

4. Induced Pluripotent Stem Cells

Takahashi and Yamanaka[40] discovered that somatic cells can be reprogrammed or "induced" into pluripotent stem cells by reintroducing only four factors (Oct 3/4, Sox2, c-Myc, and Klf4). Recently, Srivastava's group[41] has demonstrated that the combination of three different transcription factors, namely GATA-4, MEF-2c, and Tbx5, could directly reprogram fibroblasts into CM-like cells. Based on the current understanding of the molecular pathways underlying iPSC generation, the application of fewer transcription factors combined with small molecules is now used to promote pluripotency.[42,43] Pasha et al.[13] have reported that mouse SMs can be efficiently reprogrammed into iPSCs with the DNA methyltransferase inhibitor RG108 by induction of the transcription factor Oct3/4. These iPSCs share the same morphological and differentiation characteristics of ESCs. When injected in myocardium after MI, SM-derived iPSCs are associated with remarkable regeneration of myocardium and even form gap junctions with the native CMs. These SM-derived iPSCs are associated with a significant attenuation of infarct size expansion and improvement of global LV function when they are transplanted into an infarcted mouse heart.[13,14] The advantage of iPSCs is that they can be generated from a patient's own somatic cells, and therefore there is no immunological concern. Moreover, there are no ethical or moral issues involved in their generation and use. However, since undifferentiated iPSCs do exhibit teratogenicity similar to ESCs,[14,44] this challenge will need to be resolved before iPSCs will be approved for use as a human clinical therapy.

5. Bone-Marrow-Derived Stem Cells

BMSCs include hematopoietic cells, endothelial progenitor cells,[15] and mesenchymal stem cells (MSCs).[16–18] Bone marrow is an easily accessible source of autologous adult stem cells. Several investigators have reported improved cardiac function, reduced infarct size, and enhanced myocardial regeneration following transplantation of BMSCs into infarcted myocardium.[16–19,45,46] Clinical studies using autologous BMSCs have documented safety and feasibility in patients with acute MI and in those with chronic ischemia. Patients exhibited a significant improvement in the global left ventricular ejection fraction (LVEF) following intracoronary injection of autologous BMSCs.[47–53] The other stem cell types include umbilical-cord-derived stem cells,[54] adipose-tissue-derived stem cells,[55] and very small ES-like cells.[56,57] The cardiogenic potential of these cell types is being assessed currently in both in vitro and in vivo experimental animal models with encouraging results.

B. Contribution of Cytotherapy to Heart Repair

Transplantation of stem cells into the infarcted rat heart *in vivo* significantly increases cardiac function by attenuating tissue injury; inhibiting fibrotic remodeling; stimulating stem cell recruitment to, and proliferation in, the heart tissue; reducing inflammatory oxidative stress; and enhancing myocardial angiogenesis. Transplanted stem cells participate in the myocardial tissue repair process by one or more of the mechanisms including transdifferentiation of implanted cells into CM and endothelial cells[18,19,58,59] as well as through the release of biologically active factors[17,45,46,60–62] with proangiogenic and cardioprotective effects.

BM-derived MSCs are particularly attractive as therapeutic cells. These cells are easily isolated and can be greatly expanded *ex vivo* without loss of phenotype or differentiation capacity. Once transplanted, they "home" in to sites of inflammation and injury. Moreover, they are easily transfected and amenable to genetic modification *in vitro*. MSCs can be extracted from adult tissue; they do not pose the ethical concerns that ESCs do. Importantly, transplantation into an allogeneic host may not require immunosuppression.[63]

The *in vitro* capacity of mouse BM-derived MSCs to differentiate into CMs induced by 5-azacytidine was first reported in 1999 by Makino *et al.*[64] MSCs can undergo myogenic differentiation and become electromechanically coupled with host CMs.[18,65] Injected cells differentiated not only into CMs but also into endothelial cells and smooth muscle cells in animal MI models.[18,58,66] Rota *et al.*[59] have reported that transplanted stem cells form functional myocytes and regenerate the myocardium following engraftment in the infarcted myocardium. Within 12–48 h after infarction and cell implantation, donor cells are integrated structurally with resident cells. Intercellular junction and adhesion complexes are detected between transplanted stem cells and between stem cells and adjacent native CMs and fibroblasts. Donor-cell-derived CMs exhibit electrical characteristics similar to those of normal CMs, but with time (15–30 days posttransplantation), they show a prolongation of the action potential and enhanced cell shortening.[59]

Increasing evidence suggests that, in addition to transdifferentiation into cardiac phenotypes, paracrine factors from stem cells may contribute to tissue protection and repair.[17,45,46,60,62] Direct myocardial delivery of allogeneic stem cells 2 days following MI improved cardiac function before differentiation occurred.[47] Moreover, intramuscular or intracoronary injection of MSCs, or conditioned medium (CdM), or concentrated active factors secreted from MSCs are therapeutically effective in treating heart failure,[67–69] suggesting that paracrine factors of MSCs promote functional recovery of the infarcted heart. Injection of stem cells into the ischemic myocardium at the time of acute coronary artery occlusion attenuated the structural, contractile, and energetic abnormalities at 4 weeks after occlusion, resulting in preservation of global LV function, despite a low engraftment rate, and minimal differentiation of stem cells into CM and endothelial cells.[45] It has been demonstrated that implantation of BMSCs may constitute a novel safety strategy for achieving optimal therapeutic angiogenesis by secreting potent angiogenic ligands and cytokines.[70] Under physiological oxygen conditions, levels of vascular endothelial growth factor (VEGF), basic fibroblast growth factor (bFGF), insulin-like growth factor (IGF), and stromal-cell-derived factor (SDF) in CdM of MSC are significantly higher compared with CdM obtained from CM. After 4 h of anoxia, these proteins are significantly increased by 30–150% in CdM of MSC.[61] Burchfield et al.[62] reported that the paracrine factors bFGF, VEGF, and SDF-1α released from transplanted stem cells into the surrounding tissue can directly and favorably influence the healing process in the heart, including promoting neovascularization, reducing cardiac myocyte apoptosis, inflammation, and fibrosis, improving contractility, normalizing bioenergetics, and enhancing endogenous repair.[62] Moreover, Dzau and colleagues[60,67] demonstrated that the cytoprotective effects on CMs from genetically modified MSCs were largely attributable to paracrine effects and not to *de novo* regeneration by cardiomyogenic differentiation from MSCs. These results support the hypothesis that the heart repair, observed as cardioprotection and neovascularization, is at least in part mediated by paracrine factors arising from stem cells.

C. Limitation of Stem Cell Therapy

Human clinical trials examine the safety and efficacy of stem cell administration following MI. Available 5-year follow-up results from the randomized-controlled "BOOST" trial have shown that the initial improvement in LV function performance was not sustained in patients who underwent successful percutaneous coronary injection of stem cells following MI.[71,72] LVEF improvement could not be demonstrated in patients with ST-elevation MI who underwent intracoronary BMSC transplantation compared with patients in control or placebo groups at 6 months.[73,74] It is generally accepted that the capacity of stem cells to repair the ischemic myocardium depends on their transdifferentiation into myocardial phenotypes and the protection of the native CMs threatened by ischemia. Several studies have sparked intense debate over the ability of stem cells to differentiate into cardiac phenotypes. MSCs do not show high efficiency of transdifferentiation into functional CMs when transplanted into cardiac tissue.[75] No evidence for differentiation into CMs can be detected, and no engraftment is detected beyond 2 weeks after cell injections in human subjects with MI.[76] Moreover, it is reported that MSCs obtained from patients with ischemic heart disease express troponin T and

GATA-4 when cocultured with neonatal CMs. However, these MSCs do not appear to express other CM biomarkers including alpha-actinin, myosin heavy chain, or troponin I.[77] These data may explain why only limited functional improvement has been reported to date in clinical trials using MSCs after MI.

Poor survival of transplanted stem cells in the acidotic and hypoxic microenvironment of the infarcted myocardium is another of the major challenges to clinical application of stem-cell-based therapies. The highest average donor cell survival rate reported in the first hour after injection into the myocardial ischemic border is 58%.[78] In another study, $59 \pm 2\%$ of MSCs were retained in hearts 1 h after cell transplantation, only $10 \pm 6\%$ remained 1 day later, and about 1% of transplanted MSCs were found in the heart after a week.[79] Sheikh *et al.*[80] also demonstrated short-lived survival of cells following transplant, with less than 1% of cells surviving at 6 weeks after transplantation. No evidence of engraftment could be documented by immunohistochemistry, electron microscopy, or PCR at 1 month.[76]

III. Genetic Engineering of MSCs with Cytoprotective Factors

Three critical challenges MSC-mediated cardiac therapy faces are increased engraftment, long-term survival in ischemic myocardium, and transdifferentiation into cardiac cell phenotypes. To this end, various stem cell preconditioning (PC) and programming strategies have been developed to better prepare the cells for transplantation.[81] One emergent technology is the genetic engineering of MSCs achieved by modifying these cells with plasmids encoding for survival signaling molecules that improve MSCs' therapeutic potential. This line of research has generated much interest since it offers a useful strategy for boosting the therapeutic efficacy of MSCs.[82–85]

Such genetic modification of stem cells is performed *ex vivo*, where only stem cells are targeted and the entire organism is not exposed to the gene manipulation. Such an approach is safer than *in vivo* gene therapy and allows for regulated expression of the transgene in the target stem cells. Such engineering of MSCs has improved their survival and engraftment as well as their differentiation into desired cell lineages. To date, Bcl-2,[83] heme oxygenase-1 (HO-1),[84] Akt,[60,67,86] and growth factors[85,87] have each been used with the aim of improving cell survival and resistance to oxidative stress. In addition, engineering MSCs with GSK-3β (glycogen synthase kinase 3β) not only increases their survival but also induces their differentiation into cardiac phenotypes.[82] Gene modification can also contribute significantly to MSC-mediated paracrine effects[60,67] leading to an increase in CM survival and angiogenesis.

A. Transduction of Cardiac Transcription Factor GATA-4 into MSC

GATA-4 is a member of the highly conserved zinc-finger transcription factors that play a critical role in regulating differentiation, growth, and survival of a wide range of cell types.[88,89] The GATA family has six members divided into two subfamilies based on sequence similarity and expression profiles. GATA-1, GATA-2, and GATA-3 are expressed in hematopoietic cells and are involved in regulating proliferation and differentiation of blood cells.[88] GATA-4, GATA-5, and GATA-6 are expressed in cardiac tissue and endodermal derivatives.[88] The latter three proteins contain two conserved zinc fingers and their adjacent basic domains bind to (A/T)GATA(A/G) sequences.[89] GATA-4 is highly expressed in CMs throughout embryonic development, postnatal growth, and adulthood, during which it functions as a critical regulator of cardiac differentiation.[90] GATA-4 functions as a key regulator of numerous specific cardiac genes, such as atrial natriuretic peptide (ANP), B-type natriuretic peptide (BNP), α-myosin heavy chain (α-MHC), and β-myosin heavy chain (β-MHC). Depletion of GATA-4 downregulates expression of cardioregulators and upregulation of GATA-4 promotes early differentiation markers and enhanced cardiogenesis.[91]

There is increasing evidence that GATA-4 controls cell survival in addition to regulation of differentiation by several mechanisms. First, the activity of GATA-4 is regulated by preconditioning (PC), ischemia reperfusion (I/R), and nitric oxide (NO) in isolated rat hearts.[92] PC activates GATA-4 activity, while I/R and NO suppress the gene transcription of GATA-4.[92] Second, overexpression of GATA-4 protects CMs from anthracycline-induced apoptosis and cardiomyopathy,[93] whereas GATA-4 gene silencing renders anthracycline more toxic.[94,95] Third, inhibition of GATA-4 DNA-binding activity by a dominant-negative mutant of GATA-4 itself induces CM apoptosis.[96]

An additional *in vivo* cardioprotective role for GATA-4 has also been reported. Cell-based intercellular transfer of GATA-4 protein is shown to improve cardiac function in ischemic cardiomyopathy by reducing fibrosis of infarcted myocardium and by directly protecting the myocardium.[97] Recently, Rysa *et al.*[98] reported that MI significantly decreases the DNA-binding activity of GATA-4. Rescuing GATA-4 activity prevents adverse postinfarction myocardial remodeling by preventing apoptosis and promoting myocardial angiogenesis, as well as stem cell recruitment.[98] It is therefore reasonable to suggest that restoration of GATA-4 activity attenuates apoptosis, which may be an effective approach to ischemic heart repair.

Accordingly, we postulated that engineering MSCs with GATA-4 would increase cell survival and differentiation, leading to an enhancement of MSC-mediated improvement of cardiac function following transplantation of these cells into ischemic myocardium. We constructed a GATA-4 plasmid (Fig. 1) and transduced GATA-4 into MSCs obtained from male rat bone marrow using a pMSCV retroviral expression system.[99] IRES-EGFP was cloned into pMSCV vectors at XhoI and EcoRI sites, and then GATA-4 was excised from pcDNA-GATA-4[100] with HindIII and XhoI restriction enzymes and cloned into pMSCV-IRES-EGFP at BglII and SalI sites. Control cells were either not transduced—basal MSC (MSC^{bas})—or transduced with an empty vector (MSC^{Null}). Successfully transduced cell clones were acquired by selecting with puromycin (3 μg/ml). MSCs transduced with GATA-4 (MSC^{GATA-4}) were GFP immunopositive and were stained intensely for GATA-4. MSC^{GATA-4} exhibited higher levels of GATA-4 mRNA and protein. MSC^{GATA-4} retain a stem cell phenotype by expression of a stem cell marker (c-kit)[101] without significant morphological difference compared to its controls, namely, MSC^{Null} and MSC^{bas}.

B. Role of GATA-4 in MSC-Mediated Heart Repair

The local delivery of GATA-4 to the myocardial infarct border zone induces multiple effects resulting in ventricular tissue regeneration and improved global LV function.[97,98] To document the effects of MSC^{GATA-4} on the

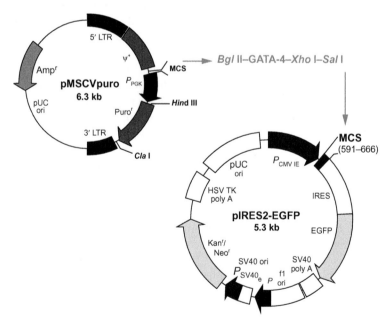

FIG. 1. Construction of the GATA-4 plasmid. (For color version of this figure, the reader is referred to the online version of this chapter.)

preservation and repair of myocardial tissue, MSC^{GATA-4} cells were directly transplanted into the ischemic myocardium of a rat following ligation of the left anterior descending coronary artery. We documented reduced CM apoptosis, improved LV function, smaller infarct size, and a thicker LV anterior wall, as well as higher myocardial perfusion in hearts transplanted with MSC^{GATA-4} than in control hearts with MSC^{Null} or with MSC^{bas}.[99] We concluded that MSC-transduced GATA-4 may represent a novel and efficient therapeutic approach for postinfarct myocardial regeneration. We have also investigated the effect of GATA-4 on MSC differentiation and survival, as well as the paracrine effects of MSC^{GATA-4}.[99,102]

1. Overexpression of GATA-4 Increases MSC Transdifferentiation into Cardiac Phenotypes

Optimally, strategies of genetic enhancement should be able to direct stem cells to differentiate into cardiac tissue cell phenotypes. The approach to combine gene transfection of stem cells to deliver GATA-4 can lead to an increase in cardiac protein expression in stem cells and to CM regeneration.[97] Moreover, it has been reported that upregulation of GATA-4 significantly increased P19 (a murine embryonic carcinoma cell line), ESC, and CSC differentiation into cardiac phenotypes.[103–105] We designed a series of *in vitro* experiments to determine whether MSC^{GATA-4} can indeed transdifferentiate efficiently into cardiac phenotypes. We first screened the cardiac gene expression in MSCs using mRNA microarray analysis. We noted upregulation of the expression of a number of cardiac genes in MSC^{GATA-4} (Table I).

In addition, using quantitative RT-PCR and Western blot methods, we documented the increased expression of several cardiac biomarkers at both mRNA and protein levels in MSC^{GATA-4} compared to MSC^{Null}; α-sarcomeric actinin (α-SA), BNP, and islet-1 were significantly increased. After 7 days of coculture with native CMs from neonatal rat ventricles, MSC^{GATA-4} cells were

TABLE I

Cardiac Gene Regulation in MSC^{GATA-4}

Gene name	Symbol	Fold increase
Myosin light chain, regulatory B	*Mylc2b*	20.18
Myosin X	*Myo10*	8.94
Myosin, heavy polypeptide 7, cardiac muscle, beta	*Myh7*	3.61
Myosin, light polypeptide kinase	*Mylk*	17.28
Tropomyosin 1, alpha	*Tpm1*	4.64
Troponin T2, cardiac	*Tnnt2*	10.09

positive to α-SA staining, showed a typical sarcomeric structure, and formed gap junctions with native CMs. FACS analysis revealed significantly more α-SA-positive cells in the MSC^{GATA-4} population than in MSC^{Null}.[102]

Our results are consistent with several reports by others obtained in different models or cell lines. First, Sachinidis *et al.*[103] reported that ES cell lines overexpressing GATA-4 show enhanced CM differentiation, whereas GATA-4-deficient ES cell lines exhibit impaired differentiation. Hu *et al.*[104] recently investigated the effects of GATA-4 on myocardial cell differentiation by constructing the vectors that overexpressed GATA-4 or silenced RNA interference. The expression of Nkx2.5 and α-MHC was upregulated in P19 cells overexpressing GATA-4 and was downregulated in those in which GATA-4 was silenced.[104] In addition, Miyamoto *et al.*[105] demonstrated that more CSCs expressing high GATA-4 differentiate into the CM lineage, compared to CSCs expressing low levels of GATA-4. Taken together, these results suggest that GATA-4 increases the potential of stem cells to transdifferentiate into cardiac tissue cell phenotypes.

GATA-4 is a zinc-finger DNA-binding protein that regulates diverse pathways associated with embryonic morphogenesis and cellular differentiation. A number of genes contain GATA regulatory sites in their promoters, including those encoding α- and β-MHC, myosin light chains, troponin C, Na^+–Ca^{2+} exchanger, angiotensin type-1α receptor, ANP, and BNP.[106] Factors involved in the GATA-4-mediated differentiation process of MSC have not been completely elucidated. Our research indicates that GATA-4 upregulates the expression of many genes in MSCs. Insulin-like growth factor binding-protein 4 (IGFBP-4) is also upregulated in MSC^{GATA-4} (9.21-fold in MSC^{GATA-4} vs. MSC^{Null}, $p < 0.05$).[102] IGFBP-4 is reported to strongly promote CM differentiation from ESCs in the late phase after embryoid body formation.[107] Recently, substrate immobilization of IGFBP-4 has been reported as a powerful tool for differentiation of ESCs into CMs.[108] To probe the relationship between IGFBP-4 and GATA-4, we decreased IGFBP-4 activity in MSC^{GATA-4} using double-stranded smart pool siRNA designed to target IGFBP-4 to determine whether IGFBP-4 plays an important role in GATA-4-mediated myocardial transdifferentiation of MSC. Nonsilencing, scrambled siRNA was used as a negative control. Our data show that reduced IGFBP-4 activity decreases MSC transdifferentiation into cardiac phenotypes.[102]

Reporter gene assays and β-catenin stabilization assays reveal that IGFBP-4 is one of the most potent canonical Wnt inhibitors.[107] The canonical Wnt protein signals are transduced by β-catenin, which enters the nucleus to modulate expression of target genes.[109] It has been shown that activation of canonical Wnt signals inhibit cardiogenesis.[110] Inhibition of canonical Wnt signaling promotes ESCs differentiation into CM and also in chick, *Xenopus*, and zebrafish embryos.[110–112] Zhu *et al.*[107] demonstrated in P19 cells that

IGFBP-4 promotes cardiogenesis by antagonizing the Wnt/β-catenin pathway through direct interactions with Frizzled and LRP5/6. Further study is needed to understand how GATA-4 upregulates IGFBP-4 resulting in MSC myocardial transdifferentiation.

Overexpression of GATA-4 upregulates many other growth factors besides IGFBP-4 in MSCs, which may also contribute to CM differentiation. Growth factors may also be able to promote *in vivo* organ-specific differentiation of ESCs.[113]

2. GATA-4 INCREASES STEM CELL SURVIVAL THROUGH UPREGULATING ANTIAPOPTOTIC PROTEINS

Several strategies leading to an increase in stem cell survival within the MI zone have demonstrated significant improvements in cardiac function. We exposed various cells to two *in vitro* oxidative stress conditions namely, anoxia (5% CO_2, 94% N_2, and 1% H_2) and hypoxia (5% CO_2, 94% N_2, and 1% O_2) cultures, to further investigate the resistance of MSCs to oxidative stress. The number of apoptotic cells was counted by FACS following annexin-V labeling. Fewer apoptotic MSCs were detected in MSC^{GATA-4} than in MSC^{Null} after exposure to hypoxia for 48 h.[99] We further evaluated the survival of MSC^{GATA-4} in ischemic myocardium based on the expression of Sry gene from male MSCs in female recipient hearts. More Sry gene was expressed in peri-infarcted and infarcted myocardium in MSC^{GATA-4}-transplanted hearts than in hearts with MSC^{Null}-transplanted cells 4 days postimplantation. We believe that this data supports the claim that overexpression of GATA-4 increases MSC survival in the low-oxygen environment of transplanted hearts.[99]

GATA-4 also regulates the expression of cell-survival signaling in stem cells.[92,94] Our previous studies indicate that cell survival mediated by GATA-4 in mobilized progenitor cells is associated with upregulation of Bcl-2.[114] Others have reported that transcription factor GATA-4 regulates cardiac Bcl-2 gene expression *in vitro* and *in vivo*.[115] We found that Bcl-2 expression was significantly higher in MSC^{GATA-4} than in MSC^{Null}. There are two GATA consensus motifs in the 5′-promoter region of the Bcl-2 gene,[116] and activation of GATA-4 by phosphorylation of Ser105 promotes the expression of Bcl-2.[115,117] Kitta *et al.*[117] suggest that GATA-4 phosphorylation also leads to the upregulation of Bcl-xL.

Bcl-2 and Bcl-xL are key antiapoptotic proteins and serve as critical regulators of cell survival pathways. Li *et al.*[83] genetically modified MSCs with Bcl-2 gene and found that Bcl-2 overexpression significantly reduced MSC apoptosis and enhanced VEGF secretion under hypoxic conditions. Transplantation of Bcl-2-MSCs into the ischemic myocardium increased MSC survival by 2.2-, 1.9-, and 1.2-fold at 4 days, 3 weeks, and 6 weeks, respectively, compared to the vector MSC group.[83] Bcl-xL promotes cell survival and increases the cloning efficiency of dissociated ESCs without altering ESC self-renewal.[118] Bcl-2 is well known for its inhibitory role in the oligomerization of proapoptotic proteins Bax and Bak, and it can also inhibit mitochondrial outer membrane permeability.[119] Therefore, the essential role of GATA-4 as a survival factor may be explained, at least in part, by its function as an upstream activator of Bcl-2 family and hence by preserving mitochondrial integrity and function.

However, it is currently unclear how upregulation of GATA-4 regulates the expression of Bcl-2. Our research suggests that GATA-4-mediated upregulation of Bcl-2 may be associated with the expression of miRNA (miR) in MSCs. miRs are very small, noncoding, posttranscriptional, regulating RNAs and play a critical role as determinants of stem cell functions. Single-stranded miRNA usually binds to specific mRNA through sequences that are imperfectly complementary to the target mRNA. The bound mRNA remains untranslated, resulting in reduced levels of the corresponding protein, or can be degraded resulting in reduced levels of the corresponding mRNA.[120] Bcl-2 is considered one of the target genes of the miR-15 family (miR-15b, miR-16, and miR-195),[121] and Bcl-2 repression by the miR-15 family leads to apoptosis in a leukemic cell line.[122] Overexpression of GATA-4 significantly downregulates the expression of miR-15 family in MSCs and may result in upregulation of Bcl-2 in MSC^{GATA-4} (Yu et al., unpublished data).

Although the genes of the Bcl-2 family appear to be important antiapoptotic targets that are regulated by GATA-4, further research is needed to identify other apoptosis and survival control genes downstream of the GATA-4 pathway. Rysa et al.[98] identified 361 genes through microarray analysis that were upregulated by more than 1.5-fold in the LV myocardium overexpressing GATA-4. We have also compared gene expression in response to GATA-4 transduction in MSCs using microarray analysis. Our data indicate that many mRNAs are regulated by overexpression of GATA-4, and certainly some of these mRNAs may be involved in GATA-4-mediated cell survival (Table II).

The expression of heat shock proteins plays an important role in refolding damaged proteins and restoring their function. Overexpression of a single heat shock protein, Hsp20, improved MSC survival under ischemic conditions.[123] VEGF, FGF-2, and IGF-1, are refolded in Hsp20-MSCs, resulting in an increase in the ratio of Bcl-2 to Bax.[123] MSCs overexpressing HO-1 also show enhanced antiapoptotic and antioxidative properties.[124] Interestingly, MSCs engineered to overexpress HO-1 improve function and decrease ventricular remodeling after transplantation in ischemically injured rat hearts by improving injected MSC survival and increasing VEGF and bFGF secretion.[84] Previous studies have also reported that direct delivery of angiogenic cytokines and growth factor genes promotes donor cell survival and engraftment.[125,126] Recently, Hi-VEGF-MSCs have been documented to be retained in the infarcted myocardium in substantial numbers associated with a remarkably reduced number of apoptotic cells the infarcted area.[87]

TABLE II

Genes Upregulated in the MSC^{GATA-4} (microarray)

Gene name	Symbol	Fold increase
Glutathione peroxidase 4	Gpx4	8.30
Glutathione S-transferase, alpha 2 (Yc2)	Gsta2	4.04
Glutathione S-transferase, mu 5	Gstm5	3.37
Glutathione S-transferase, mu 6	Gstm6	5.27
Heat shock protein 2	Hspb2	5.26
Heat shock protein 90 kDa alpha (cytosolic), class B member 1	Hsp90ab1	11.46
Heme oxygenase (decycling) 1	Hmox1	3.52
Heme oxygenase (decycling) 2	Hmox2	3.81
Mitogen-activated protein kinase 13	Mapk13	4.57
Mitogen-activated protein kinase 9	Mapk9	2.85
Mitogen-activated protein kinase kinase kinase 7 interacting protein 3	Map3k7ip3	3.07
Mitogen-activated protein kinase kinase kinase kinase 4	Map4k4	8.72
Nnerve growth factor receptor (TNFRSF16)-associated protein 1	Ngfrap1	5.36
Nerve growth factor, beta	Ngfb	5.50
Transforming growth factor, beta 1	Tgfb1	2.65

Therefore, the cytoprotection by GATA-4 would appear to be related to upregulation of prosurvival factors (e.g., Bcl-2), growth factors, heat shock proteins, and cytokines. GATA-4 appears to orchestrate molecular expression pathways through transcriptional control of apoptotic regulatory genes in coordination with other target genes that collaborate to enhance cell survival.

3. GATA-4 Enhances Paracrine Effect of MSCs

Restoration of injured myocardial tissue in order to return the heart to its normal functional dynamic range is the primary goal of stem cell-based treatment following MI. Prevention of cell loss, especially during early stages of cardiac ischemia, is a corollary objective. Paracrine factors produced by stem cells are documented as contributors to functional improvement of the infarcted heart by protecting CMs from ischemic injury.[45,46] We investigated the cardioprotective effect of MSC^{GATA-4} in primary cultured CMs exposed to hypoxia from 24 to 48 h. Hypoxic culture conditions significantly increased the numbers of annexin-V-positive apoptotic and dead (pyknotic) cells and increased the amount of LDH released from CMs. Coculture with MSC^{GATA-4} significantly reduced CM apoptosis, decreased the amount of LDH release, and increased overall CM survival compared to MSCs transduced with the empty vector. Moreover, in vivo transplantation of MSC^{GATA-4} also significantly reduced apoptotic CM in the ischemic border area.[99]

MSC^{GATA-4} provided significant cardioprotection, indicating a trophic effect; that is, the paracrine factors released from these cells act on host CMs and spare them from apoptosis. Such paracrine actions might be the result of humoral stimulation of preexisting cells by MSC^{GATA-4} and could play a pivotal role in MSC^{GATA-4}-mediated cardioprotection. To further test the paracrine hypothesis, we examined the effect of MSC^{GATA-4}-mediated cardioprotection *in vitro* using a dual-chamber culture system and using CdM obtained from various MSCs. CM survival in hypoxic culture was significantly increased in coculture with MSC^{GATA-4} or when treated with CdM obtained from MSC^{GATA-4} (CdM^{GATA-4}), compared to that cultured with MSC^{Null} or treated with CdM obtained from MSC^{Null} (CdM^{Null}).

It is well known that the beneficial effect of stem cell treatment *in vivo* is partially attributed to vasculogenesis, which by providing a tissue perfusion route prevents apoptosis of native CMs and promotes regeneration of injured myocardium. This view is in agreement with evidence that overexpression of GATA-4 increases the density of myocardial capillary and small conducting vessels and increases coronary flow.[98,127] We assessed the angiogenic effect of paracrine factors contained in culture media from MSC^{GATA-4} *in vitro* by quantifying capillary-like tube formation and spheroid sprouting of HUVECs. CdM^{GATA-4} stimulated the formation of capillary-like structures. The number of emerging capillary sprouts and the cumulative sprout lengths were greater in HUVECs treated with CdM^{GATA-4} than in those treated with CdM^{Null}.[99] Our results confirm that the paracrine factors arising from MSCs are directly responsible for MSC^{GATA-4}-mediated angiogenesis.

It is commonly believed that MSCs release many cytokines into the adjacent environment. IGF-1 and VEGF are specifically elevated in MSC culture media and are therefore associated with increased CM survival when cocultured with MSCs.[105,128] Moreover, MSCs transduced with plasmids encoding for angiogenic cytokines and growth factors, serving as carriers for therapeutic genes, may improve regional perfusion of the ischemic heart due to increased angiogenesis[79] and thus protect CMs.[67] Supporting this is evidence that coinjection of VEGF and MSCs into ischemic hearts improves cardiac function and MSC survival, compared to injection of either one alone.[79] Transplantation of MSCs engineered to express VEGF enhances neovascularization while simultaneously reducing apoptosis in the infarcted myocardium.[87] Furthermore, MSCs overexpressing the VEGF gene increase angiogenesis in the infarct region after acute MI.[129] In a similar manner, IGF-1 gene delivery has also been combined with other growth factors to promote donor cell survival, engraftment, and differentiation as part of a multimodal therapy approach.[85,125,126] Comparing control MSCs, studies with MSC^{GATA-4} have identified several genes that at least in part may account for the enhanced paracrine effects (Table II). IGF-1 and VEGF-A concentration in CdM^{GATA-4} is significantly higher than that in CdM^{Null}.

Moreover, the effect of CdM^{GATA-4} on capillary formation is diminished by VEGF-A- and IGF-1-neutralizing antibodies.[99] Overexpression of GATA-4 modulates the paracrine behavior of MSCs by enhancing the expression of cytokines to achieve the desired results. It therefore appears that increased growth factors are not only involved in MSC survival and differentiation but also act in association with the MSC^{GATA-4} construct to mediate cardioprotection and angiogenesis.

Single cytokine growth factor therapeutic regimens have been attempted with HGF, IGF, and VEGF—with encouraging results.[85,117,130,131] Our microarray results support the notion that overexpression of GATA-4 in MSCs upregulates several growth factor genes other than IGF-1 and VEGF-A, such as bone morphogenetic protein 1, β-nerve growth factor, TGF-α, etc. A combination of these bioactive factors released from transplanted stem cells might represent an integrated therapeutic strategy necessary to orchestrate comprehensive cardiac tissue repair.[46,67,132,133] At this point, we do not know whether the cytokines exert synergistic effects or whether there is cross talk among cytokines secreted from MSC^{GATA-4}. Further elucidation of the relevant pathways responsible for such benefits is needed; including those following cell therapy in injured heart tissue. The connections between GATA-4 and the measured paracrine effects remain elusive. GATA-4 has been postulated to act as an integrator of two signaling pathways: protein kinase C (PKC) and JAK–STAT. PKC phosphorylation enhances GATA-4 DNA-binding activity, and STAT-1 interacts with GATA-4 functionally and physically to synergistically activate growth-factor-inducible promoters.[134] Thus, STAT proteins appear to act as growth-factor-inducible coactivators of tissue-specific transcription factors, and overexpression of GATA-4 in MSCs may upregulate and increase secretion of an array of cytokines. Moreover, GATA factors are able to recruit STAT proteins to target promoters via GATA-binding sites.[134] The action of GATA-4 certainly involves many combinatorial interactions with other inducible transcription or cell-restricted factors, including Nkx2.5, MEF-2, SRF, nuclear factor of activated T cells, SMAD, etc.

IV. Conclusions

Engraftment of stem cells into ischemia-induced MI areas has myriad effects: prevention of large-scale myocardial necrosis, differentiation of stem cells into functional CMs that electromechanically couple with the host myocardium, and secretion of cytokines that protect native CMs and promote angiogenesis. The use of stem cells for cardiac regeneration is relatively new and incorporates a novel application of molecular genetics. Therapeutic approaches that use a single cytokine or a particular stem cell phenotype may not achieve therapeutic goals. Combinatorial approaches for stem-cell-based therapies may be necessary to overcome the limitations and shortcomings of singular approaches. Such guided cardiac tissue cell lineage differentiation could lead to cells capable of expressing the relevant paracrine "*milieu interieurc*".[135] Genetic manipulation of stem cells directed toward expression of a variety of cardiocentric factors will certainly advance our understanding of the basic mechanisms of stem cell survival and differentiation potential. MSCs engineered with GATA-4 will enhance the capacity of MSCs to differentiate into cardiac tissue cell phenotypes, improve survival of the progenitor MSCs and their offspring, and modulate paracrine role of stem cells to support their angiomyogenic potential and cardioprotective effects (Fig. 2).

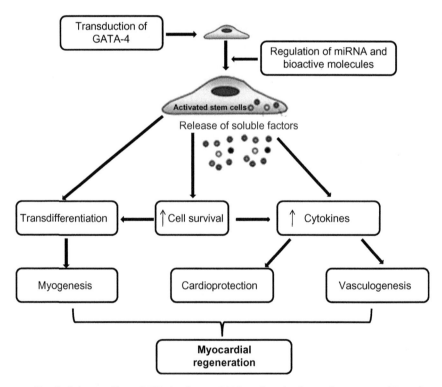

FIG. 2. Scheme of how GATA-4 enhances MSC-mediated ischemic heart repair. (For color version of this figure, the reader is referred to the online version of this chapter.)

REFERENCES

1. Tongers J, Losordo DW, Landmesser U. Stem and progenitor cell-based therapy in ischaemic heart disease: promise, uncertainties, and challenges. *Eur Heart J* 2011;**32**:1197–206.
2. Leri A, Hosoda T, Rota M, Kajstura J, Anversa P. Myocardial regeneration by exogenous and endogenous progenitor cells. *Drug Discov Today Dis Mech* 2007;**4**:197–203.
3. Beltrami AP, Barlucchi L, Torella D, et al. Adult cardiac stem cells are multipotent and support myocardial regeneration. *Cell* 2003;**114**:763–76.
4. Kajstura J, Urbanek K, Rota M, et al. Cardiac stem cells and myocardial disease. *J Mol Cell Cardiol* 2008;**45**:505–13.
5. Reinecke H, Minami E, Zhu WZ, Laflamme MA. Cardiogenic differentiation and transdifferentiation of progenitor cells. *Circ Res* 2008;**103**:1058–71.
6. Dawn B, Stein AB, Urbanek K, et al. Cardiac stem cells delivered intravascularly traverse the vessel barrier, regenerate infarcted myocardium, and improve cardiac function. *Proc Natl Acad Sci USA* 2005;**102**:3766–71.
7. Tokunaga M, Liu ML, Nagai T, et al. Implantation of cardiac progenitor cells using self-assembling peptide improves cardiac function after myocardial infarction. *J Mol Cell Cardiol* 2010;**49**:972–83.
8. Xu C, Police S, Rao N, Carpenter MK. Characterization and enrichment of cardiomyocytes derived from human embryonic stem cells. *Circ Res* 2002;**91**:501–8.
9. Asp J, Steel D, Jonsson M, et al. Cardiomyocyte clusters derived from human embryonic stem cells share similarities with human heart tissue. *J Mol Cell Biol* 2010;**2**:276–83.
10. Christoforou N, Oskouei BN, Esteso P, et al. Implantation of mouse embryonic stem cell-derived cardiac progenitor cells preserves function of infarcted murine hearts. *PLoS One* 2010;**5**:e11536.
11. Taylor DA, Atkins BZ, Hungspreugs P, et al. Regenerating functional myocardium: improved performance after skeletal myoblast transplantation. *Nat Med* 1998;**4**:929–33.
12. Menasche P, Alfieri O, Janssens S, et al. The Myoblast Autologous Grafting in Ischemic Cardiomyopathy (MAGIC) trial: first randomized placebo-controlled study of myoblast transplantation. *Circulation* 2008;**117**:1189–200.
13. Pasha Z, Haider H, Ashraf M. Efficient non-viral reprogramming of myoblasts to stemness with a single small molecule to generate cardiac progenitor cells. *PLoS One* 2011;**6**:e23667.
14. Ahmed RP, Haider HK, Buccini S, Li L, Jiang S, Ashraf M. Reprogramming of skeletal myoblasts for induction of pluripotency for tumor-free cardiomyogenesis in the infarcted heart. *Circ Res* 2011;**109**:60–70.
15. Kuliczkowski W, Derzhko R, Prajs I, Podolak-Dawidziak M, Serebruany VL. Endothelial progenitor cells and left ventricle function in patients with acute myocardial infarction: potential therapeutic considertions. *Am J Ther* 2012;**19**:44–50.
16. Liu JF, Wang BW, Hung HF, Chang H, Shyu KG. Human mesenchymal stem cells improve myocardial performance in a splenectomized rat model of chronic myocardial infarction. *J Formos Med Assoc* 2008;**107**:165–74.
17. Li L, Zhang S, Zhang Y, Yu B, Xu Y, Guan Z. Paracrine action mediate the antifibrotic effect of transplanted mesenchymal stem cells in a rat model of global heart failure. *Mol Biol Rep* 2009;**36**:725–31.
18. Quevedo HC, Hatzistergos KE, Oskouei BN, et al. Allogeneic mesenchymal stem cells restore cardiac function in chronic ischemic cardiomyopathy via trilineage differentiating capacity. *Proc Natl Acad Sci USA* 2009;**106**:14022–7.
19. Orlic D, Kajstura J, Chimenti S, et al. Bone marrow cells regenerate infarcted myocardium. *Nature* 2001;**410**:701–5.

20. Oh H, Bradfute SB, Gallardo TD, et al. Cardiac progenitor cells from adult myocardium: homing, differentiation, and fusion after infarction. *Proc Natl Acad Sci USA* 2003;**100**: 12313–8.

21. Martin CM, Meeson AP, Robertson SM, et al. Persistent expression of the ATP-binding cassette transporter, Abcg2, identifies cardiac SP cells in the developing and adult heart. *Dev Biol* 2004;**265**:262–75.

22. Nakano A, Nakano H, Chien KR. Multipotent islet-1 cardiovascular progenitors in development and disease. *Cold Spring Harb Symp Quant Biol* 2008;**73**:297–306.

23. Oyama T, Nagai T, Wada H, et al. Cardiac side population cells have a potential to migrate and differentiate into cardiomyocytes in vitro and in vivo. *J Cell Biol* 2007;**176**:329–41.

24. Moretti A, Caron L, Nakano A, et al. Multipotent embryonic isl1 + progenitor cells lead to cardiac, smooth muscle, and endothelial cell diversification. *Cell* 2006;**127**:1151–65.

25. Kehat I, Khimovich L, Caspi O, et al. Electromechanical integration of cardiomyocytes derived from human embryonic stem cells. *Nat Biotechnol* 2004;**22**:1282–9.

26. Xue T, Cho HC, Akar FG, et al. Functional integration of electrically active cardiac derivatives from genetically engineered human embryonic stem cells with quiescent recipient ventricular cardiomyocytes: insights into the development of cell-based pacemakers. *Circulation* 2005;**111**:11–20.

27. Laflamme MA, Gold J, Xu C, et al. Formation of human myocardium in the rat heart from human embryonic stem cells. *Am J Pathol* 2005;**167**:663–71.

28. Min JY, Huang X, Xiang M, et al. Homing of intravenously infused embryonic stem cell-derived cells to injured hearts after myocardial infarction. *J Thorac Cardiovasc Surg* 2006;**131**:889–97.

29. Singla DK, Lyons GE, Kamp TJ. Transplanted embryonic stem cells following mouse myocardial infarction inhibit apoptosis and cardiac remodeling. *Am J Physiol Heart Circ Physiol* 2007;**293**:H1308–14.

30. Nussbaum J, Minami E, Laflamme MA, et al. Transplantation of undifferentiated murine embryonic stem cells in the heart: teratoma formation and immune response. *FASEB J* 2007;**21**:1345–57.

31. Duckers HJ, Houtgraaf J, Hehrlein C, et al. Final results of a phase IIa, randomised, open-label trial to evaluate the percutaneous intramyocardial transplantation of autologous skeletal myoblasts in congestive heart failure patients: the SEISMIC trial. *EuroIntervention* 2011;**6**:805–12.

32. Dib N, Dinsmore J, Lababidi Z, et al. One-year follow-up of feasibility and safety of the first U.S., randomized, controlled study using 3-dimensional guided catheter-based delivery of autologous skeletal myoblasts for ischemic cardiomyopathy (CAuSMIC study). *JACC Cardiovasc Interv* 2009;**2**:9–16.

33. Farahmand P, Lai TY, Weisel RD, et al. Skeletal myoblasts preserve remote matrix architecture and global function when implanted early or late after coronary ligation into infarcted or remote myocardium. *Circulation* 2008;**118**:S130–7.

34. Fukushima S, Coppen SR, Lee J, et al. Choice of cell-delivery route for skeletal myoblast transplantation for treating post-infarction chronic heart failure in rat. *PLoS One* 2008;**3**: e3071.

35. Gavira JJ, Nasarre E, Abizanda G, et al. Repeated implantation of skeletal myoblast in a swine model of chronic myocardial infarction. *Eur Heart J* 2010;**31**:1013–21.

36. Hagege AA, Marolleau JP, Vilquin JT, et al. Skeletal myoblast transplantation in ischemic heart failure: long-term follow-up of the first phase I cohort of patients. *Circulation* 2006;**114**: I108–13.

37. Siminiak T, Fiszer D, Jerzykowska O, et al. Percutaneous trans-coronary-venous transplantation of autologous skeletal myoblasts in the treatment of post-infarction myocardial contractility impairment: the POZNAN trial. *Eur Heart J* 2005;**26**:1188–95.

38. Reinecke H, Poppa V, Murry CE. Skeletal muscle stem cells do not transdifferentiate into cardiomyocytes after cardiac grafting. *J Mol Cell Cardiol* 2002;**34**:241–9.
39. Menasche P. Stem cell therapy for heart failure: are arrhythmias a real safety concern? *Circulation* 2009;**119**:2735–40.
40. Takahashi K, Yamanaka S. Induction of pluripotent stem cells from mouse embryonic and adult fibroblast cultures by defined factors. *Cell* 2006;**126**:663–76.
41. Ieda M, Fu JD, Delgado-Olguin P, et al. Direct reprogramming of fibroblasts into functional cardiomyocytes by defined factors. *Cell* 2010;**142**:375–86.
42. Nakagawa M, Koyanagi M, Tanabe K, et al. Generation of induced pluripotent stem cells without Myc from mouse and human fibroblasts. *Nat Biotechnol* 2008;**26**:101–6.
43. Huangfu D, Osafune K, Maehr R, et al. Induction of pluripotent stem cells from primary human fibroblasts with only Oct4 and Sox2. *Nat Biotechnol* 2008;**26**:1269–75.
44. Ahmed RP, Ashraf M, Buccini S, Shujia J, Haider H. Cardiac tumorigenic potential of induced pluripotent stem cells in an immunocompetent host with myocardial infarction. *Regen Med* 2011;**6**:171–8.
45. Zeng L, Hu Q, Wang X, et al. Bioenergetic and functional consequences of bone marrow-derived multipotent progenitor cell transplantation in hearts with postinfarction left ventricular remodeling. *Circulation* 2007;**115**:1866–75.
46. Tang YL, Zhao Q, Qin X, et al. Paracrine action enhances the effects of autologous mesenchymal stem cell transplantation on vascular regeneration in rat model of myocardial infarction. *Ann Thorac Surg* 2005;**80**:229–36, discussion 236–237.
47. Medicetty S, Wiktor D, Lehman N, et al. Percutaneous adventitial delivery of allogeneic bone marrow derived stem cells via infarct related artery improves long-term ventricular function in acute myocardial infarction. *Cell Transplant* 2011. http://dx.doi.org/10.3727/096368911X603657.
48. Boonbaichaiyapruck S, Pienvichit P, Limpijarnkij T, et al. Transcoronary infusion of bone marrow derived multipotent stem cells to preserve left ventricular geometry and function after myocardial infarction. *Clin Cardiol* 2010;**33**:E10–5.
49. Wollert KC, Meyer GP, Lotz J, et al. Intracoronary autologous bone-marrow cell transfer after myocardial infarction: the BOOST randomised controlled clinical trial. *Lancet* 2004;**364**:141–8.
50. Assmus B, Rolf A, Erbs S, et al. Clinical outcome 2 years after intracoronary administration of bone marrow-derived progenitor cells in acute myocardial infarction. *Circ Heart Fail* 2010;**3**:89–96.
51. Yousef M, Schannwell CM, Kostering M, Zeus T, Brehm M, Strauer BE. The BALANCE Study: clinical benefit and long-term outcome after intracoronary autologous bone marrow cell transplantation in patients with acute myocardial infarction. *J Am Coll Cardiol* 2009;**53**:2262–9.
52. Meyer GP, Wollert KC, Lotz J, et al. Intracoronary bone marrow cell transfer after myocardial infarction: eighteen months' follow-up data from the randomized, controlled BOOST (BOne marrOw transfer to enhance ST-elevation infarct regeneration) trial. *Circulation* 2006;**113**:1287–94.
53. Martin-Rendon E, Brunskill SJ, Hyde CJ, Stanworth SJ, Mathur A, Watt SM. Autologous bone marrow stem cells to treat acute myocardial infarction: a systematic review. *Eur Heart J* 2008;**29**:1807–18.
54. Nishiyama N, Miyoshi S, Hida N, et al. The significant cardiomyogenic potential of human umbilical cord blood-derived mesenchymal stem cells in vitro. *Stem Cells* 2007;**25**:2017–24.
55. Choi YS, Dusting GJ, Stubbs S, et al. Differentiation of human adipose-derived stem cells into beating cardiomyocytes. *J Cell Mol Med* 2010;**14**:878–89.
56. Zuba-Surma EK, Guo Y, Taher H, et al. Transplantation of expanded bone marrow-derived very small embryonic-like stem cells (VSEL-SCs) improves left ventricular function and remodelling after myocardial infarction. *J Cell Mol Med* 2011;**15**:1319–28.

57. Wojakowski W, Tendera M, Kucia M, et al. Cardiomyocyte differentiation of bone marrow-derived Oct-4+CXCR4+SSEA-1+ very small embryonic-like stem cells. *Int J Oncol* 2010;**37**:237–47.

58. Kajstura J, Rota M, Whang B, et al. Bone marrow cells differentiate in cardiac cell lineages after infarction independently of cell fusion. *Circ Res* 2005;**96**:127–37.

59. Rota M, Kajstura J, Hosoda T, et al. Bone marrow cells adopt the cardiomyogenic fate in vivo. *Proc Natl Acad Sci USA* 2007;**104**:17783–8.

60. Gnecchi M, Zhang Z, Ni A, Dzau VJ. Paracrine mechanisms in adult stem cell signaling and therapy. *Circ Res* 2008;**103**:1204–19.

61. Uemura R, Xu M, Ahmad N, Ashraf M. Bone marrow stem cells prevent left ventricular remodeling of ischemic heart through paracrine signaling. *Circ Res* 2006;**98**:1414–21.

62. Burchfield JS, Dimmeler S. Role of paracrine factors in stem and progenitor cell mediated cardiac repair and tissue fibrosis. *Fibrogenesis Tissue Repair* 2008;**1**:4.

63. DelaRosa O, Lombardo E. Modulation of adult mesenchymal stem cells activity by toll-like receptors: implications on therapeutic potential. *Mediators Inflamm* 2010;**2010**:865601.

64. Makino S, Fukuda K, Miyoshi S, et al. Cardiomyocytes can be generated from marrow stromal cells in vitro. *J Clin Invest* 1999;**103**:697–705.

65. Price MJ, Chou CC, Frantzen M, et al. Intravenous mesenchymal stem cell therapy early after reperfused acute myocardial infarction improves left ventricular function and alters electrophysiologic properties. *Int J Cardiol* 2006;**111**:231–9.

66. Krause DS, Theise ND, Collector MI, et al. Multi-organ, multi-lineage engraftment by a single bone marrow-derived stem cell. *Cell* 2001;**105**:369–77.

67. Gnecchi M, He H, Noiseux N, et al. Evidence supporting paracrine hypothesis for Akt-modified mesenchymal stem cell-mediated cardiac protection and functional improvement. *FASEB J* 2006;**20**:661–9.

68. Shabbir A, Zisa D, Suzuki G, Lee T. Heart failure therapy mediated by the trophic activities of bone marrow mesenchymal stem cells: a noninvasive therapeutic regimen. *Am J Physiol Heart Circ Physiol* 2009;**296**:H1888–97.

69. Nguyen BK, Maltais S, Perrault LP, et al. Improved function and myocardial repair of infarcted heart by intracoronary injection of mesenchymal stem cell-derived growth factors. *J Cardiovasc Transl Res* 2010;**3**:547–58.

70. Kamihata H, Matsubara H, Nishiue T, et al. Implantation of bone marrow mononuclear cells into ischemic myocardium enhances collateral perfusion and regional function via side supply of angioblasts, angiogenic ligands, and cytokines. *Circulation* 2001;**104**:1046–52.

71. Meyer GP, Wollert KC, Lotz J, et al. Intracoronary bone marrow cell transfer after myocardial infarction: 5-year follow-up from the randomized-controlled BOOST trial. *Eur Heart J* 2009;**30**:2978–84.

72. Schaefer A, Zwadlo C, Fuchs M, et al. Long-term effects of intracoronary bone marrow cell transfer on diastolic function in patients after acute myocardial infarction: 5-year results from the randomized-controlled BOOST trial—an echocardiographic study. *Eur J Echocardiogr* 2010;**11**:165–71.

73. Srimahachota S, Boonyaratavej S, Rerkpattanapipat P, et al. Intra-coronary bone marrow mononuclear cell transplantation in patients with ST-elevation myocardial infarction: a randomized controlled study. *J Med Assoc Thai* 2011;**94**:657–63.

74. Traverse JH, McKenna DH, Harvey K, et al. Results of a phase 1, randomized, double-blind, placebo-controlled trial of bone marrow mononuclear stem cell administration in patients following ST-elevation myocardial infarction. *Am Heart J* 2010;**160**:428–34.

75. Noiseux N, Gnecchi M, Lopez-Ilasaca M, et al. Mesenchymal stem cells overexpressing Akt dramatically repair infarcted myocardium and improve cardiac function despite infrequent cellular fusion or differentiation. *Mol Ther* 2006;**14**:840–50.

76. Agbulut O, Mazo M, Bressolle C, et al. Can bone marrow-derived multipotent adult progenitor cells regenerate infarcted myocardium? *Cardiovasc Res* 2006;**72**:175–83.

77. Koninckx R, Hensen K, Daniels A, et al. Human bone marrow stem cells co-cultured with neonatal rat cardiomyocytes display limited cardiomyogenic plasticity. *Cytotherapy* 2009;**11**: 778–92.

78. Muller-Ehmsen J, Whittaker P, Kloner RA, et al. Survival and development of neonatal rat cardiomyocytes transplanted into adult myocardium. *J Mol Cell Cardiol* 2002;**34**:107–16.

79. Pons J, Huang Y, Takagawa J, et al. Combining angiogenic gene and stem cell therapies for myocardial infarction. *J Gene Med* 2009;**11**:743–53.

80. Sheikh AY, Huber BC, Narsinh KH, et al. In vivo functional and transcriptional profiling of bone marrow stem cells after transplantation into ischemic myocardium. *Arterioscler Thromb Vasc Biol* 2012;**32**:92–102.

81. Haider H, Mustafa A, Feng Y, Ashraf M. Genetic modification of stem cells for improved therapy of the infarcted myocardium. *Mol Pharm* 2011;**8**:1446–57.

82. Cho J, Zhai P, Maejima Y, Sadoshima J. Myocardial injection with GSK-3beta-overexpressing bone marrow-derived mesenchymal stem cells attenuates cardiac dysfunction after myocardial infarction. *Circ Res* 2011;**108**:478–89.

83. Li W, Ma N, Ong LL, et al. Bcl-2 engineered MSCs inhibited apoptosis and improved heart function. *Stem Cells (Dayton Ohio)* 2007;**25**:2118–27.

84. Zeng B, Ren X, Lin G, et al. Paracrine action of HO-1-modified mesenchymal stem cells mediates cardiac protection and functional improvement. *Cell Biol Int* 2008;**32**:1256–64.

85. Haider H, Jiang S, Idris NM, Ashraf M. IGF-1-overexpressing mesenchymal stem cells accelerate bone marrow stem cell mobilization via paracrine activation of SDF-1alpha/CXCR4 signaling to promote myocardial repair. *Circ Res* 2008;**103**:1300–8.

86. Mangi AA, Noiseux N, Kong D, et al. Mesenchymal stem cells modified with Akt prevent remodeling and restore performance of infarcted hearts. *Nat Med* 2003;**9**:1195–201.

87. Kim SH, Moon HH, Kim HA, Hwang KC, Lee M, Choi D. Hypoxia-inducible vascular endothelial growth factor-engineered mesenchymal stem cells prevent myocardial ischemic injury. *Mol Ther* 2011;**19**:741–50.

88. Patient RK, McGhee JD. The GATA family (vertebrates and invertebrates). *Curr Opin Genet Dev* 2002;**12**:416–22.

89. Molkentin JD. The zinc finger-containing transcription factors GATA-4, -5, and -6. Ubiquitously expressed regulators of tissue-specific gene expression. *J Biol Chem* 2000;**275**:38949–52.

90. Oka T, Xu J, Molkentin JD. Re-employment of developmental transcription factors in adult heart disease. *Semin Cell Dev Biol* 2007;**18**:117–31.

91. Grepin C, Nemer G, Nemer M. Enhanced cardiogenesis in embryonic stem cells overexpressing the GATA-4 transcription factor. *Development* 1997;**124**:2387–95.

92. Suzuki YJ, Nagase H, Day RM, Das DK. GATA-4 regulation of myocardial survival in the preconditioned heart. *J Mol Cell Cardiol* 2004;**37**:1195–203.

93. Li L, Takemura G, Li Y, et al. Preventive effect of erythropoietin on cardiac dysfunction in doxorubicin-induced cardiomyopathy. *Circulation* 2006;**113**:535–43.

94. Aries A, Paradis P, Lefebvre C, Schwartz RJ, Nemer M. Essential role of GATA-4 in cell survival and drug-induced cardiotoxicity. *Proc Natl Acad Sci USA* 2004;**101**:6975–80.

95. Kobayashi S, Volden P, Timm D, Mao K, Xu X, Liang Q. Transcription factor GATA4 inhibits doxorubicin-induced autophagy and cardiomyocyte death. *J Biol Chem* 2010;**285**: 793–804.

96. Liang Q, De Windt LJ, Witt SA, Kimball TR, Markham BE, Molkentin JD. The transcription factors GATA4 and GATA6 regulate cardiomyocyte hypertrophy in vitro and in vivo. *J Biol Chem* 2001;**276**:30245–53.

97. Bian J, Popovic ZB, Benejam C, Kiedrowski M, Rodriguez LL, Penn MS. Effect of cell-based intercellular delivery of transcription factor GATA4 on ischemic cardiomyopathy. *Circ Res* 2007;**100**:1626–33.
98. Rysa J, Tenhunen O, Serpi R, et al. GATA-4 is an angiogenic survival factor of the infarcted heart. *Circ Heart Fail* 2010;**3**:440–50.
99. Li H, Zuo S, He Z, et al. Paracrine factors released by GATA-4 overexpressed mesenchymal stem cells increase angiogenesis and cell survival. *Am J Physiol Heart Circ Physiol* 2010;**299**: H1772–81.
100. Dai YS, Markham BE. p300 Functions as a coactivator of transcription factor GATA-4. *J Biol Chem* 2001;**276**:37178–85.
101. Xu M, Wani M, Dai Y-S, et al. Differentiation of bone marrow stromal cells into the cardiac phenotype requires intercellular communication with myocytes. *Circulation* 2004;**110**:2658–65.
102. Li H, Zuo S, Pasha Z, et al. GATA-4 promotes myocardial transdifferentiation of mesenchymal stromal cells via up-regulating IGFBP-4. *Cytotherapy* 2011;**13**:1057–65.
103. Sachinidis A, Fleischmann BK, Kolossov E, Wartenberg M, Sauer H, Hescheler J. Cardiac specific differentiation of mouse embryonic stem cells. *Cardiovasc Res* 2003;**58**:278–91.
104. Hu DL, Chen FK, Liu YQ, et al. GATA-4 promotes the differentiation of P19 cells into cardiac myocytes. *Int J Mol Med* 2010;**26**:365–72.
105. Miyamoto S, Kawaguchi N, Ellison GM, Matsuoka R, Shin'oka T, Kurosawa H. Characterization of long-term cultured c-kit+ cardiac stem cells derived from adult rat hearts. *Stem Cells Dev* 2010;**19**:105–16.
106. Charron F, Paradis P, Bronchain O, Nemer G, Nemer M. Cooperative interaction between GATA-4 and GATA-6 regulates myocardial gene expression. *Mol Cell Biol* 1999;**19**:4355–65.
107. Zhu W, Shiojima I, Ito Y, et al. IGFBP-4 is an inhibitor of canonical Wnt signalling required for cardiogenesis. *Nature* 2008;**454**:345–9.
108. Minato A, Ise H, Goto M, Akaike T. Cardiac differentiation of embryonic stem cells by substrate immobilization of insulin-like growth factor binding protein 4 with elastin-like polypeptides. *Biomaterials* 2012;**33**:515–23.
109. Moon RT, Kohn AD, De Ferrari GV, Kaykas A. WNT and beta-catenin signalling: diseases and therapies. *Nat Rev Genet* 2004;**5**:691–701.
110. Naito AT, Shiojima I, Akazawa H, et al. Developmental stage-specific biphasic roles of Wnt/beta-catenin signaling in cardiomyogenesis and hematopoiesis. *Proc Natl Acad Sci USA* 2006;**103**:19812–7.
111. Ueno S, Weidinger G, Osugi T, et al. Biphasic role for Wnt/beta-catenin signaling in cardiac specification in zebrafish and embryonic stem cells. *Proc Natl Acad Sci USA* 2007;**104**: 9685–90.
112. Foley A, Mercola M. Heart induction: embryology to cardiomyocyte regeneration. *Trends Cardiovasc Med* 2004;**14**:121–5.
113. Kofidis T, de Bruin JL, Yamane T, et al. Stimulation of paracrine pathways with growth factors enhances embryonic stem cell engraftment and host-specific differentiation in the heart after ischemic myocardial injury. *Circulation* 2005;**111**:2486–93.
114. Dai Y, Ashraf M, Zuo S, et al. Mobilized bone marrow progenitor cells serve as donors of cytoprotective genes for cardiac repair. *J Mol Cell Cardiol* 2008;**44**:607–17.
115. Kobayashi S, Lackey T, Huang Y, et al. Transcription factor gata4 regulates cardiac BCL2 gene expression in vitro and in vivo. *FASEB J* 2006;**20**:800–2.
116. Morisco C, Seta K, Hardt SE, Lee Y, Vatner SF, Sadoshima J. Glycogen synthase kinase 3beta regulates GATA4 in cardiac myocytes. *J Biol Chem* 2001;**276**:28586–97.
117. Kitta K, Day RM, Kim Y, Torregroza I, Evans T, Suzuki YJ. Hepatocyte growth factor induces GATA-4 phosphorylation and cell survival in cardiac muscle cells. *J Biol Chem* 2003; **278**:4705–12.

118. Bai H, Chen K, Gao YX, et al. Bcl-xL enhances single-cell survival and expansion of human embryonic stem cells without affecting self-renewal. *Stem Cell Res* 2012;**8**:26–37.
119. Chipuk JE, Green DR. How do BCL-2 proteins induce mitochondrial outer membrane permeabilization? *Trends Cell Biol* 2008;**18**:157–64.
120. Lim LP, Lau NC, Garrett-Engele P, et al. Microarray analysis shows that some microRNAs downregulate large numbers of target mRNAs. *Nature* 2005;**433**:769–73.
121. Yang J, Cao Y, Sun J, Zhang Y. Curcumin reduces the expression of Bcl-2 by upregulating miR-15a and miR-16 in MCF-7 cells. *Med Oncol* 2010;**27**:1114–8.
122. Cimmino A, Calin GA, Fabbri M, et al. miR-15 and miR-16 induce apoptosis by targeting BCL2. *Proc Natl Acad Sci USA* 2005;**102**:13944–9.
123. Wang X, Zhao T, Huang W, et al. Hsp20-engineered mesenchymal stem cells are resistant to oxidative stress via enhanced activation of Akt and increased secretion of growth factors. *Stem Cells* 2009;**27**:3021–31.
124. Tsubokawa T, Yagi K, Nakanishi C, et al. Impact of anti-apoptotic and anti-oxidative effects of bone marrow mesenchymal stem cells with transient overexpression of heme oxygenase-1 on myocardial ischemia. *Am J Physiol Heart Circ Physiol* 2010;**298**:H1320–9.
125. Kanemitsu N, Tambara K, Premaratne GU, et al. Insulin-like growth factor-1 enhances the efficacy of myoblast transplantation with its multiple functions in the chronic myocardial infarction rat model. *J Heart Lung Transplant* 2006;**25**:1253–62.
126. Kofidis T, de Bruin JL, Yamane T, et al. Insulin-like growth factor promotes engraftment, differentiation, and functional improvement after transfer of embryonic stem cells for myocardial restoration. *Stem Cells* 2004;**22**:1239–45.
127. Heineke J, Auger-Messier M, Xu J, et al. Cardiomyocyte GATA4 functions as a stress-responsive regulator of angiogenesis in the murine heart. *J Clin Invest* 2007;**117**:3198–210.
128. Sadat S, Gehmert S, Song YH, et al. The cardioprotective effect of mesenchymal stem cells is mediated by IGF-I and VEGF. *Biochem Biophys Res Commun* 2007;**363**:674–9.
129. Gao F, He T, Wang H, et al. A promising strategy for the treatment of ischemic heart disease: mesenchymal stem cell-mediated vascular endothelial growth factor gene transfer in rats. *Can J Cardiol* 2007;**23**:891–8.
130. Losordo DW, Vale PR, Symes JF, et al. Gene therapy for myocardial angiogenesis: initial clinical results with direct myocardial injection of phVEGF165 as sole therapy for myocardial ischemia. *Circulation* 1998;**98**:2800–4.
131. Suzuki G, Lee TC, Fallavollita JA, Canty Jr. JM. Adenoviral gene transfer of FGF-5 to hibernating myocardium improves function and stimulates myocytes to hypertrophy and reenter the cell cycle. *Circ Res* 2005;**96**:767–75.
132. Takahashi M, Li TS, Suzuki R, et al. Cytokines produced by bone marrow cells can contribute to functional improvement of the infarcted heart by protecting cardiomyocytes from ischemic injury. *Am J Physiol Heart Circ Physiol* 2006;**291**:H886–93.
133. Herrmann JL, Abarbanell AM, Wang Y, et al. Transforming growth factor-alpha enhances stem cell-mediated postischemic myocardial protection. *Ann Thorac Surg* 2011;**92**:1719–25.
134. Wang J, Paradis P, Aries A, et al. Convergence of protein kinase C and JAK-STAT signaling on transcription factor GATA-4. *Mol Cell Biol* 2005;**25**:9829–44.
135. Millard RW, Wang Y. Milieu interieur: the search for myocardial arteriogenic signals. *J Am Coll Cardiol* 2009;**53**:2148–9.

Progenitor Cell Mobilization and Recruitment: SDF-1, CXCR4, α4-integrin, and c-kit

MIN CHENG* AND
GANGJIAN QIN[†]

*Department of Cardiology, Union Hospital, Tongji Medical College, Huazhong University of Science and Technology, Wuhan, Hubei, PR China

[†]Feinberg Cardiovascular Research Institute, Department of Medicine – Cardiology, Northwestern University Feinberg School of Medicine, Chicago, Illinois, USA

Progenitor cell retention and release are largely governed by the binding of stromal-cell-derived factor 1 (SDF-1) to CXC chemokine receptor 4 (CXCR4) and by α4-integrin signaling. Both of these pathways are dependent on c-kit activity: the mobilization of progenitor cells in response to either CXCR4 antagonism or α4-integrin blockade is impaired by the loss of c-kit kinase activity; and c-kit–kinase inactivation blocks the retention of CXCR4-positive progenitor cells in the bone marrow. SDF-1/CXCR4 and α4-integrin signaling are also crucial for the retention of progenitor cells in the ischemic region, which may explain, at least in part, why clinical trials of progenitor cell therapy have failed to display the efficacy observed in preclinical investigations. The lack of effectiveness is often attributed to poor retention of the transplanted cells and, to date, most of the trial protocols have mobilized cells with injections of granulocyte colony-stimulating factor (G-CSF), which activates extracellular proteases that irreversibly cleave cell-surface adhesion molecules, including α4-integrin and CXCR4. Thus, the retention of G-CSF-mobilized cells in the ischemic region may be impaired, and the mobilization of agents that reversibly disrupt SDF-1/CXCR4 binding, such as

AMD3100, may improve patient response. Efforts to supplement SDF-1 levels in the ischemic region may also improve progenitor cell recruitment and the effectiveness of stem cell therapy.

I. Introduction

Over the last decade, a compelling body of evidence has accumulated to suggest that progenitor cells of bone marrow origin, such as endothelial progenitor cells (EPCs) and mesenchymal stem cells (MSCs), play a significant role in postnatal physiological and pathophysiological vasculogenesis[1-7] and could provide a promising new therapeutic approach for the treatment of ischemic disease.[8-15] These cells form the structural components of the new vasculature, mediate favorable cell–cell contacts, and release growth factors that contribute to vessel growth and protect against cell death in the ischemic tissue.[14,16,17] Furthermore, abnormally low levels of peripheral blood EPCs are closely associated with risk factors for cardiovascular disease, cardiovascular events, and mortality.[18,19]

Currently, most clinical trials of cell therapy for the treatment of ischemic heart disease have used progenitor cells of bone marrow origin,[20-22] which are usually administered via intracoronary infusion or transplanted directly into the ischemic region. In general, the trials have found evidence of therapeutic benefit, but with only modest efficacy,[21-26] and the absence of more definitive results is often attributed to poor retention and survival of the transplanted cells.[21,22,27] Because increases in circulating progenitor cell levels are expected to enhance the number of cells recruited to the ischemic tissue,[28-31] techniques that promote progenitor cell mobilization are being rigorously investigated.[32-36] The effectiveness of this strategy has been demonstrated in numerous preclinical studies[30,31,35-38] and has led to frequent investigations of progenitor-cell-mobilizing agents in early clinical trials.[28,29,39-50] Granulocyte colony-stimulating factor (G-CSF) has been the most commonly used mobilizing agent, but the results from these trials have not met the expectations, despite substantial increases in peripheral blood progenitor cell counts.[28,29,44,46,48,51,52] Thus, a better understanding of how progenitor cells interact with the microenvironment in the bone marrow and in the ischemic region could lead to the development of more effective cell-based therapies.

II. Progenitor Cell Mobilization

The mobilization of progenitor cells from bone marrow to the peripheral circulation is highly regulated under both normal physiological conditions and stress.[53,54] In adult bone tissue, progenitor cells are retained predominantly in specialized microenvironments near the endosteum (i.e., the osteoblast niche), where they interact with spindle-shaped, N-cadherin-expressing osteo-blasts,[55,56] and in the sinusoids (i.e., the vascular niche), where they interact with SDF-1-expressing reticular cells.[57–59] Many different cell types, matrix proteins, and soluble factors cooperatively regulate the self-renewal, differentiation, and maintenance of progenitor cells[55–57,60–65]; however, the bulk of experimental evidence suggests that progenitor cell retention and release are largely governed by two pathways, one of which is dependent on stromal-cell-derived factor 1 (SDF-1, also called CXC chemokine ligand 12 [CXCL12]) and the SDF-1 receptor CXC chemokine receptor 4 (CXCR4), and the other on $\alpha4\beta1$-integrin (also called very late antigen-4 [VLA-4]).[57,59,60,66–69] Initially, SDF-1/CXCR4 and $\alpha4\beta1$-integrin signaling appear to proceed independently; for example, the $\alpha4\beta1$-integrin antagonist Groβ can mobilize progenitor cells in mice transplanted with CXCR4-knockout bone marrow.[70] However, results from our recent studies suggest that c-kit, a receptor tyrosine kinase that binds stem cell factor (SCF), is an integral downstream component of both pathways.[71]

A. SDF-1/CXCR4

CXCR4 is a G protein-coupled receptor composed of 352 amino acids with seven transmembrane helices[72–74] and is broadly expressed by both mononuclear cells and progenitor cells in the bone marrow.[72–78] The ligand for CXCR4, SDF-1, is a secreted or membrane-bound protein that is abundantly expressed by osteoblasts, endothelial cells, and a subset of reticular cells in the osteoblast and vascular niches.[57,79–81] SDF-1/CXCR4 signaling induces the directional migration of cells and is involved in many biological processes, including cardiovascular organogenesis, hematopoiesis, immune response, and cancer metastasis. Interactions between SDF-1 and CXCR4 are crucial for maintaining populations of hematopoietic stem cells (HSCs) in adult animals,[57,66,82–87] and mice that lack either SDF-1 or CXCR4 exhibit nearly identical phenotypes characterized by late gestational lethality and defects in bone marrow colonization, B-cell lymphopoiesis, blood vessel formation, and cardiac septum formation.[83,85,88–90] Thus, the SDF-1/CXCR4 axis appears to have a fundamental role in both vasculogenesis and cardiogenesis.

The roles of SDF-1 and CXCR4 in bone marrow progenitor cell retention and release are well established.[66] Selective antagonism of CXCR4 with the pharmacological agent AMD3100 rapidly and potently mobilizes bone marrow progenitor cells in both animals and humans,[86,91–93] and systemically injected bone marrow progenitor cells accumulate predominantly in subdomains of bone marrow microvessels that are rich in SDF-1 expression.[68,94] Notably, both SDF-1 and CXCR4 expression are upregulated by relatively low oxygen tension (hypoxia) in discrete regions of the bone marrow and by the activation of hypoxia inducible factor 1 (HIF-1).[87,95–100]

B. α4-integrin

Integrins are heterodimeric transmembrane receptors composed of non-covalently joined α and β subunits and have the remarkable ability to transmit both incoming and outgoing signals across the cell membrane.[101,102] Integrins usually induce signaling pathways by acting synergistically with growth-factor receptors to regulate cell shape, adhesion, migration, proliferation, and differentiation; but both can also function independently.[103,104] The α4-integrins, α4β1 and α4β7, bind to vascular cell adhesion molecule 1 (VCAM-1), which is expressed on the surface of endothelial and stromal cells[105–109] and to fibronectin in the extracellular matrix. These binding interactions are crucial for the adhesion of progenitor cells to the microenvironment and, consequently, to progenitor cell retention and recruitment.[61,64,65] The expression of α4-integrin is downregulated during progenitor cell mobilization,[110,111] and the cleavage of VCAM-1 and α4-integrin is a critical step during cytokine-induced bone marrow progenitor cell mobilization.[112] In adult mice, deletion of α4-integrin persistently alters the distribution of progenitor cells,[108,113,114] and antibody-mediated α4-integrin blockade mobilizes progenitor cells in both animals and humans.[62,115,116] The level of α4-integrin expression on mobilized peripheral blood progenitor cells is inversely correlated with bone marrow homing and predicts the rate of engraftment in patients who have received autologous progenitor cell transplantation.[117] Furthermore, we have shown that the transient blockade of α4-integrin activity leads to higher peripheral blood EPC levels, greater EPC-mediated neovascularization, and less adverse cardiac remodeling after myocardial infarction, and that α4-integrin antibodies can release bone marrow EPCs from immobilized VCAM-1 or bone marrow stromal cells.[31] Thus, α4-integrin antibodies appear to mobilize EPCs from the bone marrow by disrupting VCAM-1:α4-integrin binding.

C. c-kit

c-kit (also called CD117) is a type III receptor tyrosine kinase expressed predominantly in bone marrow stem/progenitor cells[118] and has recently been identified as a marker for EPC and cardiac progenitor cell identity.[119–121] The ligand for c-kit, namely, SCF, is expressed in bone marrow endothelial cells and stromal cells as either a membrane-bound protein or a soluble one.[122,123] Dimers of SCF bind to c-kit, which triggers c-kit homodimerization and the phosphorylation of specific c-kit tyrosine residues.[124] The pattern of c-kit phosphorylation determines which signaling event is activated and can induce both positive and negative pathways.[124–126] SCF/c-kit signaling is essential for embryonic hematopoiesis,[127,128] and mutations that lead to the loss of c-kit (i.e., the W mutation), c-kit kinase activity (e.g., the *W42* mutation), or SCF (the *Sl* mutation)[129] cause severe macrocytic anemia and death *in utero* or during the perinatal period. Notably, defects in c-kit activity are also associated with impaired vascular development and angiogenesis,[130–132] and cancer therapies that target c-kit are cardiotoxic.[133,134] c-kit also supports progenitor cell maintenance[135–137] and is a crucial component of cardiac regeneration.[138] After myocardial infarction, c-kit-positive bone marrow cells are recruited to the ischemic myocardium and facilitate cardiac repair by differentiating into cardiac cell lineages and by expressing angiogenic cytokines.[35,131]

The mobilization of progenitor cells in response to α4-integrin blockade is markedly blunted in c-kit$^{W/W-V}$ mutant mice, which are defective in c-kit kinase activity but have normal levels of c-kit expression and SCF binding at the cell surface.[131,139–141] Thus, the kinase activity of c-kit appears to be crucial for progenitor cell mobilization, but the mechanism by which c-kit participates in the retention and release of progenitor cells is unclear. In the bone marrow, membrane-bound SCF can be cleaved by SDF-1 to form soluble SCF, which subsequently activates c-kit and leads to progenitor cell mobilization,[59] but the functional blockade of c-kit (with the c-kit–neutralizing antibody ACK2) has also been shown to mobilize bone marrow HSCs to the peripheral circulation and to enhance the engraftment of systemically injected donor bone marrow cells.[142] Thus, both the activation and blockade of c-kit activity have been associated with progenitor cell mobilization.[59,142,143] Furthermore, c-kit is the only known receptor for SCF, but SCF is not always required for c-kit activity,[144–146] and neither SCF nor an SCF-neutralizing antibody are potent mobilizers,[147,148] so SCF-binding alone cannot adequately explain the role of c-kit in progenitor cell mobilization.[142,147,148]

D. SDF-1/CXCR4–c-kit Signaling

The kinetics of bone marrow progenitor cell mobilization induced by a c-kit-neutralizing antibody (i.e., ACK2) and by antagonism of CXCR4 with the pharmacological CXCR4 antagonist AMD3100 are similar, so we investigated whether c-kit is involved in CXCR4-mediated bone marrow progenitor cell trafficking. Peripheral blood progenitor cell levels significantly increased, and bone marrow progenitor cell levels significantly declined after AMD3100 was injected into wild-type mice, but not after it was injected into c-kit kinase-defective (c-kit$^{W/W-V}$) mice.[71] To determine which specific subpopulations of bone marrow cells were affected by the c-kit kinase deficiency, we developed a short-term, *in vivo* bone marrow clearance/repopulation assay. AMD3100 was administered to wild-type and c-kit$^{W/W-V}$ mice, and then labeled bone marrow mononuclear cells were injected into the peripheral circulation and allowed to repopulate the bone marrow. Significantly fewer systemically administered CXCR4-expressing progenitor cells were observed in the bone marrow of c-kit$^{W/W-V}$ mice than in the bone marrow of wild-type mice.

AMD3100 also failed to mobilize bone marrow progenitor cells that expressed a constitutively active c-kit kinase (c-kit^{D816V}) mutation. Thus, both the loss and the constitutive activation of c-kit kinase activity impaired AMD3100-induced bone marrow progenitor cell mobilization, which may seem contradictory. However, bone marrow progenitor cell levels were lower in c-kit kinase-defective mice than in wild-type mice before mobilization, and, after mobilization, systemically administered CXCR4-positive progenitor cells could not repopulate the bone marrow of c-kit kinase-defective mice. These observations suggest that c-kit kinase inactivation blocks the retention of CXCR4-positive progenitor cells and, consequently, that the cells susceptible to AMD3100-induced mobilization are (in effect) already mobilized.

In isolated bone marrow mononuclear cells, SDF-1/CXCR4 signaling upregulates, and the antagonism or genetic deletion of CXCR4 downregulates, c-kit phosphorylation. These results, as well as the lack of AMD3100-induced progenitor cell mobilization in mice with bone marrow cells that expressed a constitutively active c-kit mutant, suggest that CXCR4-mediated mobilization requires c-kit deactivation. Thus, AMD3100 and G-CSF appear to induce progenitor cell mobilization through fundamentally different mechanisms, because G-CSF-induced mobilization requires an increase in c-kit activation.[59] Furthermore, G-CSF-induced mobilization occurs 3–5 days after administration and is accompanied by an increase in the number of progenitor cells present in the perivascular niche,[149] whereas AMD3100-induced mobilization occurs within a few hours and, consequently, is unlikely to be preceded by the perivascular accumulation of progenitor cells. The two agents also appear to mobilize different subpopulations of progenitor cells, and more cells are mobilized when G-CSF and AMD3100 are combined than when G-CSF is administered alone.[150–155] Collectively, these observations would suggest that G-CSF and other slow-acting agents increase c-kit phosphorylation by upregulating SCF, which (by itself) has only a modest effect on mobilization but potently promotes progenitor cell proliferation; if so, G-CSF-induced mobilization may be delayed until an adequate surplus of progenitor cells is available for release to the peripheral blood.[149] Conversely, fast-acting agents, such as AMD3100, may mobilize progenitor cells directly by reducing c-kit phosphorylation in the perivascular niche.[156] This hypothesis is also supported by recent evidence that progenitor cells can be rapidly mobilized by the administration of a c-kit neutralizing antibody.[142]

E. α4-integrin–c-kit Signaling

Because c-kit also appears to participate in progenitor cell mobilization through an α4-integrin-mediated mechanism,[140] and interactions between α4-integrin and VCAM-1 support the adhesion and retention of mononuclear cells in the bone marrow,[31] we performed a series of *in vitro* experiments to determine whether the phosphorylation state of c-kit is altered by α4-integrin-mediated adhesion. Wild-type bone marrow mononuclear cells were applied to VCAM-1-coated or uncoated plates, allowed to adhere for 15 min, incubated with or without AMD3100 for another 15 min, and then c-kit phosphorylation at tyrosine 719 was evaluated. Phosphorylated c-kit levels were notably higher in adherent wild-type cells (i.e., cells from VCAM-1–coated plates) than in nonadherent wild-type cells (i.e., cells from uncoated plates), and treatment with an α4-integrin-blocking antibody reduced phosphorylated c-kit levels, whereas treatment with the CXCR4 ligand SDF-1 markedly increased c-kit phosphorylation. Furthermore, both SDF-1 and SCF induced c-kit phosphorylation, but c-kit levels were highest when the cells were incubated with both factors, and AMD3100 treatment suppressed SDF-1-induced, but not SCF-induced, c-kit phosphorylation. Collectively, these observations suggest that α4-integrin-mediated mononuclear cell adhesion is associated with an increase in phosphorylated c-kit levels, and that in adherent mononuclear cells, SDF-1 upregulates, and AMD3100 downregulates, c-kit phosphorylation. Thus, SDF-1- and SCF-induced c-kit activation may occur independently and regulate different cellular activities (Fig. 1).

III. Progenitor Cell Recruitment and Retention

Mobilized progenitor cells are recruited from the peripheral circulation to the ischemic region, where they become incorporated into the growing vasculature.[14,157] Several of the intermediate steps during progenitor cell recruitment, including chemotaxis, transendothelial migration, and adhesion to single layers of mature endothelial cells and integrin, are regulated by SDF-1/CXCR4 binding,[30,99,158–160] but the mechanisms and downstream components of SDF-1/CXCR4 signaling at the injury site are poorly understood. SDF-1 expression is significantly elevated in the plasma of patients with acute myocardial infarction,[161] and the expression of both CXCR4 and SDF-1 is elevated in ischemic myocardium,[158,162,163] whereas the blockade of SDF-1/CXCR4 signaling diminishes progenitor cell recruitment,[99,138,162,164] and impairments in CXCR4 signaling contribute to the reduced angiogenic potency of EPCs from patients with coronary artery disease and related conditions such as aging and diabetes.[164–167] SDF-1 and CXCR4 may also participate in vascular remodeling by recruiting smooth muscle progenitor cells[168] and protect cardiomyocytes against ischemia/reperfusion damage by activating the antiapoptotic kinases Akt and extracellular-regulated kinase.[169]

Hypoxia induces SDF-1 expression at the injury site,[163] where platelets are an important source of SDF-1 expression,[170,171] and we have recently shown that hypoxic preconditioning enhances the recruitment of cardiosphere-derived

Retention   **Mobilization**

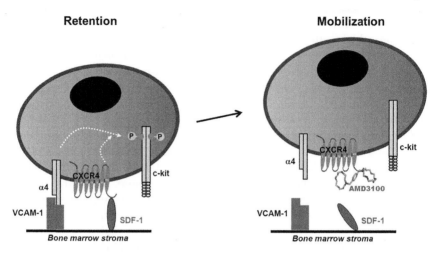

FIG. 1. c-kit is a common component of two signaling pathways that regulate progenitor cell trafficking. Progenitor cell retention and release are largely governed by two pathways, one of which is dependent on the binding of SDF-1 to CXCR4 and the other on α4-integrin/VCAM-1 binding. Both interactions lead to the phosphorylation of c-kit, which is crucial for the retention of progenitor cells in the bone marrow. AMD3100 disrupts the SDF-1/CXCR4 interaction, which reduces c-kit phosphorylation and mobilizes progenitor cells from the bone marrow. SCF also increases phosphorylated c-kit levels by binding directly to c-kit, but disruption of the SCF/c-kit interaction does not appear to induce progenitor cell mobilization. Furthermore, AMD3100 suppresses SDF-1-induced, but not SCF-induced, c-kit phosphorylation, and phosphorylated c-kit levels are higher when cells are incubated with both SDF-1 and SCF than with either individual factor. Thus, SDF-1 and SCF appear to regulate c-kit phosphorylation independently and likely coordinate different cellular activities. (For color version of this figure, the reader is referred to the online version of this chapter.)

Lin-negative, c-kit-positive progenitor (CLK) cells (i.e., cardiac progenitor cells) by inducing CXCR4 expression.[138] CXCR4 expression is much lower in CLK cells than in bone marrow mononuclear cells under normoxic conditions, but increases significantly in response to hypoxia. The increase is accompanied by elevations in SDF-1 expression and preceded by the upregulation of HIF-1α, whereas the siRNA-mediated inactivation of HIF-1α abolishes CXCR4 upregulation. Hypoxic treatment also increased the migration of isolated CLK cells toward SDF-1 in a CXCR4-dependent manner, and hypoxic preconditioning was associated with a 2.5-fold increase in the recruitment of systemically injected CLK cells to the ischemic myocardium of mice after surgically induced myocardial infarction. The recruited cells expressed cardiac troponin I, von Willebrand factor, and smooth muscle actin, indicating that CLK cells can differentiate into cardiomyocytes, endothelial cells, and vascular smooth muscle cells, respectively.

IV. Therapeutic Implications

Both the release of progenitor cells from the bone marrow to the peripheral blood and the recruitment and retention of progenitor cells in ischemic tissue are regulated by interactions between SDF-1 and CXCR4.[37,66,68,75,85,91,99,158,162,163,172–174] The interaction must be disrupted before progenitor cells can be mobilized from the bone marrow to the peripheral circulation and restored to enable retention of the mobilized cells in the ischemic tissue. Thus, the effectiveness of progenitor cell therapy is crucially dependent on how mobilization is induced and on the level of SDF-1 expression in the ischemic region at the time of cell administration. To date, G-CSF is the most frequently used mobilizing agent in clinical trials of progenitor cell therapy,[28,29,44,46,48,51,52] but the efficacy results from many of these trials have been disappointing, perhaps because G-CSF mobilizes progenitor cells by activating extracellular proteases that irreversibly cleave cell-surface adhesion molecules, including α4-integrin, VCAM-1, and CXCR4.[92,175] Thus, the retention of G-CSF-mobilized cells in the ischemic region may be impaired, and mobilizing agents that reversibly disrupt SDF-1/CXCR4 binding, such as AMD3100, may improve the effectiveness of cell therapy[37] (Fig. 2).

Cardiac SDF-1 expression is upregulated within minutes to an hour after myocardial infarction but declines 4–7 days later.[158] Thus, if progenitor cells are administered days after, or even years after (i.e., in patients with established coronary disease), the infarct event, retention of the administered cells is likely to be poor in the ischemic region. Thus, several approaches to increase SDF-1 levels in the ischemic region are currently being investigated. Locally delivered SDF-1 protein increased vascular growth in the injured limbs of mice after surgically induced hind-limb ischemia,[176] and SDF-1 also improved cardiac function after ischemic myocardial injury by increasing progenitor cell recruitment and angiogenesis and by reducing scar formation.[177] However, local SDF-1 delivery is limited by the rapid diffusion of the administered protein and by the activity of proteases in the inflammatory environment of the injury. Furthermore, the induction of SDF-1/CXCR4 signaling also stimulates the production of matrix metalloproteases (MMPs),[178–181] including MMP-2, which (in concert with several exopeptidases) cleaves SDF-1 to produce a neurotoxin that has been implicated in some forms of dementia.[163,182] To overcome this limitation, Segers et al. designed an SDF-1 variant that retains the chemotactic properties of the native molecule but is resistant to MMP-2 and exopeptidase cleavage. Nanofiber-mediated delivery of this construct, S-SDF-1 (also called S4V), promoted progenitor cell recruitment and improved cardiac function in a murine model of myocardial infarction.[183] Furthermore, sustained SDF-1 release has been achieved by covalently linking it to a polyethylene glycol fibrin patch.[174] When the patch was applied to the

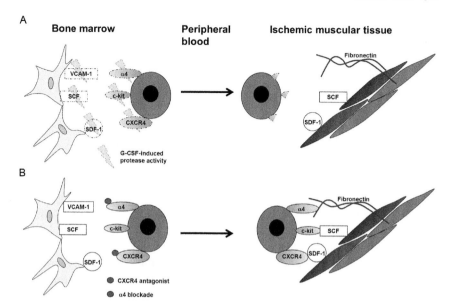

FIG. 2. Mechanisms of therapeutic progenitor cell mobilization. (A) Growth factors (e.g., G-CSF) mobilize progenitor cells by activating extracellular proteases that irreversibly cleave cell-surface adhesion molecules, including α4-integrin, c-kit, and CXCR4, which are crucial for progenitor cell retention in the ischemic region. (B) Receptor antagonists, such as the CXCR4 antagonist AMD3100 or the α4-integrin-blocking antibody Natalizumab, mobilize progenitor cells by reversibly blocking interactions that bind progenitor cells to the bone marrow substrate without cleaving the adhesion molecules. Thus, the use of reversible antagonists, rather than growth factors, for therapeutic progenitor cell mobilization may increase the number of mobilized cells retained in the ischemic region. (For color version of this figure, the reader is referred to the online version of this chapter.)

surface of infarcted mouse hearts, SDF-1 continued to be released for 28 days, and the treatment was associated with greater numbers of incorporated c-kit-positive cells (i.e., progenitor cells) and with improvements in left ventricular function.

Local SDF-1 expression has also been increased through the administration of genetically engineered MSCs.[184,185] Intravenous injections of either unmodified MSCs or MSCs that overexpressed SDF-1 to rats after acute myocardial infarction were associated with improvements in cardiac function, and the beneficial effects appeared to evolve primarily through the preservation of preexisting cardiomyocytes rather than the generation of new cardiomyocytes within the infarct zone.[184] Vascular density, cardiomyocyte survival, and cardiac myosin-positive area were greater in animals treated with the modified cells than in those treated with unmodified cells, and SDF-1 overexpression also increased the number of small cardiac myosin-expressing cells that had not differentiated into mature cardiac myocytes, but were capable of depolarizing and, consequently, may have contributed to improvements in contractile function.[185]

V. Summary

Progenitor cell retention and release are largely governed by SDF-1/ CXCR4 and α4-integrin signaling. The initial steps of these two pathways appear to proceed independently, but both regulate c-kit phosphorylation,[71] and the mobilization of progenitor cells in response to either CXCR4 antagonism or α4-integrin blockade is impaired by the loss of c-kit kinase activity.[131,139–141] Furthermore, bone marrow progenitor cell levels are lower in c-kit kinase-defective mice than in wild-type mice before mobilization, and systemically administered CXCR4-positive progenitor cells cannot repopulate the bone marrow in the absence of c-kit kinase activity.[71] Collectively, these observations suggest that c-kit-kinase inactivation blocks the retention of CXCR4-positive progenitor cells in the bone marrow, and that c-kit could function as a final common mediator of fundamental importance to the regulation of bone marrow progenitor cell trafficking.

SDF-1/CXCR4 and α4-integrin signaling are also crucial for the retention of progenitor cells in the ischemic region, which may explain, at least in part, why clinical trials of progenitor cell therapy have failed to display the efficacy observed in preclinical investigations. The lack of effectiveness is often attributed to poor retention of the transplanted cells,[21,22,27] and, to date, most of the trial protocols have mobilized cells via G-CSF administration, which activates extracellular proteases that irreversibly cleave cell-surface adhesion molecules, including α4-integrin and CXCR4.[92,175] Thus, the retention of G-CSF-mobilized cells in the ischemic region may be impaired, and mobilizing agents that reversibly disrupt SDF-1/CXCR4 binding, such as AMD3100, may improve patient response.[37] Efforts to supplement SDF-1 levels in the ischemic region may also improve progenitor cell recruitment and the effectiveness of stem cell therapy.

ACKNOWLEDGMENT

We thank W. Kevin Meisner, PhD, for editorial assistance.

References

1. Asahara T, Murohara T, Sullivan A, Silver M, van der Zee R, Li T, et al. Isolation of putative progenitor endothelial cells for angiogenesis. *Science* 1997;**275**:964–7.
2. Shi Q, Rafii S, Wu MH, Wijelath ES, Yu C, Ishida A, et al. Evidence for circulating bone marrow-derived endothelial cells. *Blood* 1998;**92**:362–7.
3. Urbich C, Dimmeler S. Endothelial progenitor cells: characterization and role in vascular biology. *Circ Res* 2004;**95**:343–53.
4. Jiang Y, Jahagirdar BN, Reinhardt RL, Schwartz RE, Keene CD, Ortiz-Gonzalez XR, et al. Pluripotency of mesenchymal stem cells derived from adult marrow. *Nature* 2002;**418**:41–9.
5. Williams AR, Hare JM. Mesenchymal stem cells: biology, pathophysiology, translational findings, and therapeutic implications for cardiac disease. *Circ Res* 2011;**109**:923–40.
6. Asahara T, Masuda H, Takahashi T, Kalka C, Pastore C, Silver M, et al. Bone marrow origin of endothelial progenitor cells responsible for postnatal vasculogenesis in physiological and pathological neovascularization. *Circ Res* 1999;**85**:221–8.
7. Schuleri KH, Amado LC, Boyle AJ, Centola M, Saliaris AP, Gutman MR, et al. Early improvement in cardiac tissue perfusion due to mesenchymal stem cells. *Am J Physiol Heart Circ Physiol* 2008;**294**:H2002–11.
8. Losordo DW, Henry TD, Davidson C, Sup Lee J, Costa MA, Bass T, et al. Intramyocardial, autologous CD34+ cell therapy for refractory angina. *Circ Res* 2011;**109**:428–36.
9. Losordo DW, Schatz RA, White CJ, Udelson JE, Veereshwarayya V, Durgin M, et al. Intra-myocardial transplantation of autologous CD34+ stem cells for intractable angina: a phase I/IIa double-blind, randomized controlled trial. *Circulation* 2007;**115**:3165–72.
10. Hare JM, Traverse JH, Henry TD, Dib N, Strumpf RK, Schulman SP, et al. A randomized, double-blind, placebo-controlled, dose-escalation study of intravenous adult human mesenchymal stem cells (prochymal) after acute myocardial infarction. *J Am Coll Cardiol* 2009;**54**:2277–86.
11. Lunde K, Solheim S, Aakhus S, Arnesen H, Abdelnoor M, Egeland T, et al. Intracoronary injection of mononuclear bone marrow cells in acute myocardial infarction. *N Engl J Med* 2006;**355**:1199–209.
12. Assmus B, Honold J, Schachinger V, Britten MB, Fischer-Rasokat U, Lehmann R, et al. Transcoronary transplantation of progenitor cells after myocardial infarction. *N Engl J Med* 2006;**355**:1222–32.
13. Tateishi-Yuyama E, Matsubara H, Murohara T, Ikeda U, Shintani S, Masaki H, et al. Therapeutic angiogenesis for patients with limb ischaemia by autologous transplantation of bone-marrow cells: a pilot study and a randomised controlled trial. *Lancet* 2002;**360**:427–35.
14. Dimmeler S, Zeiher AM, Schneider MD. Unchain my heart: the scientific foundations of cardiac repair. *J Clin Invest* 2005;**115**:572–83.
15. Isner JM, Asahara T. Angiogenesis and vasculogenesis as therapeutic strategies for postnatal neovascularization. *J Clin Invest* 1999;**103**:1231–6.
16. Hatzistergos KE, Quevedo H, Oskouei BN, Hu Q, Feigenbaum GS, Margitich IS, et al. Bone marrow mesenchymal stem cells stimulate cardiac stem cell proliferation and differentiation. *Circ Res* 2010;**107**:913–22.
17. Gnecchi M, Zhang Z, Ni A, Dzau VJ. Paracrine mechanisms in adult stem cell signaling and therapy. *Circ Res* 2008;**103**:1204–19.
18. Werner N, Kosiol S, Schiegl T, Ahlers P, Walenta K, Link A, et al. Circulating endothelial progenitor cells and cardiovascular outcomes. *N Engl J Med* 2005;**353**:999–1007.

19. Hill JM, Zalos G, Halcox JP, Schenke WH, Waclawiw MA, Quyyumi AA, et al. Circulating endothelial progenitor cells, vascular function, and cardiovascular risk. *N Engl J Med* 2003;**348**:593–600.
20. Segers VF, Lee RT. Stem-cell therapy for cardiac disease. *Nature* 2008;**451**:937–42.
21. Chavakis E, Koyanagi M, Dimmeler S. Enhancing the outcome of cell therapy for cardiac repair: progress from bench to bedside and back. *Circulation* 2010;**121**:325–35.
22. Wollert KC, Drexler H. Cell therapy for the treatment of coronary heart disease: a critical appraisal. *Nat Rev Cardiol* 2010;**7**:204–15.
23. Lipinski MJ, Biondi-Zoccai GG, Abbate A, Khianey R, Sheiban I, Bartunek J, et al. Impact of intracoronary cell therapy on left ventricular function in the setting of acute myocardial infarction: a collaborative systematic review and meta-analysis of controlled clinical trials. *J Am Coll Cardiol* 2007;**50**:1761–7.
24. Rosenzweig A. Cardiac cell therapy—mixed results from mixed cells. *N Engl J Med* 2006;**355**:1274–7.
25. Reffelmann T, Konemann S, Kloner RA. Promise of blood- and bone marrow-derived stem cell transplantation for functional cardiac repair: putting it in perspective with existing therapy. *J Am Coll Cardiol* 2009;**53**:305–8.
26. Abdel-Latif A, Bolli R, Tleyjeh IM, Montori VM, Perin EC, Hornung CA, et al. Adult bone marrow-derived cells for cardiac repair: a systematic review and meta-analysis. *Arch Intern Med* 2007;**167**:989–97.
27. Hofmann M, Wollert KC, Meyer GP, Menke A, Arseniev L, Hertenstein B, et al. Monitoring of bone marrow cell homing into the infarcted human myocardium. *Circulation* 2005;**111**:2198–202.
28. Zohlnhofer D, Ott I, Mehilli J, Schomig K, Michalk F, Ibrahim T, et al. Stem cell mobilization by granulocyte colony-stimulating factor in patients with acute myocardial infarction: a randomized controlled trial. *JAMA* 2006;**295**:1003–10.
29. Ripa RS, Jorgensen E, Wang Y, Thune JJ, Nilsson JC, Sondergaard L, et al. Stem cell mobilization induced by subcutaneous granulocyte-colony stimulating factor to improve cardiac regeneration after acute ST-elevation myocardial infarction: result of the double-blind, randomized, placebo-controlled stem cells in myocardial infarction (STEMMI) trial. *Circulation* 2006;**113**:1983–92.
30. Chavakis E, Aicher A, Heeschen C, Sasaki K, Kaiser R. El Makhfi N, *et al.* Role of beta2-integrins for homing and neovascularization capacity of endothelial progenitor cells. *J Exp Med* 2005;**201**:63–72.
31. Qin G, Ii M, Silver M, Wecker A, Bord E, Ma H, et al. Functional disruption of alpha4 integrin mobilizes bone marrow-derived endothelial progenitors and augments ischemic neovascularization. *J Exp Med* 2006;**203**:153–63.
32. Adams GB, Scadden DT. A niche opportunity for stem cell therapeutics. *Gene Ther* 2008;**15**:96–9.
33. Aicher A, Kollet O, Heeschen C, Liebner S, Urbich C, Ihling C, et al. The Wnt antagonist Dickkopf-1 mobilizes vasculogenic progenitor cells via activation of the bone marrow endos-teal stem cell niche. *Circ Res* 2008;**103**:796–803.
34. Zaruba MM, Theiss HD, Vallaster M, Mehl U, Brunner S, David R, et al. Synergy between CD26/DPP-IV inhibition and G-CSF improves cardiac function after acute myocardial infarction. *Cell Stem Cell* 2009;**4**:313–23.
35. Orlic D, Kajstura J, Chimenti S, Limana F, Jakoniuk I, Quaini F, et al. Mobilized bone marrow cells repair the infarcted heart, improving function and survival. *Proc Natl Acad Sci USA* 2001;**98**:10344–9.
36. Iwakura A, Shastry S, Luedemann C, Hamada H, Kawamoto A, Kishore R, et al. Estradiol enhances recovery after myocardial infarction by augmenting incorporation of bone marrow-

derived endothelial progenitor cells into sites of ischemia-induced neovascularization via endothelial nitric oxide synthase-mediated activation of matrix metalloproteinase-9. *Circulation* 2006;**113**:1605–14.

37. Jujo K, Hamada H, Iwakura A, Thorne T, Sekiguchi H, Clarke T, et al. CXCR4 blockade augments bone marrow progenitor cell recruitment to the neovasculature and reduces mortality after myocardial infarction. *Proc Natl Acad Sci USA* 2010;**107**:11008–13.

38. Roncalli J, Renault MA, Tongers J, Misener S, Thorne T, Kamide C, et al. Sonic hedgehog-induced functional recovery after myocardial infarction is enhanced by AMD3100-mediated progenitor-cell mobilization. *J Am Coll Cardiol* 2011;**57**:2444–52.

39. Voors AA, Belonje AM, Zijlstra F, Hillege HL, Anker SD, Slart RH, et al. A single dose of erythropoietin in ST-elevation myocardial infarction. *Eur Heart J* 2010;**31**:2593–600.

40. Belonje AM, Voors AA, van Gilst WH, Anker SD, Slart RH, Tio RA, et al. Effects of erythropoietin after an acute myocardial infarction: rationale and study design of a prospective, randomized, clinical trial (HEBE III). *Am Heart J* 2008;**155**:817–22.

41. Mancini DM, Katz SD, Lang CC, LaManca J, Hudaihed A, Androne AS. Effect of erythropoietin on exercise capacity in patients with moderate to severe chronic heart failure. *Circulation* 2003;**107**:294–9.

42. Taniguchi N, Nakamura T, Sawada T, Matsubara K, Furukawa K, Hadase M, et al. Erythropoietin prevention trial of coronary restenosis and cardiac remodeling after ST-elevated acute myocardial infarction (EPOC-AMI): a pilot, randomized, placebo-controlled study. *Circ J* 2010;**74**:2365–71.

43. Nakamura R, Takahashi A, Yamada T, Miyai N, Irie H, Kinoshita N, et al. Erythropoietin in patients with acute coronary syndrome and its cardioprotective action after percutaneous coronary intervention. *Circ J* 2009;**73**:1920–6.

44. Engelmann MG, Theiss HD, Hennig-Theiss C, Huber A, Wintersperger BJ, Werle-Ruedinger AE, et al. Autologous bone marrow stem cell mobilization induced by granulocyte colony-stimulating factor after subacute ST-segment elevation myocardial infarction undergoing late revascularization: final results from the G-CSF-STEMI (Granulocyte Colony-Stimulating Factor ST-Segment Elevation Myocardial Infarction) trial. *J Am Coll Cardiol* 2006;**48**:1712–21.

45. Valgimigli M, Rigolin GM, Cittanti C, Malagutti P, Curello S, Percoco G, et al. Use of granulocyte-colony stimulating factor during acute myocardial infarction to enhance bone marrow stem cell mobilization in humans: clinical and angiographic safety profile. *Eur Heart J* 2005;**26**:1838–45.

46. Ince H, Petzsch M, Kleine HD, Schmidt H, Rehders T, Korber T, et al. Preservation from left ventricular remodeling by front-integrated revascularization and stem cell liberation in evolving acute myocardial infarction by use of granulocyte-colony-stimulating factor (FIRSTLINE-AMI). *Circulation* 2005;**112**:3097–106.

47. Kang HJ, Kim HS, Koo BK, Kim YJ, Lee D, Sohn DW, et al. Intracoronary infusion of the mobilized peripheral blood stem cell by G-CSF is better than mobilization alone by G-CSF for improvement of cardiac function and remodeling: 2-year follow-up results of the Myocardial Regeneration and Angiogenesis in Myocardial Infarction with G-CSF and Intra-Coronary Stem Cell Infusion (MAGIC Cell) 1 trial. *Am Heart J* 2007;**153**:237 e1–8.

48. Ellis SG, Penn MS, Bolwell B, Garcia M, Chacko M, Wang T, et al. Granulocyte colony stimulating factor in patients with large acute myocardial infarction: results of a pilot dose-escalation randomized trial. *Am Heart J* 2006;**152**:1051 e9–14.

49. Takano H, Hasegawa H, Kuwabara Y, Nakayama T, Matsuno K, Miyazaki Y, et al. Feasibility and safety of granulocyte colony-stimulating factor treatment in patients with acute myocardial infarction. *Int J Cardiol* 2007;**122**:41–7.

50. Engelmann MG, Theiss HD, Theiss C, Huber A, Wintersperger BJ, Werle-Ruedinger AE, et al. G-CSF in patients suffering from late revascularized ST elevation myocardial infarction: analysis on the timing of G-CSF administration. *Exp Hematol* 2008;**36**:703–9.

51. Hill JM, Syed MA, Arai AE, Powell TM, Paul JD, Zalos G, et al. Outcomes and risks of granulocyte colony-stimulating factor in patients with coronary artery disease. *J Am Coll Cardiol* 2005;**46**:1643–8.

52. Srinivas G, Anversa P, Frishman WH. Cytokines and myocardial regeneration: a novel treatment option for acute myocardial infarction. *Cardiol Rev* 2009;**17**:1–9.

53. Wright DE, Wagers AJ, Gulati AP, Johnson FL, Weissman IL. Physiological migration of hematopoietic stem and progenitor cells. *Science* 2001;**294**:1933–6.

54. Takahashi T, Kalka C, Masuda H, Chen D, Silver M, Kearney M, et al. Ischemia- and cytokine-induced mobilization of bone marrow-derived endothelial progenitor cells for neovascularization. *Nat Med* 1999;**5**:434–8.

55. Xie Y, Yin T, Wiegraebe W, He XC, Miller D, Stark D, et al. Detection of functional haematopoietic stem cell niche using real-time imaging. *Nature* 2009;**457**:97–101.

56. Zhang J, Niu C, Ye L, Huang H, He X, Tong WG, et al. Identification of the haematopoietic stem cell niche and control of the niche size. *Nature* 2003;**425**:836–41.

57. Sugiyama T, Kohara H, Noda M, Nagasawa T. Maintenance of the hematopoietic stem cell pool by CXCL12-CXCR4 chemokine signaling in bone marrow stromal cell niches. *Immunity* 2006;**25**:977–88.

58. Kiel MJ, Yilmaz OH, Iwashita T, Terhorst C, Morrison SJ. SLAM family receptors distinguish hematopoietic stem and progenitor cells and reveal endothelial niches for stem cells. *Cell* 2005;**121**:1109–21.

59. Heissig B, Hattori K, Dias S, Friedrich M, Ferris B, Hackett NR, et al. Recruitment of stem and progenitor cells from the bone marrow niche requires MMP-9 mediated release of kit-ligand. *Cell* 2002;**109**:625–37.

60. Papayannopoulou T, Scadden DT. Stem-cell ecology and stem cells in motion. *Blood* 2008;**111**:3923–30.

61. Oostendorp RA, Dormer P. VLA-4-mediated interactions between normal human hematopoietic progenitors and stromal cells. *Leuk Lymphoma* 1997;**24**:423–35.

62. Vermeulen M, Le Pesteur F, Gagnerault MC, Mary JY, Sainteny F, Lepault F. Role of adhesion molecules in the homing and mobilization of murine hematopoietic stem and progenitor cells. *Blood* 1998;**92**:894–900.

63. Papayannopoulou T, Priestley GV, Nakamoto B, Zafiropoulos V, Scott LM, Harlan JM. Synergistic mobilization of hemopoietic progenitor cells using concurrent beta1 and beta2 integrin blockade or beta2-deficient mice. *Blood* 2001;**97**:1282–8.

64. Papayannopoulou T, Priestley GV, Nakamoto B, Zafiropoulos V, Scott LM. Molecular pathways in bone marrow homing: dominant role of alpha(4)beta(1) over beta(2)-integrins and selectins. *Blood* 2001;**98**:2403–11.

65. Peled A, Kollet O, Ponomaryov T, Petit I, Franitza S, Grabovsky V, et al. The chemokine SDF-1 activates the integrins LFA-1, VLA-4, and VLA-5 on immature human CD34(+) cells: role in transendothelial/stromal migration and engraftment of NOD/SCID mice. *Blood* 2000;**95**:3289–96.

66. Peled A, Petit I, Kollet O, Magid M, Ponomaryov T, Byk T, et al. Dependence of human stem cell engraftment and repopulation of NOD/SCID mice on CXCR4. *Science* 1999;**283**:845–8.

67. Wright DE, Bowman EP, Wagers AJ, Butcher EC, Weissman IL. Hematopoietic stem cells are uniquely selective in their migratory response to chemokines. *J Exp Med* 2002;**195**:1145–54.

68. Lo Celso C, Fleming HE, Wu JW, Zhao CX, Miake-Lye S, Fujisaki J, et al. Live-animal tracking of individual haematopoietic stem/progenitor cells in their niche. *Nature* 2009;**457**: 92–6.

69. Kopp HG, Avecilla ST, Hooper AT, Rafii S. The bone marrow vascular niche: home of HSC differentiation and mobilization. *Physiology (Bethesda)* 2005;**20**:349–56.

70. Christopher MJ, Liu F, Hilton MJ, Long F, Link DC. Suppression of CXCL12 production by bone marrow osteoblasts is a common and critical pathway for cytokine-induced mobilization. *Blood* 2009;**114**:1331–9.

71. Cheng M, Zhou J, Wu M, Boriboun C, Thorne T, Liu T, et al. CXCR4-mediated bone marrow progenitor cell maintenance and mobilization are modulated by c-kit activity. *Circ Res* 2010;**107**:1083–93.

72. Wu B, Chien EY, Mol CD, Fenalti G, Liu W, Katritch V, et al. Structures of the CXCR4 chemokine GPCR with small-molecule and cyclic peptide antagonists. *Science* 2010;**330**:1066–71.

73. Loetscher M, Geiser T, O'Reilly T, Zwahlen R, Baggiolini M, Moser B. Cloning of a human seven-transmembrane domain receptor, LESTR, that is highly expressed in leukocytes. *J Biol Chem* 1994;**269**:232–7.

74. Nomura H, Nielsen BW, Matsushima K. Molecular cloning of cDNAs encoding a LD78 receptor and putative leukocyte chemotactic peptide receptors. *Int Immunol* 1993;**5**:1239–49.

75. Nie Y, Han YC, Zou YR. CXCR4 is required for the quiescence of primitive hematopoietic cells. *J Exp Med* 2008;**205**:777–83.

76. Rossi D, Zlotnik A. The biology of chemokines and their receptors. *Annu Rev Immunol* 2000;**18**:217–42.

77. Murphy PM, Baggiolini M, Charo IF, Hebert CA, Horuk R, Matsushima K, et al. International union of pharmacology. XXII. Nomenclature for chemokine receptors. *Pharmacol Rev* 2000;**52**:145–76.

78. Murdoch C, Finn A. Chemokine receptors and their role in inflammation and infectious diseases. *Blood* 2000;**95**:3032–43.

79. Aiuti A, Webb IJ, Bleul C, Springer T, Gutierrez-Ramos JC. The chemokine SDF-1 is a chemoattractant for human CD34+ hematopoietic progenitor cells and provides a new mechanism to explain the mobilization of CD34+ progenitors to peripheral blood. *J Exp Med* 1997;**185**:111–20.

80. Dorshkind K. Regulation of hemopoiesis by bone marrow stromal cells and their products. *Annu Rev Immunol* 1990;**8**:111–37.

81. Fuchs E, Tumbar T, Guasch G. Socializing with the neighbors: stem cells and their niche. *Cell* 2004;**116**:769–78.

82. Ma Q, Jones D, Springer TA. The chemokine receptor CXCR4 is required for the retention of B lineage and granulocytic precursors within the bone marrow microenvironment. *Immunity* 1999;**10**:463–71.

83. Nagasawa T, Hirota S, Tachibana K, Takakura N, Nishikawa S, Kitamura Y, et al. Defects of B-cell lymphopoiesis and bone-marrow myelopoiesis in mice lacking the CXC chemokine PBSF/SDF-1. *Nature* 1996;**382**:635–8.

84. Kawabata K, Ujikawa M, Egawa T, Kawamoto H, Tachibana K, Iizasa H, et al. A cell-autonomous requirement for CXCR4 in long-term lymphoid and myeloid reconstitution. *Proc Natl Acad Sci USA* 1999;**96**:5663–7.

85. Zou YR, Kottmann AH, Kuroda M, Taniuchi I, Littman DR. Function of the chemokine receptor CXCR4 in haematopoiesis and in cerebellar development. *Nature* 1998;**393**:595–9.

86. Shepherd RM, Capoccia BJ, Devine SM, Dipersio J, Trinkaus KM, Ingram D, et al. Angiogenic cells can be rapidly mobilized and efficiently harvested from the blood following treatment with AMD3100. *Blood* 2006;**108**:3662–7.

87. Staller P, Sulitkova J, Lisztwan J, Moch H, Oakeley EJ, Krek W. Chemokine receptor CXCR4 downregulated by von Hippel-Lindau tumour suppressor pVHL. *Nature* 2003;**425**: 307–11.

88. Tachibana K, Hirota S, Iizasa H, Yoshida H, Kawabata K, Kataoka Y, et al. The chemokine receptor CXCR4 is essential for vascularization of the gastrointestinal tract. *Nature* 1998;**393**:591–4.

89. Ara T, Tokoyoda K, Okamoto R, Koni PA, Nagasawa T. The role of CXCL12 in the organ-specific process of artery formation. *Blood* 2005;**105**:3155–61.

90. Ara T, Itoi M, Kawabata K, Egawa T, Tokoyoda K, Sugiyama T, et al. A role of CXC chemokine ligand 12/stromal cell-derived factor-1/pre-B cell growth stimulating factor and its receptor CXCR4 in fetal and adult T cell development in vivo. *J Immunol* 2003;**170**:4649–55.

91. Broxmeyer HE, Orschell CM, Clapp DW, Hangoc G, Cooper S, Plett PA, et al. Rapid mobilization of murine and human hematopoietic stem and progenitor cells with AMD3100, a CXCR4 antagonist. *J Exp Med* 2005;**201**:1307–18.

92. Larochelle A, Krouse A, Metzger M, Orlic D, Donahue RE, Fricker S, et al. AMD3100 mobilizes hematopoietic stem cells with long-term repopulating capacity in nonhuman primates. *Blood* 2006;**107**:3772–8.

93. Liles WC, Broxmeyer HE, Rodger E, Wood B, Hubel K, Cooper S, et al. Mobilization of hematopoietic progenitor cells in healthy volunteers by AMD3100, a CXCR4 antagonist. *Blood* 2003;**102**:2728–30.

94. Sipkins DA, Wei X, Wu JW, Runnels JM, Cote D, Means TK, et al. In vivo imaging of specialized bone marrow endothelial microdomains for tumour engraftment. *Nature* 2005;**435**:969–73.

95. Hirota K, Semenza GL. Regulation of angiogenesis by hypoxia-inducible factor 1. *Crit Rev Oncol Hematol* 2006;**59**:15–26.

96. Schioppa T, Uranchimeg B, Saccani A, Biswas SK, Doni A, Rapisarda A, et al. Regulation of the chemokine receptor CXCR4 by hypoxia. *J Exp Med* 2003;**198**:1391–402.

97. Zagzag D, Krishnamachary B, Yee H, Okuyama H, Chiriboga L, Ali MA, et al. Stromal cell-derived factor-1alpha and CXCR4 expression in hemangioblastoma and clear cell-renal cell carcinoma: von Hippel-Lindau loss-of-function induces expression of a ligand and its receptor. *Cancer Res* 2005;**65**:6178–88.

98. Vanbervliet B, Bendriss-Vermare N, Massacrier C, Homey B, de Bouteiller O, Briere F, et al. The inducible CXCR3 ligands control plasmacytoid dendritic cell responsiveness to the constitutive chemokine stromal cell-derived factor 1 (SDF-1)/CXCL12. *J Exp Med* 2003;**198**:823–30.

99. Ceradini DJ, Kulkarni AR, Callaghan MJ, Tepper OM, Bastidas N, Kleinman ME, et al. Progenitor cell trafficking is regulated by hypoxic gradients through HIF-1 induction of SDF-1. *Nat Med* 2004;**10**:858–64.

100. Mohyeldin A, Garzon-Muvdi T, Quinones-Hinojosa A. Oxygen in stem cell biology: a critical component of the stem cell niche. *Cell Stem Cell* 2010;**7**:150–61.

101. Hemler ME, Lobb RR. The leukocyte beta 1 integrins. *Curr Opin Hematol* 1995;**2**:61–7.

102. Shattil SJ, Kim C, Ginsberg MH. The final steps of integrin activation: the end game. *Nat Rev Mol Cell Biol* 2010;**11**:288–300.

103. Assoian RK, Schwartz MA. Coordinate signaling by integrins and receptor tyrosine kinases in the regulation of G1 phase cell-cycle progression. *Curr Opin Genet Dev* 2001;**11**:48–53.

104. Liu S, Thomas SM, Woodside DG, Rose DM, Kiosses WB, Pfaff M, et al. Binding of paxillin to alpha4 integrins modifies integrin-dependent biological responses. *Nature* 1999;**402**:676–81.

105. Liu S, Rose DM, Han J, Ginsberg MH. Alpha4 integrins in cardiovascular development and diseases. *Trends Cardiovasc Med* 2000;**10**:253–7.

106. Katayama Y, Hidalgo A, Peired A, Frenette PS. Integrin alpha4beta7 and its counterreceptor MAdCAM-1 contribute to hematopoietic progenitor recruitment into bone marrow following transplantation. *Blood* 2004;**104**:2020–6.
107. Arroyo AG, Yang JT, Rayburn H, Hynes RO. Alpha4 integrins regulate the proliferation/differentiation balance of multilineage hematopoietic progenitors in vivo. *Immunity* 1999;**11**:555–66.
108. Scott LM, Priestley GV, Papayannopoulou T. Deletion of alpha4 integrins from adult hematopoietic cells reveals roles in homeostasis, regeneration, and homing. *Mol Cell Biol* 2003;**23**:9349–60.
109. Ulyanova T, Priestley GV, Nakamoto B, Jiang Y, Papayannopoulou T. VCAM-1 ablation in nonhematopoietic cells in MxCre+ VCAM-1f/f mice is variable and dictates their phenotype. *Exp Hematol* 2007;**35**:565–71.
110. Prosper F, Stroncek D, McCarthy JB, Verfaillie CM. Mobilization and homing of peripheral blood progenitors is related to reversible downregulation of alpha4 beta1 integrin expression and function. *J Clin Invest* 1998;**101**:2456–67.
111. Wagers AJ, Allsopp RC, Weissman IL. Changes in integrin expression are associated with altered homing properties of Lin(-/lo)Thy1.1(lo)Sca-1(+)c-kit(+) hematopoietic stem cells following mobilization by cyclophosphamide/granulocyte colony-stimulating factor. *Exp Hematol* 2002;**30**:176–85.
112. Levesque JP, Takamatsu Y, Nilsson SK, Haylock DN, Simmons PJ. Vascular cell adhesion molecule-1 (CD106) is cleaved by neutrophil proteases in the bone marrow following hematopoietic progenitor cell mobilization by granulocyte colony-stimulating factor. *Blood* 2001;**98**:1289–97.
113. Priestley GV, Ulyanova T, Papayannopoulou T. Sustained alterations in biodistribution of stem/progenitor cells in Tie2Cre+ alpha4(f/f) mice are hematopoietic cell autonomous. *Blood* 2007;**109**:109–11.
114. Arroyo AG, Yang JT, Rayburn H, Hynes RO. Differential requirements for alpha4 integrins during fetal and adult hematopoiesis. *Cell* 1996;**85**:997–1008.
115. Zohren F, Toutzaris D, Klarner V, Hartung HP, Kieseier B, Haas R. The monoclonal anti-VLA-4 antibody natalizumab mobilizes CD34+ hematopoietic progenitor cells in humans. *Blood* 2008;**111**:3893–5.
116. Papayannopoulou T, Nakamoto B. Peripheralization of hemopoietic progenitors in primates treated with anti-VLA4 integrin. *Proc Natl Acad Sci USA* 1993;**90**:9374–8.
117. Hartz B, Volkmann T, Irle S, Loechelt C, Neubauer A, Brendel C. alpha4 integrin levels on mobilized peripheral blood stem cells predict rapidity of engraftment in patients receiving autologous stem cell transplantation. *Blood* 2011;**118**:2362–5.
118. Besmer P, Murphy JE, George PC, Qiu FH, Bergold PJ, Lederman L, et al. A new acute transforming feline retrovirus and relationship of its oncogene v-kit with the protein kinase gene family. *Nature* 1986;**320**:415–21.
119. Beltrami AP, Barlucchi L, Torella D, Baker M, Limana F, Chimenti S, et al. Adult cardiac stem cells are multipotent and support myocardial regeneration. *Cell* 2003;**114**:763–76.
120. Tallini YN, Greene KS, Craven M, Spealman A, Breitbach M, Smith J, et al. c-kit expression identifies cardiovascular precursors in the neonatal heart. *Proc Natl Acad Sci USA* 2009;**106**: 1808–13.
121. Wu SM, Fujiwara Y, Cibulsky SM, Clapham DE, Lien CL, Schultheiss TM, et al. Developmental origin of a bipotential myocardial and smooth muscle cell precursor in the mammalian heart. *Cell* 2006;**127**:1137–50.
122. Flanagan JG, Leder P. The kit ligand: a cell surface molecule altered in steel mutant fibroblasts. *Cell* 1990;**63**:185–94.
123. Broudy VC. Stem cell factor and hematopoiesis. *Blood* 1997;**90**:1345–64.

124. Yuzawa S, Opatowsky Y, Zhang Z, Mandiyan V, Lax I, Schlessinger J. Structural basis for activation of the receptor tyrosine kinase KIT by stem cell factor. *Cell* 2007;**130**:323–34.
125. Arakawa T, Yphantis DA, Lary JW, Narhi LO, Lu HS, Prestrelski SJ, et al. Glycosylated and unglycosylated recombinant-derived human stem cell factors are dimeric and have extensive regular secondary structure. *J Biol Chem* 1991;**266**:18942–8.
126. Blume-Jensen P, Jiang G, Hyman R, Lee KF, O'Gorman S, Hunter T. Kit/stem cell factor receptor-induced activation of phosphatidylinositol 3'-kinase is essential for male fertility. *Nat Genet* 2000;**24**:157–62.
127. McCulloch EA, Siminovitch L, Till JE, Russell ES, Bernstein SE. The cellular basis of the genetically determined hemopoietic defect in anemic mice of genotype Sl-Sld. *Blood* 1965;**26**:399–410.
128. Barker JE. Sl/Sld hematopoietic progenitors are deficient in situ. *Exp Hematol* 1994;**22**: 174–7.
129. Tan JC, Nocka K, Ray P, Traktman P, Besmer P. The dominant W42 spotting phenotype results from a missense mutation in the c-kit receptor kinase. *Science* 1990;**247**:209–12.
130. Heissig B, Werb Z, Rafii S, Hattori K. Role of c-kit/Kit ligand signaling in regulating vasculogenesis. *Thromb Haemost* 2003;**90**:570–6.
131. Fazel S, Cimini M, Chen L, Li S, Angoulvant D, Fedak P, et al. Cardioprotective c-kit+ cells are from the bone marrow and regulate the myocardial balance of angiogenic cytokines. *J Clin Invest* 2006;**116**:1865–77.
132. Ayach BB, Yoshimitsu M, Dawood F, Sun M, Arab S, Chen M, et al. Stem cell factor receptor induces progenitor and natural killer cell-mediated cardiac survival and repair after myocardial infarction. *Proc Natl Acad Sci USA* 2006;**103**:2304–9.
133. Kerkela R, Grazette L, Yacobi R, Iliescu C, Patten R, Beahm C, et al. Cardiotoxicity of the cancer therapeutic agent imatinib mesylate. *Nat Med* 2006;**12**:908–16.
134. Heinrich MC, Griffith DJ, Druker BJ, Wait CL, Ott KA, Zigler AJ. Inhibition of c-kit receptor tyrosine kinase activity by STI 571, a selective tyrosine kinase inhibitor. *Blood* 2000;**96**: 925–32.
135. Zsebo KM, Williams DA, Geissler EN, Broudy VC, Martin FH, Atkins HL, et al. Stem cell factor is encoded at the Sl locus of the mouse and is the ligand for the c-kit tyrosine kinase receptor. *Cell* 1990;**63**:213–24.
136. Williams DE, Eisenman J, Baird A, Rauch C, Van Ness K, March CJ, et al. Identification of a ligand for the c-kit proto-oncogene. *Cell* 1990;**63**:167–74.
137. Huang E, Nocka K, Beier DR, Chu TY, Buck J, Lahm HW, et al. The hematopoietic growth factor KL is encoded by the Sl locus and is the ligand of the c-kit receptor, the gene product of the W locus. *Cell* 1990;**63**:225–33.
138. Tang YL, Zhu W, Cheng M, Chen L, Zhang J, Sun T, et al. Hypoxic preconditioning enhances the benefit of cardiac progenitor cell therapy for treatment of myocardial infarction by inducing CXCR4 expression. *Circ Res* 2009;**104**:1209–16.
139. Roberts AW, Foote S, Alexander WS, Scott C, Robb L, Metcalf D. Genetic influences determining progenitor cell mobilization and leukocytosis induced by granulocyte colony-stimulating factor. *Blood* 1997;**89**:2736–44.
140. Papayannopoulou T, Priestley GV, Nakamoto B. Anti-VLA4/VCAM-1-induced mobilization requires cooperative signaling through the kit/mkit ligand pathway. *Blood* 1998;**91**:2231–9.
141. Nocka K, Tan JC, Chiu E, Chu TY, Ray P, Traktman P, et al. Molecular bases of dominant negative and loss of function mutations at the murine c-kit/white spotting locus: W37, Wv, W41 and W. *EMBO J* 1990;**9**:1805–13.
142. Czechowicz A, Kraft D, Weissman IL, Bhattacharya D. Efficient transplantation via antibody-based clearance of hematopoietic stem cell niches. *Science* 2007;**318**:1296–9.

143. Fazel SS, Chen L, Angoulvant D, Li SH, Weisel RD, Keating A, et al. Activation of c-kit is necessary for mobilization of reparative bone marrow progenitor cells in response to cardiac injury. *FASEB J* 2008;**22**:930–40.
144. Ikuta K, Weissman IL. Evidence that hematopoietic stem cells express mouse c-kit but do not depend on steel factor for their generation. *Proc Natl Acad Sci USA* 1992;**89**:1502–6.
145. Tsujimura T, Furitsu T, Morimoto M, Isozaki K, Nomura S, Matsuzawa Y, et al. Ligand-independent activation of c-kit receptor tyrosine kinase in a murine mastocytoma cell line P-815 generated by a point mutation. *Blood* 1994;**83**:2619–26.
146. Mol CD, Lim KB, Sridhar V, Zou H, Chien EY, Sang BC, et al. Structure of a c-kit product complex reveals the basis for kinase transactivation. *J Biol Chem* 2003;**278**:31461–4.
147. Briddell RA, Hartley CA, Smith KA, McNiece IK. Recombinant rat stem cell factor synergizes with recombinant human granulocyte colony-stimulating factor in vivo in mice to mobilize peripheral blood progenitor cells that have enhanced repopulating potential. *Blood* 1993;**82**:1720–3.
148. Andrews RG, Briddell RA, Knitter GH, Opie T, Bronsden M, Myerson D, et al. In vivo synergy between recombinant human stem cell factor and recombinant human granulocyte colony-stimulating factor in baboons enhanced circulation of progenitor cells. *Blood* 1994;**84**: 800–10.
149. Li Z, Li L. Understanding hematopoietic stem-cell microenvironments. *Trends Biochem Sci* 2006;**31**:589–95.
150. Fruehauf S, Veldwijk MR, Seeger T, Schubert M, Laufs S, Topaly J, et al. A combination of granulocyte-colony-stimulating factor (G-CSF) and plerixafor mobilizes more primitive peripheral blood progenitor cells than G-CSF alone: results of a European phase II study. *Cytotherapy* 2009;**11**:992–1001.
151. DiPersio JF, Micallef IN, Stiff PJ, Bolwell BJ, Maziarz RT, Jacobsen E, et al. Phase III prospective randomized double-blind placebo-controlled trial of plerixafor plus granulocyte colony-stimulating factor compared with placebo plus granulocyte colony-stimulating factor for autologous stem-cell mobilization and transplantation for patients with non-Hodgkin's lymphoma. *J Clin Oncol* 2009;**27**:4767–73.
152. Flomenberg N, Devine SM, Dipersio JF, Liesveld JL, McCarty JM, Rowley SD, et al. The use of AMD3100 plus G-CSF for autologous hematopoietic progenitor cell mobilization is superior to G-CSF alone. *Blood* 2005;**106**:1867–74.
153. Fowler CJ, Dunn A, Hayes-Lattin B, Hansen K, Hansen L, Lanier K, et al. Rescue from failed growth factor and/or chemotherapy HSC mobilization with G-CSF and plerixafor (AMD3100): an institutional experience. *Bone Marrow Transplant* 2009;**43**:909–17.
154. Blum A, Childs RW, Smith A, Patibandla S, Zalos G, Samsel L, et al. Targeted antagonism of CXCR4 mobilizes progenitor cells under investigation for cardiovascular disease. *Cytotherapy* 2009;**11**:1016–9.
155. Donahue RE, Jin P, Bonifacino AC, Metzger ME, Ren J, Wang E, et al. Plerixafor (AMD3100) and granulocyte colony-stimulating factor (G-CSF) mobilize different CD34+ cell populations based on global gene and microRNA expression signatures. *Blood* 2009;**114**:2530–41.
156. Zampetaki A, Kirton JP, Xu Q. Vascular repair by endothelial progenitor cells. *Cardiovasc Res* 2008;**78**:413–21.
157. Vandervelde S, van Luyn MJ, Tio RA, Harmsen MC. Signaling factors in stem cell-mediated repair of infarcted myocardium. *J Mol Cell Cardiol* 2005;**39**:363–76.
158. Askari AT, Unzek S, Popovic ZB, Goldman CK, Forudi F, Kiedrowski M, et al. Effect of stromal-cell-derived factor 1 on stem-cell homing and tissue regeneration in ischaemic cardiomyopathy. *Lancet* 2003;**362**:697–703.

159. Chavakis E, Carmona G, Urbich C, Gottig S, Henschler R, Penninger JM, et al. Phosphati-dylinositol-3-kinase-gamma is integral to homing functions of progenitor cells. *Circ Res* 2008;**102**:942–9.

160. Burger JA, Kipps TJ. CXCR4: a key receptor in the crosstalk between tumor cells and their microenvironment. *Blood* 2006;**107**:1761–7.

161. Chang LT, Yuen CM, Sun CK, Wu CJ, Sheu JJ, Chua S, et al. Role of stromal cell-derived factor-1alpha, level and value of circulating interleukin-10 and endothelial progenitor cells in patients with acute myocardial infarction undergoing primary coronary angioplasty. *Circ J* 2009;**73**:1097–104.

162. Abbott JD, Huang Y, Liu D, Hickey R, Krause DS, Giordano FJ. Stromal cell-derived factor-1alpha plays a critical role in stem cell recruitment to the heart after myocardial infarction but is not sufficient to induce homing in the absence of injury. *Circulation* 2004;**110**:3300–5.

163. Penn MS. Importance of the SDF-1:CXCR4 axis in myocardial repair. *Circ Res* 2009;**104**: 1133–5.

164. Walter DH, Haendeler J, Reinhold J, Rochwalsky U, Seeger F, Honold J, et al. Impaired CXCR4 signaling contributes to the reduced neovascularization capacity of endothelial pro-genitor cells from patients with coronary artery disease. *Circ Res* 2005;**97**:1142–51.

165. Egan CG, Lavery R, Caporali F, Fondelli C, Laghi-Pasini F, Dotta F, et al. Generalised reduction of putative endothelial progenitors and CXCR4-positive peripheral blood cells in type 2 diabetes. *Diabetologia* 2008;**51**:1296–305.

166. Oh BJ, Kim DK, Kim BJ, Yoon KS, Park SG, Park KS, et al. Differences in donor CXCR4 expression levels are correlated with functional capacity and therapeutic outcome of angio-genic treatment with endothelial colony forming cells. *Biochem Biophys Res Commun* 2010;**398**:627–33.

167. Tendera M, Wojakowski W, Ruzyllo W, Chojnowska L, Kepka C, Tracz W, et al. Intracoronary infusion of bone marrow-derived selected CD34+CXCR4+ cells and non-selected mononu-clear cells in patients with acute STEMI and reduced left ventricular ejection fraction: results of randomized, multicentre myocardial regeneration by intracoronary infusion of selected population of stem cells in acute myocardial infarction (REGENT) trial. *Eur Heart J* 2009;**30**:1313–21.

168. Zernecke A, Schober A, Bot I, von Hundelshausen P, Liehn EA, Mopps B, et al. SDF-1alpha/CXCR4 axis is instrumental in neointimal hyperplasia and recruitment of smooth muscle progenitor cells. *Circ Res* 2005;**96**:784–91.

169. Hu X, Dai S, Wu WJ, Tan W, Zhu X, Mu J, et al. Stromal cell derived factor-1 alpha confers protection against myocardial ischemia/reperfusion injury: role of the cardiac stromal cell derived factor-1 alpha CXCR4 axis. *Circulation* 2007;**116**:654–63.

170. Massberg S, Konrad I, Schurzinger K, Lorenz M, Schneider S, Zohlnhoefer D, et al. Platelets secrete stromal cell-derived factor 1alpha and recruit bone marrow-derived progenitor cells to arterial thrombi in vivo. *J Exp Med* 2006;**203**:1221–33.

171. Stellos K, Langer H, Daub K, Schoenberger T, Gauss A, Geisler T, et al. Platelet-derived stromal cell-derived factor-1 regulates adhesion and promotes differentiation of human CD34 + cells to endothelial progenitor cells. *Circulation* 2008;**117**:206–15.

172. De Falco E, Porcelli D, Torella AR, Straino S, Iachininoto MG, Orlandi A, et al. SDF-1 involvement in endothelial phenotype and ischemia-induced recruitment of bone marrow progenitor cells. *Blood* 2004;**104**:3472–82.

173. Seeger FH, Rasper T, Koyanagi M, Fox H, Zeiher AM, Dimmeler S. CXCR4 expression determines functional activity of bone marrow-derived mononuclear cells for therapeutic neovascularization in acute ischemia. *Arterioscler Thromb Vasc Biol* 2009;**29**:1802–9.

174. Zhang G, Nakamura Y, Wang X, Hu Q, Suggs LJ, Zhang J. Controlled release of stromal cell-derived factor-1 alpha in situ increases c-kit+ cell homing to the infarcted heart. *Tissue Eng* 2007;**13**:2063–71.

175. Hill JM, Bartunek J. The end of granulocyte colony-stimulating factor in acute myocardial infarction? Reaping the benefits beyond cytokine mobilization. *Circulation* 2006;**113**:1926–8.

176. Yamaguchi J, Kusano KF, Masuo O, Kawamoto A, Silver M, Murasawa S, et al. Stromal cell-derived factor-1 effects on ex vivo expanded endothelial progenitor cell recruitment for ischemic neovascularization. *Circulation* 2003;**107**:1322–8.

177. Saxena A, Fish JE, White MD, Yu S, Smyth JW, Shaw RM, et al. Stromal cell-derived factor-1alpha is cardioprotective after myocardial infarction. *Circulation* 2008;**117**:2224–31.

178. Fernandis AZ, Prasad A, Band H, Klosel R, Ganju RK. Regulation of CXCR4-mediated chemotaxis and chemoinvasion of breast cancer cells. *Oncogene* 2004;**23**:157–67.

179. Janowska-Wieczorek A, Marquez LA, Dobrowsky A, Ratajczak MZ, Cabuhat ML. Differential MMP and TIMP production by human marrow and peripheral blood CD34(+) cells in response to chemokines. *Exp Hematol* 2000;**28**:1274–85.

180. Samara GJ, Lawrence DM, Chiarelli CJ, Valentino MD, Lyubsky S, Zucker S, et al. CXCR4-mediated adhesion and MMP-9 secretion in head and neck squamous cell carcinoma. *Cancer Lett* 2004;**214**:231–41.

181. Spiegel A, Kollet O, Peled A, Abel L, Nagler A, Bielorai B, et al. Unique SDF-1-induced activation of human precursor-B ALL cells as a result of altered CXCR4 expression and signaling. *Blood* 2004;**103**:2900–7.

182. McQuibban GA, Butler GS, Gong JH, Bendall L, Power C, Clark-Lewis I, et al. Matrix metalloproteinase activity inactivates the CXC chemokine stromal cell-derived factor-1. *J Biol Chem* 2001;**276**:43503–8.

183. Segers VF, Tokunou T, Higgins LJ, MacGillivray C, Gannon J, Lee RT. Local delivery of protease-resistant stromal cell derived factor-1 for stem cell recruitment after myocardial infarction. *Circulation* 2007;**116**:1683–92.

184. Zhang M, Mal N, Kiedrowski M, Chacko M, Askari AT, Popovic ZB, et al. SDF-1 expression by mesenchymal stem cells results in trophic support of cardiac myocytes after myocardial infarction. *FASEB J* 2007;**21**:3197–207.

185. Unzek S, Zhang M, Mal N, Mills WR, Laurita KR, Penn MS. SDF-1 recruits cardiac stem cell-like cells that depolarize in vivo. *Cell Transplant* 2007;**16**:879–86.

Genetically Manipulated Progenitor/Stem Cells Restore Function to the Infarcted Heart Via the SDF-1α/CXCR4 Signaling Pathway

YIGANG WANG AND
KRISTIN LUTHER

Department of Pathology and Laboratory Medicine, College of Medicine, University of Cincinnati, Cincinnati, Ohio, USA

Progenitor/stem cells are a viable option to replace myocytes lost subsequent to myocardial infarction (MI). The stromal cell-derived factor-1α/CXC-chemokine receptor type 4 (SDF-1α/CXCR4) axis plays an important role in numerous biological processes including hematopoiesis, cardiogenesis, vasculogenesis, and neuronal development, as well as endothelial progenitor cell trafficking. The secretion of chemoattractants such as SDF-1α at the site of injury creates an environment facilitating the homing of circulating CXCR4 positive and other stem cells (such as mast/stem cell growth factor receptor kit-positive (c-kit$^+$) and c-kit$^+$/GATA binding protein 4 positive (GATA4$^+$) cells) for organ regeneration and tissue repair. SDF-1α is also secreted by hematopoietic progenitor/stem cells and is involved in the autocrine/paracrine regulation of their development and survival. Hypoxic preconditioning activates SDF-1α/CXCR4 signaling and upregulates several vascular/angiogenic factors that cause mobilization of progenitor cells. The SDF-1α/CXCR4 signaling pathway can thus be effectively exploited for cell-based therapy.

SDF-1α/CXCR4 signaling is important for endogenous processes, including organogenesis and hematopoiesis, as well as for response to tissue injury. The secretion of SDF-1α acts as a chemoattractant to facilitate the homing of circulating CXCR4 positive cells as well as other progenitor/stem cells to the

Progress in Molecular Biology
and Translational Science, Vol. 111
http://dx.doi.org/10.1016/B978-0-12-398459-3.00012-5

site of injury to initiate organ regeneration and repair. Both SDF-1α and CXCR4 are constitutively expressed in a variety of tissues and cell types and play a pivotal role in cell mobilization, proliferation, and survival. In the case of cardiovascular diseases, and particularly myocardial infarction, this signaling axis is implicated in many of these processes and is also the mainstay of cells utilizing paracrine mechanisms to enhance cell survival, promote angiogenesis, and stimulate differentiation.[1] Despite these advancements, much remains to be discovered in order for new treatments to be better able to regenerate tissue and recover function. In this review chapter, we provide a perspective on the current and future applications of progenitor/stem cell therapies that take advantage of this signaling axis, and the implications of these promising treatment alternatives.

I. Importance of SDF-1α and CXCR4 Interaction in Ischemic Hearts

SDF-1α is a member of the chemokine CXC subfamily initially cloned from the murine bone marrow stromal cell lines ST-2 and PA6,[1] and purified from supernatant from the murine MS-5 cell line[2] in the early 1990s. Chemoattractant cytokines, or chemokines, are small proteins secreted from areas of inflammation; they function through binding to their specific receptor on leukocytes or other cell types, resulting in chemotaxis and activation.[3] SDF-1α mediates its effects through a specific G protein-coupled receptor, CXCR4.[4] Both SDF-1α and CXCR4 are constitutively expressed in a variety of tissues and cell types and play a pivotal role in cell migration, proliferation, and survival.[4] Expression of CXCR4 allows for efficient SDF-1α-induced progenitor/stem cell mobilization during development of the heart and following injury, thus representing an endogenous repair mechanism present in the myocardium. SDF-1α/CXCR4 interactions are important for the recruitment of CXCR4-positive progenitor/stem cells to the heart after myocardial infarction (MI), mediating their homing to the site of injury.[5] Numerous types of progenitor/stem cells can respond to SDF-1α release by migrating to the ischemic myocardium, including endothelial progenitor cells (EPCs), bone-marrow-derived cells,[6] mesenchymal stem cells (MSCs), and endogenous cardiac stem cells (CSCs).[7,8] Thus, SDF-1α appears to have an important role in cardioprotection and regeneration following ischemia.[9] EPCs protect the ischemic area by releasing paracrine factors[10] or through regeneration.[8] The bone-marrow-derived cells participate actively in the repair process[6] and enhance survival via paracrine effects.[4] Endogenous CSC-like cells depolarize *in vivo* and thus may contribute to increased contractile function even in the absence of maturation into mature cardiomyocytes.[11,12] Through binding to CXCR4, SDF-1α acts to repopulate the ischemic zone and bring about revascularization.

II. Role of SDF-1α/CXCR4 as Therapeutic Targets in Heart Disease

A functional role of SDF-1α/CXCR4 in nonhematopoietic tissues is supported by the fact that SDF-1[−/−] and CXCR4[−/−] mice exhibit a cardiac ventricular septal defect, deformation of the large vessels supplying the gastrointestinal tract, and suppressed hematopoiesis in the bone marrow.[13,14] Additionally, B-cell lymphopoiesis and myelopoiesis are both severely affected. Because of the embryonic lethality associated with loss-of-function mutations, no known pathologies involve complete SDF-1α/CXCR4 deficiencies.

The secretion of chemoattractants such as SDF-1α in or around injured tissue is a crucial process that creates an environment facilitating the homing of circulating progenitor cells for organ regeneration and tissue repair.[5] The SDF-1α/CXCR4 axis seems to be particularly important in progenitor cell chemotaxis, homing, engraftment, and retention in the damaged myocardium.[6] We[15] and others[16] have shown that SDF-1α is significantly expressed in myocardial tissue after MI and is most highly expressed in the peri-infarct zone, the area to which progenitor/stem cells are recruited. SDF-1α is also secreted by hematopoietic progenitor/stem cells and is involved in autocrine/paracrine regulation of their development and survival. Specifically, as summarized in Fig. 1, SDF-1α signals through binding to CXCR4 and initiating various cascades, such as activation of calcium flux and focal adhesion components including focal adhesion kinase (FAK) and proline-rich kinase-2, as well as the adaptor proteins paxillin, noncatalytic region of tyrosine kinase adaptor protein (Nck), adaptor molecule Crk, Crk-like protein, and Crk-associated substrate. Additionally, phospholipase C-γ is stimulated, cleaving phosphatidylinositol 4,5-bisphosphate and activating protein kinase C (PKC).[4] Through these interactions, SDF-1α is known to induce phosphorylation of mitogen-activated protein kinase (MAPK) p42/44 and RAC-alpha serine/threonine-protein kinase (Akt) and to upregulate vascular endothelial growth factor (VEGF) after coronary occlusion *in vivo*.[4] MAPK and Akt pathways are widely accepted as being associated with cell survival and proliferation, while VEGF is associated with vasculogenesis and angiogenesis. SDF-1α, together with other cytokines, enhances the survival and proliferation of both human cluster of differentiation molecule 34-positive (CD34[+]) cells and murine progenitor/stem cells.[10] Signaling through the Gα12/13–Rho axis has also been implicated in

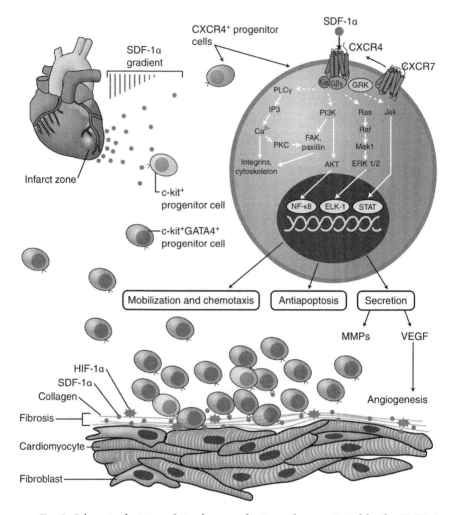

FIG. 1. Schematic depiction of signaling transduction pathways activated by the SDF-1α/
CXCR4 interaction. SDF-1α binding to the CXCR4 activates G protein-coupled receptor kinases,
which are responsible for many signaling transduction pathways in cells. The PI3K-stimulated
activation of extracellular signal-regulated kinase (ERK)-1/2 has been implicated in SDF-1α-
induced chemotaxis, cell proliferation, and regulation of integrin activity. Increased nuclear translocation of nuclear factor-κB (NF-κB) and DNA-binding activity occur after SDF-1α treatment.
SDF-1α promotes the association of CXCR4 with Jak, followed by tyrosine phosphorylation.
Phosphorylated Jaks induce the activation and nuclear translocation of STAT. Thus, activation of
these pathways regulates (a) mobilization via phosphorylation of focal adhesion proteins, such as
FAK and paxilin, by both PI3K and PKC, (b) chemotactic responses via release of autocrine/
paracrine factors, (c) adhesion, (d) secretion of anti-fibrotic enzymes such as matrix metalloproteases (MMPs) for enhancing progenitor cell engraftment, and (e) secretion of angiopoietic factors
by the progenitor cells. In vivo, an ischemic environment leads to apoptosis, but pretreatment of

SDF-1α-induced migration.[17] While CXCR4 is considered to be the main target of SDF-1α, there is additional binding to CXC-chemokine receptor type 7 (CXCR7), which appears to stimulate MAPK, phosphoinositide 3-kinase (PI3K), and Janus kinase/signal transducer and activator of transcription (Jak–STAT) signaling. It is reported that CXCR7 forms functional heterodimers with CXCR4 to regulate SDF-1α-mediated G protein-coupled signal pathway activation; these heterodimers have novel properties that contribute to the functional plasticity of SDF-1α. While the effects of CXCR7 signaling have yet to be fully defined, this receptor is clearly important in the heart: it is highly expressed in cardiomyocytes of wild-type mice, and CXCR7-deficient mice die within 1 week of birth because of cardiovascular defects.[18]

Transplanted or endogenous progenitor/stem cells can help to repair the damaged myocardium by stimulating angiogenesis,[8] that is, the development of new blood vessels from preexisting vessels. During embryonic development, blood vessels are constructed via three overlapping processes: vasculogenesis, angiogenesis, and arteriogenesis. In vasculogenesis, mesodermal cells become hemangioblasts, in a process mediated by fibroblast growth factor-2, and cluster together into aggregations known as blood islands.[19] The inner cells differentiate into hematopoietic stem cells, while the outer cells become blood vessel precursors. These angioblasts multiply, and through VEGF signaling, become endothelial cells (ECs). The ECs form tubes that give rise to a capillary network or plexus. Additionally, differentiated vascular cells including pericytes and smooth muscle cells (SMCs) are recruited through angiopoietin-1 and platelet-derived growth factor signaling, and migrate to cover the new vessels. Angiogenesis involves the process of remodeling and shaping the network of endothelial tubes into distinct capillary beds through VEGF-mediated processes, involving the splitting of capillaries and sprouting of new capillaries. An overlapping process known as arteriogenesis also occurs at this time, and is mediated by VEGF, blood flow, and shear stress.[20] Intercellular contacts between ECs are loosened, allowing the capillaries to be remodeled into larger diameter vessels such as arterioles, venules, arteries, and veins. Extracellular matrix degradation at localized points continues to allow for the formation of new capillaries. Thus organs such as the heart are vascularized during development.

---

progenitor cells with hypoxic preconditioning or inducing overexpression of prosurvival genes such as Akt and hypoxia-inducible factor-1α (HIF-1α) can reduce cell death in subsequent ischemic conditions. In addition, c-kit⁺ and c-kit⁺/GATA4⁺ cells are also mobilized from the bone marrow and heart, respectively, to the site of injury by the treatment with SDF-1α, and their presence contributes to healing. (See Color Insert.)

Until the discovery of EPCs in 1997,[21] it was believed that any new blood vessel growth was the result of local rearrangements and proliferation in vessel walls. With the characterization of EPCs came the realization that there are circulating progenitor cells capable of contributing to neovascularization through processes related to vasculogenesis.[19] The implication of this for progenitor/stem cell therapy is clear: through promoting the homing of progenitor/stem cells and/or increasing their number, ischemic tissues can be resupplied with oxygen and nutrients, aiding in their recovery. Additionally, any therapy that promotes angiogenic activities among the progenitor population, such as tube formation, will likely contribute to recovery after MI. During the past decade, the regenerative and angiogenic potential of many varieties of progenitor/stem cells has been revealed, in both endogenous and genetically modified cell populations.

The endothelium constitutes a single layer of cells that lines the walls of all blood vessels and acts as both a barrier and point of interaction between the bloodstream and the rest of the body. It actively responds to both physical and chemical stimuli. For instance, in response to shear stress or acetylcholine, healthy endothelium triggers vasodilation by releasing nitric oxide. Endothelial dysfunction is associated with many health problems, including heart disease, atherosclerosis, and diabetes.[22]

Researchers have performed manipulations of the SDF-1α/CXCR4 signaling axis to induce greater therapeutic responses and improve functional outcomes following MI. Cells overexpressing SDF-1α display an increased capacity for cellular growth, protection against interleukin-4-induced apoptosis,[4] and prevention of γ-irradiation damage,[23] possibly through Akt-mediated effects. In the ischemic heart, direct injections of SDF-1α into the peri-infarct zone improved function, as did overexpression of SDF-1α with adenoviral delivery; AMD3100, an inhibitor of CXCR4, was shown to abolish these benefits.[9] The increased level of SDF-1α secretion following MI is transient, and is not sustained for more than 7 days.[15] This is due to the fact that it is inactivated by dipeptidyl peptidase IV (DPP-IV).[24] Consequently, efforts have also been made to prolong the effect of SDF-1α in the infarcted heart by inhibiting DPP-IV using diprotin A. This treatment stabilizes and prolongs the active myocardial SDF-1α level, which was shown to enhance progenitor/stem cell mobilization to the ischemic myocardium.[25] Additionally, SDF-1α is also thought to activate EPC integrins involved in the extravasation process. Local SDF-1α administration induces the accumulation of transplanted EPCs at sites of ischemia and enhances neovascularization after EPC transplantation.

CXCR4 expression is regulated by various factors including cytokines, chemokines, stromal cells, adhesion molecules, and proteolytic enzymes[4] and increased by transcription factors related to stress/hypoxia and tissue damage, such as NF-κB,[4] HIF-1α,[26] glucocorticoids,[27] lysophosphatidylcholine,[5] transforming growth factor-ß1 (TGF-ß1),[28] VEGF,[4] interferon-α,[29] and several interleukins (IL-2, IL-4, and IL-7).[17] Thus, it is likely that stress-related conditions in the heart may upregulate CXCR4 expression in progenitor/stem cells.

Manipulations have been undertaken to increase CXCR4 expression to improve outcomes after MI. For example, increased CXCR4 expression through adenoviral vectors enhances vascularization in the damaged myocardium.[6] In human progenitor/stem cells, CXCR4 can be upregulated by short-term ($\sim$ 4 h) *in vitro* treatment with cytokines.[30] This enhances their *in vitro* migration in response to an SDF-1α gradient. SDF-1α/CXCR4 interactions are also involved in hematopoietic stem cell functions, such as secretion of paracrine factors.[4,31] MSCs overexpressing CXCR4 release antifibrotic enzymes (MMP-2 and -9) under hypoxic conditions,[15] which can help these cells in crossing the basement membrane, thus leading to more efficient migration, reduced scar formation, and improvement in cardiac function.

III. SDF-1α/CXCR4 as Therapeutic Targets in Vascular Diseases

SDF-1α's effect of attracting progenitor/stem cells to areas of tissue damage is not always desirable. Long-term overexpression of SDF-1α has been reported to cause chronic allograft deterioration associated with development of transplant arteriosclerosis.[6] In transplant arteriosclerosis, the contribution of hematopoietic progenitor cells (HSCs) to neointimal lesion formation has been established in animal models and also in humans.[32,33] The mobilization and recruitment of HSCs in heterotopic abdominal heart transplant is mediated by SDF-1α in allografts in rats.[6] SDF-1α expression is also increased in injured vasculature, contributing to the recruitment of circulating progenitor cells into the neointima during intimal hyperplasia,[33] which may accelerate neointimal thickening leading to obstructive vascular disease, such as restenosis after percutaneous interventions in coronary or peripheral artery disease and transplant arteriosclerosis. SDF-1α levels appear to be important in response to wire injury to vessels, which induces apoptosis of damaged SMCs in the intima and recruits SMC progenitors.[34] SDF-1α blockage was associated with reduction of this cell recruitment and thus reduced vessel wall thickness following injury. A similarly diminished effect was exhibited in mice that did not express CXCR4.[33] Moreover, a recent study of single-nucleotide polymorphisms demonstrated that a SDF-1α gene variation has an influence on SDF-1α level and circulating EPC number, such that the plasma SDF-1α level is a predictor of EPC number and likelihood of atherosclerosis and other cardiovascular diseases.[35] In apolipoprotein E-deficient (apoE$^{-/-}$) mice, which are prone to atherosclerosis, blocking SDF-1α after wire injury resulted in a significant reduction of the neointimal area by inhibiting the accumulation of bone-marrow-derived SMC progenitors in the neointima.[34] The effect of SDF-1α on SMC progenitor recruitment was shown to be dependent on the expression of CXCR4 in bone marrow cells because neointimal hyperplasia and SMC content are diminished in apoE$^{-/-}$ mice after bone marrow reconstitution with fetal HSCs from CXCR4$^{-/-}$ mice.[34] Consequently, patients with atherosclerosis or transplant vasculitis may benefit from the administration of CXCR4 antagonists, with the possibility of reduced SMC progenitor cell infiltration and reduced neointimal hypertrophy.

In contrast to the above reports, Walter et al.[36] suggest that upregulation of CXCR4-mediated signaling may be beneficial in patients with cardiovascular diseases. In this study, EPCs from patients with coronary artery disease were found to have similar surface expression of CXCR4 as compared to healthy volunteers, but basal Jak activation was reduced, and less responsive to SDF-1α, leading to reduced capacity for EPC homing and neovascularization. Bone-marrow-derived EPCs participate in angiogenesis either by incorporating into the neovasculature[37] or by secreting proangiogenic factors.[25] Consistent with this notion, blockade of CXCR4 by either monoclonal antibody or AMD3100 partially inhibited blood flow recovery of ischemic hearts and VEGF-mediated incremental revascularization.[31] It remains to be determined whether the benefits of prolonged SDF-1α signaling to the heart would outweigh the potential negative effects to the circulatory system in patients with comorbid atherosclerosis or transplant vasculopathy.

IV. Role of SDF-1α/CXCR4 in Cell-Based Therapy

While cardiomyocytes are considered to be terminally differentiated, and the heart's potential to regenerate is known to be limited, research has revealed that SDF-1α/CXCR4-mediated signaling represents an endogenous repair mechanism, making possible the homing of bone-marrow-derived progenitor/stem cells. Progenitor/stem cell-based therapy seeks to exploit these pathways to improve outcomes after MI and in heart failure by repairing and replacing damaged tissue. This concept has proved effective in animal models, and preclinical and clinical studies are currently underway.

The first cells to be transplanted in a clinical setting were skeletal myoblasts, which conferred a slight improvement in left ventricular (LV) function, but their implementation was associated with an increased risk of arrhythmia.[38] Nonhematopoietic MSCs have also been transplanted in several clinical trials, and have been shown to be not only safe (no increased risk of arrythmia) but also effective in improving LV function after MI.[39] Other clinical trials have investigated the effects of transplanting hematopoietic progenitor/stem cells (HSCs) and/or EPCs. These are thought to contribute to recovery through antiapoptotic signaling and neovascularization, rather than transdifferentiation.[39] The efficacy of mononuclear cells from bone marrow and peripheral blood, along with a mixture of HSCs, MSCs, and EPCs, have been investigated and compared; both types were found to be effective.[40] However, there has been less evidence that these cells actually differentiate into cardiomyocytes, or that they remain a long-term component of the recovering heart.

The most likely cells to be capable of regeneration through differentiation into cardiomyocytes are cardiac progenitor/stem cells and inducible pluripotent stem (iPS) cells. Clinical trials have not yet been conducted, but they are in the planning phase.[39] In particular, iPS cells possess a promising capacity for differentiation without immunogenic or ethical problems. However, one barrier to using iPS cells is the risk of teratoma formation.[41] These cells are derived from adult cells reprogrammed with factors that can lead to mutation and oncogene expression. To minimize this, specific cell types, such as purified cardiomyocytes, SMCs, or EPCs derived from embryonic stem or iPS cells, are purified for transplantation based on the following methods in animal experiments.

(1) Fluorescence-assisted cell sorting (FACS) or magnetic microbeads are used for the isolation of specific cell types based on their unique markers.
(2) Another approach for the isolation of specific cell types is genetic selection using lentiviral vectors with cell-type-specific promoters to identify these cells. For example, a population of iPS cells can be enriched for cardiomyocytes by using vectors encoding markers and a gene for puromycin resistance under the control of the cardiomyocyte-specific promoter for Na^+/Ca^{2+} Exchanger 1. The cells can be expanded in a medium containing puromycin, to ensure that noncardiomyocytes are removed, and all the cells become cardiomyocyte precursors or cardiomyocytes.[42]
(3) It has been reported that tumors often form from reprogrammed iPS cells that remain octamer-binding protein 4 (Oct4) or myc proto-oncogene protein (c-Myc) positive. Thus, scientists aim to remove Oct4- or c-Myc-positive cells before implantation by using FACS or magnetic beads techniques.
(4) In the future, it would be ideal to generate iPS cells by reprogramming somatic cells through nonviral methods.

Another population of progenitor/stem cells that has been investigated is very small embryonic-like stem cells (VSELs).[43] These rare cells are found in adult human bone marrow, umbilical cord blood, and peripheral blood. As their name implies, they are characterized by their small size (averaging 4 μm) and distinctive morphology, that is, a relatively large nucleus, open-type chromatin, and numerous mitochondria. Even without reprogramming, they express early embryonic markers, as well as CXCR4, and have been shown to be pluripotent, capable of differentiating into all three germ layers, including the mesoderm, from which cardiomyocytes are derived.[43] Accordingly, a murine *ex vivo* expansion and differentiation protocol developed by Wojakowski *et al.*,[44] involving coculture of the isolated VSELs with myoblasts and in cardiogenic medium, generated cardiomyocytes. The cells derived using this treatment were shown to improve LV function and reduce remodeling in mice.[45]

Another direction for future clinical trials is to transplant progenitor/stem cells genetically modified in such a way that they can take advantage of the heart's endogenous homing signals through the SDF-1α/CXCR4 pathway. The delivery of SDF-1α-overexpressing MSCs to the injured tissue suggests a novel strategy whereby the recruitment of CXCR4[+] progenitor cells as well as c-kit[+] cells from bone marrow may be enhanced so as to facilitate the repair process. In a clinical situation, SDF-1α plays a significant role in hematopoietic progenitor cell mobilization.[46] Pegylated fibrin patch-based SDF-1α delivery, or augmentation through SDF-1α overexpressing MSCs, improves the rate of c-kit[+] cell homing and increases LV function after MI.[47] This is of potential importance because c-kit[+] cells play a significant role in cardiac repair and restoration of heart function after myocardial injury.[47] Bone-marrow-derived c-kit[+] cells produce high levels of VEGF, which in turn stimulates their differentiation into ECs.[48] The observation that c-kit-deficient cardiac progenitor cells in mast-cell-deficient (W/WV) mutant mice fail to show cell differentiation during aging or after injury and that c-kit-positive cells are also associated with enhanced angiogenesis highlights the importance of an increased number of c-kit-positive cells in repair processes.[47]

Even without experimental manipulation, SDF-1α level increases after MI, so another approach to cell-based therapy is to transplant MSCs that overexpress CXCR4 (MSC[CXCR4]).[49] This has been shown to enhance LV function and attenuate remodeling through improved homing and revascularization. Given that overexpression of both CXCR4 and SDF-1α improves the outcome individually, one would expect that combining these two approaches (i.e., enhancing SDF-1α level in the recipient heart and overexpressing CXCR4 in the transplanted MSCs) would be even more effective. To this end, we have successfully implemented the use of diprotin A to further improve the migration of transplanted MSC[CXCR4].[50] Recall that the level of SDF-1α released from the heart increases after MI, but falls back to baseline within about 7 days; this occurs as SDF-1α is cleaved and inactivated by DPP-IV, a process that can be inhibited by diprotin A. In a rat model of MI, treatment with diprotin A infusions for 1 week prior to grafting a monolayer of MSC[CXCR4] resulted in decreased fibrotic area, increased capillary density in the cell sheet and the border zone, decreased apoptosis, and improved heart function as measured by echocardiography.

V. Genetically Manipulated Cell Patch for Repair of Infarcted Myocardium

Despite recent advances in pharmacological and surgical approaches to rescue injured myocardium, ischemic heart disease remains the leading cause of heart failure and death in developed nations.[51] The most common methods of cell delivery for myocardial therapy are intravenous[8] or by direct intramyocardial injection into an infarcted area.[52] It is difficult, however, to control the deposition of grafted cells using these methods.[53,54] Recent progress in myocardial cell patch techniques offers a solution to this problem and a potentially beneficial strategy for tissue engineering aimed at heart regeneration to reverse deleterious tissue effects following MI.

Repair of the infarcted myocardium by genetically manipulated cell patch application is currently limited by the inability of most structural biodegradable scaffold substrate materials to cultivate local tissue replacement.[55] Polymers, the largest class of engineered biomaterials used today for myocardial tissue reconstruction, have many disadvantages such as poor elasticity, poor attachment, biodegradation, poor ECM formation, and biotoxicity. Natural internal biomaterials, such as peritoneum, omentum, diaphragm, or pericardium, represent an emerging substrate choice, and have been demonstrated to promote wound healing and stimulate revascularization of ischemic tissues.[56,57] O'Shaughnessy was the first, in 1937, to report cardio-omentopexy procedures in which pedicled omental grafts were attached to the surface of the ischemic heart through the diaphragm in humans.[58] However, the use of omentopexy was abandoned as a treatment because this procedure alone failed to produce sufficient myocardial revascularization.[55] The method was soon thereafter replaced by coronary artery bypass grafting, and coronary revascularization was subsequently accomplished through percutaneous transluminal coronary angioplasty.

However, native biomaterials used as substrates become an attractive option when used in combination with progenitor/stem cells, and have demonstrated complementary outcome features including improved cardiac contractility (via direct myogenesis or due to paracrine effects from progenitor/stem cells), enhanced tissue nutrition (via angiogenesis), and enhanced cell survival (via antiapoptosis). These benefits combine to reduce myocardial remodeling, limit the infarct size, and improve the heart's mechanical performance.

Recently, we developed and reported a method of generating genetically manipulated cell patches with MSC^{CXCR4} using a homologous peritoneum substrate that we applied after MI to repair scarred myocardium. The detailed procedure of biological substrate preparation and application to the infarcted area is shown in Fig. 2 (A–D), and has been previously reported.[59] At 4 weeks

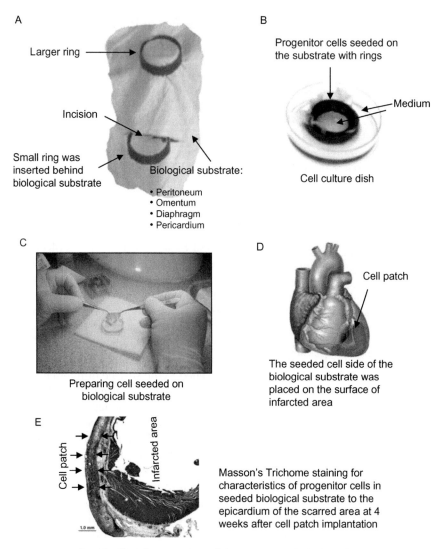

A

Larger ring

Incision

Small ring was
inserted behind
biological substrate

Biological substrate:
• Peritoneum
• Omentum
• Diaphragm
• Pericardium

B

Progenitor cells seeded on
the substrate with rings

Medium

Cell culture dish

C

Preparing cell seeded on
biological substrate

D

Cell patch

The seeded cell side of the
biological substrate was
placed on the surface of
infarcted area

E

Cell patch

Infarcted area

1.0 mm

Masson's Trichome staining for
characteristics of progenitor cells in
seeded biological substrate to the
epicardium of the scarred area at 4
weeks after cell patch implantation

FIG. 2. Biological cell patch preparation and characteristics. The biological substrate is clipped by two overlapping plastic rings (A); the cut substrate with ring is shifted to a cell culture dish, and the progenitor cells are seeded on the patch (B); monolayered progenitor cell-seeded patch at day 9 and ready to use (C). The cell patch is grafted to the surface scarred area of myocardium (D). A section of the biological cell patch (arrow) is applied to the epicardium of the scarred area at 4 weeks after cell patch implantation treated with Masson's trichrome staining (E.) (See Color Insert.)

after MSC-based biological substrate transplantation, the cell patch has become tightly adherent to the surface of the recipient heart as shown in Fig. 2E. The combination of CXCR4-overexpressing MSCs and the use of homologous peritoneum synergistically ameliorated cardiac function and dilating remodeling 5 weeks after MI. We evaluated the mechanism underlying this synergy and found the following:

(1) MSC^{CXCR4} seeded on peritoneum promoted angiogenesis when applied to the epicardial surface of minimally vascular dyskinetic scar tissue of the left ventricle following MI.

(2) The MSC^{CXCR4} patch using homologous peritoneum appeared to attenuate post-MI ventricular remodeling via increased anterior LV wall thickness and reduced LV collagen deposition and fibrosis.

(3) Growth factors released from MSC^{CXCR4} enhanced the implanted cells' survival and maturation.

Therefore, the incorporation of MSC^{CXCR4} into a peritoneal cell patch created a synergistic setting that promoted *in vivo* myogenesis in association with angiogenesis. These processes are critical for improving blood flow to supply nutrition to the damaged heart and to enable restoration of heart function after MI.

The new approach of transplanting genetically manipulated progenitor/stem cells supported by a homologous peritoneal patch addresses the limitations of preceding strategies. In developing this new approach to myocardial tissue repair, we have utilized a biocompatible tissue substrate that:

(1) serves as an effective reservoir for cell delivery;

(2) provides nutrition from outside the adjacent/adherent epicardium (newly formed vessels in cell patch) and angiogenesis within the infarcted areas;

(3) dampens systolic aneurysmal bulging over the ischemic area and prevents aneurysmal stretching of peri-ischemic myocardium;

(4) strengthens the myocardial wall and reduces LV collagen deposition, preventing or reversing further LV remodeling;

(5) manipulates genes so that progenitor cells release cytoprotective factors, which exert beneficial effects on angiogenesis, progenitor cell survival, and migration into the scarred myocardium, and

(6) allows progenitor cells to become embedded in the matrix of the underlying myocardium, resulting in a stronger, more adherent bond with the homologous peritoneum patch itself. The transplanted cells become incorporated within the healing heart, and the peritoneum remains as a permanent supportive structure.

In addition to CXCR4, another manipulation that has been shown to be beneficial is the overexpression of adenylyl cyclase 6 (AC6) in the recipient heart. After MI, fibroblasts proliferate and synthesize the extracellular matrix, forming a scar and leading to heart failure and arrhythmia. AC6 catalyzes the conversion of adenosine triphosphate to cyclic adenosine monophosphate (cAMP), which activates protein kinase A. In a recent study,[41] we found that elevation of cAMP, mediated through AC6 overexpression, leads to antifibrotic effects through the inhibition of TGF-β, ERK 1/2, and Smad pathways. Synthesis of collagen types I and III were significantly reduced. The use of forskolin, a plant-derived compound, mimics the effects of AC6 overexpression *in vitro*, though its *in vivo* effects have not been thoroughly evaluated. Similarly, the naturally occurring hormone Relaxin has been shown to reduce activation of profibrotic pathways and promote MMP9-induced collagen degradation in mice.[60] Thus, another potentially beneficial adjuvant treatment would be to combine Relaxin therapy with the genetic modifications in progenitor/stem cell transplants.

Based on successful tissue engineering techniques used to replace or restore damaged heart function in mouse, rat, rabbit, canine, and pig models,[61-63] applications such as this are moving ever closer to the clinical mainstream. A critical challenge in this discipline will be acquiring sufficient quantities of cells for clinical use. The ideal cells or combination of cells must have several key properties:

(1) they must be available in sufficient quantities;
(2) they must be able to combine in such a way that they function synergistically: for example, a patch might include cardiomyocytes (to restore heart contractility), ECs and SMCs (to build new blood vessels, which supply nutrition to ischemic tissue and carry away metabolic wastes), and fibroblasts (to provide support to the cell structure) in the appropriate ratio;
(3) they must be capable of being harvested with minimal donor site morbidity (from autologous or homologous sources), and be compatible with the recipient's immune system.

Additionally, further investigation of the outcomes of SDF-1α/CXCR4 signaling manipulations should be undertaken prior to widespread clinical implementation. Some studies in experimental models have yielded negative results, including elevated apoptosis and inflammation, increased scar size, and deteriorated contractility. In an *in vitro* study on adult cardiomyocytes, the decrease in contractility coincided with a decrease in calcium transients. CXCR4 overexpression increased this effect, and inhibition of CXCR4 abolished it.[64] When SDF-1α was directly injected into the peri-infarct myocardium in a porcine model of MI, there was no beneficial effect on perfusion or scar size, and, despite increased vessel density and reduced collagen levels, function actually deteriorated compared to non-SDF-1α-treated controls.[3] Similarly, adenovirus-mediated overexpression of CXCR4 in a rat ischemic model led to increased apoptosis, inflammation, and scar size, and administration of the inhibitor AMD3100 had the opposite effect: reduced infarct size, improved LV function, and decrease in hypertrophy.[65] It is clear that the effect of these manipulations still needs to be investigated systematically.

Tissue engineering has progressed in discovering intrinsic healing mechanisms and possible ways for enhancing these pathways through *ex vivo* progenitor/stem cell expansion and the use of genetic manipulation.[66] These facets of regenerative medicine have started to be combined with the administration of ESCs or iPS cell-derived cells, such as cardiomyocytes, SMCs, and EPCs.

As with the use of progenitor/stem cell suspensions, human clinical trials are now beginning to appear investigating the use of patches as progenitor/stem cell reservoirs. Considering biological substrates for cell patches, a suitably sized greater omentum or peritoneum sample can be easily harvested in humans; that is the reason we chose homologous peritoneum as the culture substrate. In addition, many laboratories have reported supporting results indicating that a cell patch is not only suitable for treatment of acute MI but also effective for use in heart tissue after chronic LV remodeling.[67] The benefits are considered to include increased LV wall thickness at the infarct region, attenuated LV dilation, and improved heart function.[68] Unfortunately, the current surgical procedure is invasive and requires thoracotomy. This issue reduces enthusiasm and potential significance for some prospective users. However, a novel device for minimally invasive transplantation of cell patches using endoscopic methods with video-assisted thoracoscopic instrumentation is now available and offers an alternative minimally invasive approach to applying such tissue patches to regions of MI.[69] The ultimate goal of cell patch application is to provide biocompatible, non-immunogenic cardiovascular cells with morphological and functional properties similar to those of natural myocardium to repair MI.

VI. Conclusions

In summary, the role of SDF-1α/CXCR4 in regulating bone-marrow-derived progenitor/stem cell engraftment, mobilization, vascular remodeling, and neovascularization is well accepted. Alternative strategies that direct stimulation of the natural ligand SDF-1α to attain these goals could be used to stimulate CXCR4 to promote activation/mobilization of progenitor/stem cells. Thus, CXCR4 is potentially an important therapeutic target in the treatment of the aforementioned diseases.[51] Understanding the molecular mechanisms regulating CXCR4 expression and activation in various CXCR4-positive cells is crucial in the development of therapeutic strategies for the treatment of patients with cardiovascular diseases. For example, hypoxic preconditioning[7,70] of progenitor/stem cells causes upregulation of SDF-1α/CXCR4, inducing progenitor/stem cell homing. Hypoxic preconditioning of progenitor/stem cells also increases upregulation of vascular growth factors,[70] which may stimulate SDF-1α, resulting in the mobilization and adhesion of progenitor cells, which would be highly beneficial in the treatment of ischemic diseases. Thus this phenomenon could be manipulated clinically for progenitor/stem cell-based therapy.

On the other hand, by recruiting progenitor cells (which give rise to EPCs and SMCs), SDF-1α may also contribute to accelerated neointima formation leading to obstructive vascular disease. Furthermore, neointimal hyperplasia and SMC accumulation is a significant factor in the failure of venous grafts and stenosis, arteriovenous fistulas, and grafts as vascular access to hemodialysis. Caution must be exercised in targeting these proteins because of the "double-edged sword" nature of the biological effects of SDF-1α/CXCR4. Blocking the SDF-1α/CXCR4 axis by using specific CXCR4 antagonists[71] or small-molecule antagonists for blocking SDF-1α[5] might be a valuable strategy to treat transplant vasculopathy or restenosis. This therapeutic option is compromised by the yet-unknown effects of SDF-1α in the pathogenesis of native atherosclerosis. For instance, clinical data suggest that SDF-1α may support plaque stability of native atherosclerotic lesions,[72] although very little is known about the effect of SDF-1α on atherosclerotic lesion progression. This problem needs to be further elucidated with experimental data on the specific role of SDF-1α expression and CXCR4 function under various disease conditions.

Cardiomyocytes that are lost after MI can be replaced with progenitor/stem cells. It is likely that this phenomenon occurs naturally to some extent, as inflammatory responses increase chemoattractant secretion and lead to stem and progenitor cell mobilization; this endogenous healing mechanism can be enhanced therapeutically. Many different types of cells, including nonhematopoietic MSCs, hematopoietic MSCs, EPCs, and iPS cells, have been investigated in preclinical and clinical trials, and have shown promising results. There are difficulties associated with the deposition of progenitor/stem cell suspensions; transplanting cells grown in a homologous substrate such as peritoneum may provide an effective reservoir of progenitor/stem cells, as well as providing structural support to the damaged myocardium. Much research is still needed before these experimental treatments can become mainstream first-line therapies. For instance, questions remain about the best type of cell to transplant, optimal transplantation timing, and medications to be used concurrently (such as the DPP-IV inhibitor diprotin A) to enhance the outcome.

ACKNOWLEDGMENTS

We thank Christian Paul and Wei Huang for providing technical assistance.

References

1. Nagasawa T, Kikutani H, Kishimoto T. Molecular cloning and structure of a pre-B-cell growth-stimulating factor. *Proc Natl Acad Sci USA* 1994;**91**:2305–9.
2. Tashiro K, Tada H, Heilker R, Shirozu M, Nakano T, Honjo T. Signal sequence trap: a cloning strategy for secreted proteins and type I membrane proteins. *Science* 1993;**261**:600–3.
3. Koch KC, Schaefer WM, Liehn EA, Rammos C, Mueller D, Schroeder J, et al. Effect of catheter-based transendocardial delivery of stromal cell-derived factor 1alpha on left ventricular function and perfusion in a porcine model of myocardial infarction. *Basic Res Cardiol* 2006;**101**:69–77.
4. Kucia M, Jankowski K, Reca R, Wysoczynski M, Bandura L, Allendorf DJ, et al. CXCR4-SDF-1 signaling, locomotion, chemotaxis and adhesion. *J Mol Histol* 2004;**35**:233–45.
5. Abbott JD, Huang Y, Liu D, Hickey R, Krause DS, Giordano FJ. Stromal cell-derived factor-1alpha plays a critical role in stem cell recruitment to the heart after myocardial infarction but is not sufficient to induce homing in the absence of injury. *Circulation* 2004;**110**:3300–5.
6. Wang Y, Haider H, Ahmad N, Zhang D, Ashraf M. Evidence for ischemia induced host-derived bone marrow cell mobilization into cardiac allografts. *J Mol Cell Cardiol* 2006;**41**:478–87.
7. Tang YL, Zhu W, Cheng M, Chen L, Zhang J, Sun T, et al. Hypoxic preconditioning enhances the benefit of cardiac progenitor cell therapy for treatment of myocardial infarction by inducing CXCR4 expression. *Circ Res* 2009;**104**:1209–16.
8. Zhao T, Zhang D, Millard RW, Ashraf M, Wang Y. Stem cell homing and angiomyogenesis in transplanted hearts are enhanced by combined intramyocardial SDF-1α delivery and endogenous cytokine signaling. *Am J Physiol Heart Circ Physiol* 2009;**296**:H976–86.
9. Saxena A, Fish JE, White MD, Yu S, Smyth JW, Shaw RM, et al. Stromal cell-derived factor-1alpha is cardioprotective after myocardial infarction. *Circulation* 2008;**117**:2224–31.
10. Kijowski J, Baj-Krzyworzeka M, Majka M, Reca R, Marquez LA, Christofidou-Solomidou M, et al. The SDF-1-CXCR4 axis stimulates VEGF secretion and activates integrins but does not affect proliferation and survival in lymphohematopoietic cells. *Stem Cells* 2001;**19**:453–66.
11. Unzek S, Zhang M, Mal N, Mills WR, Laurita KR, Penn MS. SDF-1 recruits cardiac stem cell like cells that depolarize in vivo. *Cell Transplant* 2007;**16**(9):879–86.
12. Zhang M, Mal N, Kiedrowski M, Chacko M, Askari AT, Popovic ZB, et al. SDF-1 expression by mesenchymal stem cells results in trophic support of cardiomyocytes after myocardial infarction. *FASEB J* 2007;**21**:3197–207.
13. Zou YR, Kottmann AH, Kuroda M, Taniuchi I, Littman DR. Function of the chemokine receptor CXCR4 in haematopoiesis and in cerebellar development. *Nature* 1998;**393**:595–9.
14. Nagasawa T, Hirota S, Tachibana K, Takakura N, Nishikawa S, Kitamura Y, et al. Defects of B-cell lymphopoiesis and bone-marrow myelopoiesis in mice lacking the CXC chemokine PBSF/SDF-1. *Nature* 1996;**382**:635–8.
15. Zhang D, Fan GC, Zhou X, Zhao T, Pasha Z, Xu M, et al. Over-expression of CXCR4 on mesenchymal stem cells augments myoangiogenesis in the infarcted myocardium. *J Mol Cell Cardiol* 2008;**44**:281–92.
16. Pittenger MF, Martin BJ. Mesenchymal stem cells and their potential as cardiac therapeutics. *Circ Res* 2004;**95**:9–20.
17. Tan W, Martin D, Gutkind JS. The Galpha13-Rho signaling axis is required for SDF-1-induced migration through CXCR4. *J Biol Chem* 2006;**281**:39542–9.
18. Gerrits H, van Ingen Schenau DS, Bakker NE, van Disseldorp AJ, Strik A, Hermens LS, et al. Early postnatal lethality and cardiovascular defects in CXCR7-deficient mice. *Genesis* 2008;**46**:235–45.

19. Gilbert SF. *Developmental biology*. 8th ed. Sunderland, MA: Sinauer Associates Inc; 2006. pp. 483–488.

20. Kovacic JC, Moore J, Herbert A, Ma D, Boehm M, Graham RM. Endothelial progenitor cells, angioblasts, and angiogenesis—old terms reconsidered from a current perspective. *Trends Cardiovasc Med* 2008;**18**(2):45–51.

21. Asahara T, Murohara T, Sullivan A, et al. Isolation of putative progenitor endothelial cells for angiogenesis. *Science* 1997;**275**:964–7.

22. Vanhoutte PM, Shimokawa H, Tang EH, Feletou M. Endothelial dysfunction and vascular disease. *Acta Physiol (Oxf)* 2009;**196**(2):193–222.

23. Herodin F, Bourin P, Mayol JF, Lataillade JJ, Drouet M. Short-term injection of antiapoptotic cytokine combinations soon after lethal gamma-irradiation promotes survival. *Blood* 2003;**101**:2609–16.

24. Zaruba MM, Theiss HD, Vallaster M, Mehl U, Brunner S, David R, et al. Synergy between CD26/DPP-IV inhibition and G-CSF improves cardiac function after acute myocardial infarction. *Cell Stem Cell* 2009;**4**:313–23.

25. Salcedo R, Wasserman K, Young HA, Grimm MC, Howard OM, Anver MR, et al. Vascular endothelial growth factor and basic fibroblast growth factor induce expression of CXCR4 on human endothelial cells. In vivo neovascularization induced by stromal-derived factor-1alpha. *Am J Pathol* 1999;**154**:1125–35.

26. Mangi AA, Noiseux N, Kong D, He H, Rezvani M, Ingwall JS, et al. Mesenchymal stem cells modified with Akt prevent remodeling and restore performance of infarcted hearts. *Nat Med* 2003;**9**:1195–201.

27. Maurel A, Azarnoush K, Sabbah L, Vignier N, Le Lorc'h M, Mandet C, et al. Can cold or heat shock improve skeletal myoblast engraftment in infarcted myocardium? *Transplantation* 2005;**80**:660–5.

28. Wysoczynski M, Reca R, Ratajczak J, Kucia M, Shirvaikar N, Honczarenko M, et al. Incorporation of CXCR4 into membrane lipid rafts primes homing-related responses of hematopoietic progenitor/stem cells to an SDF-1 gradient. *Blood* 2005;**105**:40–8.

29. Kahn J, Byk T, Jansson-Sjostrand L, Petit I, Shivtiel S, Nagler A, et al. Overexpression of CXCR4 on human CD34+ progenitors increases their proliferation, migration, and NOD/SCID repopulation. *Blood* 2004;**103**:2942–9.

30. Peled A, Petit I, Kollet O, Magid M, Ponomaryov T, Byk T, et al. Dependence of human stem cell engraftment and repopulation of NOD/SCID mice on CXCR4. *Science* 1999;**283**:845–8.

31. Dar A, Kollet O, Lapidot T. Mutual, reciprocal SDF-1/CXCR4 interactions between hematopoietic and bone marrow stromal cells regulate human stem cell migration and development in NOD/SCID chimeric mice. *Exp Hematol* 2006;**34**:967–75.

32. Sakihama H, Masunaga T, Yamashita K, Hashimoto T, Inobe M, Todo S, et al. Stromal cell-derived factor-1 and CXCR4 interaction is critical for development of transplant arteriosclerosis. *Circulation* 2004;**110**:2924–30.

33. Sata M, Saiura A, Kunisato A, Tojo A, Okada S, Tokuhisa T, et al. Hematopoietic stem cells differentiate into vascular cells that participate in the pathogenesis of atherosclerosis. *Nat Med* 2002;**8**:403–9.

34. Zernecke A, Schober A, Bot I, von Hundelshausen P, Liehn EA, Mopps B, et al. SDF-1alpha/CXCR4 axis is instrumental in neointimal hyperplasia and recruitment of smooth muscle progenitor cells. *Circ Res* 2005;**96**:784–91.

35. Xiao Q, Ye S, Oberhollenzer F, Mayr A, Jahangiri M, Willeit J, et al. SDF1 gene variation is associated with circulating SDF1alpha level and endothelial progenitor cell number. the Bruneck Study. *PLoS One* 2008;**3**:e4061.

36. Walter DH, Haendeler J, Reinhold J, Rochwalsky U, Seeger F, Honold J, et al. Impaired CXCR4 signaling contributes to the reduced neovascularization capacity of endothelial progenitor cells from patients with coronary artery disease. *Circ Res* 2005;**97**:1142–51.

37. Kaminski A, Ma N, Donndorf P, Lindenblatt N, Feldmeier G, Ong LL, et al. Endothelial NOS is required for SDF-1/CXCR4-mediated peripheral endothelial adhesion of c-kit+ bone marrow stem cells. *Lab Invest* 2008;**88**:58–69.

38. Menasche P, Hagege A, Vilquin JT, Desnon M, Abergel E, Pouzet B, et al. Autologous skeletal myoblast transplantation for severe postinfarction left ventricular dysfunction. *J Am Coll Cardiol* 2003;**41**:1078–83.

39. Ghadge S, Muhlstedt S, Ozcelik C, Bader M. SDF-1α as a therapeutic stem cell homing factor in myocardial infarction. *Pharmacol Ther* 2011;**129**:97–108.

40. Shachinger V, Erbs S, Elasser A, Haberbosch W, Hambrecht R, Holschermann H, et al. Improved clinical outcome after intracoronary administration of bone-marrow-derived progenitor cells in acute myocardial infarction: final 1-year results of the REPAIR-AMI trial. *Eur Heart J* 2006;**27**:2775–83.

41. Dai B, Huang W, Xu M, Millard RW, Gao MH, Hammond K, et al. Reduced collagen deposition in infarcted myocardium facilitates induced pluripotent stem cell engraftment and angiomyogenesis for improvement of left ventricular function. *J Am Coll Cardiol* 2011;**58**(20):2118–27.

42. Kita-Matsuo H, Barcova M, Prigozhina N, Salomonis N, Wei K, Jacot JG, et al. Lentiviral vectors and protocols for creation of stable hESC lines for fluorescent tracking and drug resistance selection of cardiomyocytes. *PLoS One* 2009;**4**(4):e5046.

43. Wojakowski W, Kucia M, Zuba-Surma E, Jadczyk T, Ksiazek B, Ratajczak MZ, et al. Very small embryonic-like stem cells in cardiovascular repair. *Pharmacol Ther* 2011;**129**:21–8.

44. Wojakowski W, Tendera M, Kucia M, Zuba-Surma E, Milewski K, Wallace-Bradley D, et al. Cardiomyocyte differentiation of bone marrow-derived Oct-4+CXCR4+SSEA-1+ very small embryonic-like stem cells. *Int J Oncol* 2010;**37**(2):237–47.

45. Zuba-Surma EK, Guo Y, Taher H, Sanganalmath SK, Hunt G, Vincent RJ, et al. Transplantation of expanded bone marrow-derived very small embryonic-like stem cells (VSEL-SCs) improves left ventricular function and remodelling after myocardial infarction. *J Cell Mol Med* 2011;**15**:1319–28.

46. Benboubker L, Watier H, Carion A, Georget MT, Desbois I, Colombat P, et al. Association between the SDF1-3'A allele and high levels of CD34(+) progenitor cells mobilized into peripheral blood in humans. *Br J Haematol* 2001;**113**:247–50.

47. Zhang G, Nakamura Y, Wang X, Hu Q, Suggs LJ, Zhang J. Controlled release of stromal cell-derived factor-1 alpha in situ increases c-kit+ cell homing to the infarcted heart. *Tissue Eng* 2007;**13**:2063–71.

48. Fazel S, Cimini M, Liwen Chen L, Li S, Angoulvant D, Fedak P, et al. Cardioprotective c-kit+ cells are from the bone marrow and regulate the myocardial balance of angiogenic cytokines. *J Clin Invest* 2006;**116**:1865–77.

49. Cheng Z, Ou L, Zhou X, Li F, Jia X, et al. Targeted migration of mesenchymal stem cells modified with CXCR4 gene to infarcted myocardium improves cardiac performance. *Mol Ther* 2008;**16**:571–9.

50. Zhang D, Huang W, Dai B, Zhao T, Ashraf A, Millard RW, et al. Genetically manipulated progenitor cell sheet with diprotin A improves myocardial function and repair of infarcted hearts. *Am J Physiol Heart Circ Physiol* 2010;**299**:H1339–47.

51. Almsherqi ZA, McLachlan CS, Slocinska MB, Sluse FE, Navet R, Kocherginsky N, et al. Reduced cardiac output is associated with decreased mitochondrial efficiency in the non-ischemic ventricular wall of the acute myocardial-infarcted dog. *Cell Res* 2006;**16**:297–305.

52. Pasha Z, Wang Y, Sheikh R, Zhang D, Zhao T, Ashraf M. Preconditioning enhances cell survival and differentiation of stem cells during transplantation in infarcted myocardium. *Cardiovasc Res* 2008;**77**:134–42.

53. Grossman PM, Han Z, Palasis M, Barry JJ, Lederman RJ. Incomplete retention after direct myocardial injection. *Catheter Cardiovasc Interv* 2002;**55**:392–7.

54. Hofmann M, Wollert KC, Meyer GP, Menke A, Arseniev L, Hertenstein B, et al. Monitoring of bone marrow cell homing into the infarcted human myocardium. *Circulation* 2005;**111**:2198–202.

55. Jawad H, Ali NN, Lyon AR, Chen QZ, Harding SE, Boccaccini AR. Myocardial tissue engineering. A review. *J Tissue Eng Regen Med* 2007;**1**:327–42.

56. Taheri SA, Ashraf H, Merhige M, Miletich RS, Satchidanand S, Malik C, et al. Myoangiogenesis after cell patch cardiomyoplasty and omentopexy in a patient with ischemic cardiomyopathy. *Tex Heart Inst J* 2005;**32**(4):598–601.

57. O'Leary DP. Use of the greater omentum in colorectal surgery. *Dis Colon Rectum* 1999;**42**:533–9.

58. O'Shaughnessy L. Surgical treatment of cardiac ischemia. *Lancet* 1937;**232**:185–94.

59. Huang W, Zhang D, Millard RW, Wang T, Zhao T, Fan GC, et al. Gene manipulated peritoneal cell patch repairs infarcted myocardium. *J Mol Cell Cardiol* 2010;**48**:702–12.

60. Samuel CS, Cendrawan S, Gao XM, Ming Z, Zhao C, Kiriazis H, et al. Relaxin remodels fibrotic healing following myocardial infarction. *Lab Invest* 2011;**91**(5):675–90.

61. Venugopal JR, Prabhakaran MP, Mukherjee S, Ravichandran R, Dan K, Ramakrishna S. Biomaterial strategies for alleviation of myocardial infarction. *J R Soc Interface* 2012;**9**(66):1–19.

62. Ravichandran R, Venugopal JR, Sundarrajan S, Mukherjee S, Ramakrishna S. Poly(Glycerol sebacate)/gelatin core/shell fibrous structure for regeneration of myocardial infarction. *Tissue Eng Part A* 2011;**17**(9–10):1363–73.

63. Godier-Furnémont AF, Martens TP, Koeckert MS, Wan L, Parks J, Arai K, et al. Composite scaffold provides a cell delivery platform for cardiovascular repair. *Proc Natl Acad Sci USA* 2011;**108**(19):7974–9.

64. Pyo RT, Sui J, Dhume A, Palomeque J, Blaxall BC, Diaz G, et al. CXCR4 modulates contractility in adult cardiac myocytes. *J Mol Cell Cardiol* 2006;**41**:834–44.

65. Chen J, Chemaly E, Liang L, Kho C, Lee A, Park J, et al. Effects of CXCR4 gene transfer on cardiac function after ischemia–reperfusion injury. *Am J Pathol* 2010;**176**:1705–15.

66. Kreutziger KL, Murry CE. Engineered human cardiac tissue. *Pediatr Cardiol* 2011;**32**(3):334–41.

67. Bhatia SK. Tissue engineering for clinical applications. *Biotechnol J* 2010;**5**(12):1309–23.

68. Cui J, Li J, Mathison M, Tondato F, Mulkey SP, Micko C, et al. A clinically relevant large-animal model for evaluation of tissue-engineered cardiac surgical patch materials. *Cardiovasc Revasc Med* 2005;**6**:113–20.

69. Maeda M, Yamato M, Kanzaki M, Iseki H, Okano T. Thoracoscopic cell sheet transplantation with a novel device. *J Tissue Eng Regen Med* 2009;**3**:255–9.

70. Haider HKh, Ashraf M. Strategies to promote donor cell survival: combining preconditioning approach with stem cell transplantation. *J Mol Cell Cardiol* 2008;**45**:554–66.

71. Coffield VM, Jiang Q, Su L. A genetic approach to inactivating chemokine receptors using a modified viral protein. *Nat Biotechnol* 2003;**21**:1321–7.

72. Damas JK, Waehre T, Yndestad A, Ueland T, Muller F, Eiken HG, et al. Stromal cell-derived factor-1alpha in unstable angina: potential antiinflammatory and matrix-stabilizing effects. *Circulation* 2002;**106**:36–42.

Genetic Modification of Stem Cells for Cardiac, Diabetic, and Hemophilia Transplantation Therapies

M. Ian Phillips* and
Yaoliang Tang[†]

*Keck Graduate Institute, Claremont, California, USA

[†]Department of Internal Medicine, University of Cincinnati, Cincinnati, Ohio, USA

Gene modification of stem cells prior to their transplantation enhances their survival and increases their function in cell therapy. Like the famous Trojan horse, the gene-modified cell has to gain entrance into the host's walls and survive to deliver its transgene products. Using cellular, molecular, and gene manipulation techniques, the transplanted cell can be protected in a hostile environment from immune rejection, inflammation, hypoxia, and apoptosis. Genetic engineering to modify cells involves construction of functional gene sequences and their insertion into stem cells. The modifications can be simple reporter genes or complex cassettes with gene switches, cell-specific pro-moters, and multiple transgenes. We discuss methods to deliver and construct gene cassettes with viral and nonviral delivery, siRNA, and conditional Cre/Lox P. We review the current uses of gene-modified stem cells in cardiovascular disease, diabetes, and hemophilia.

Progress in Molecular Biology
and Translational Science, Vol. 111
http://dx.doi.org/10.1016/B978-0-12-398459-3.00013-7

285

I. Introduction

Stem cells could provide new therapies for diseases for which there is no cure or treatment except transplantation of organs such as the heart, kidney, liver, lung, and pancreas. However, experience with transplanting simple stem cells from bone marrow or embryonic (and induced pluripotent) stem cells shows a rapid mortality of the injected cells.[1-3] Also, cells do not arrive at the injury site in sufficient numbers to be effective. What is required is gene modification of stem cells so that they can survive longer and increase in concentration at the target site. Gene modification of stem cells is the Trojan horse approach to making stem cells more effective. The idea of entering a cell with the help of the cell and then attacking it from the inside is even more ancient than this story. It is the process used by viruses, plasmids, bacteria, and parasites. By using viruses and plasmids as the Trojan horse, we can modify genes or introduce new ones to make the cell die or survive longer, secrete proteins or switch off genes, differentiate or not differentiate.

Stem cells replicate throughout life so long as a few of them do not differentiate. The daughter cells that differentiate go on to become adult cells with specific functions in the body. Adult stem cells are generally multipotent and can transform into tissues that are produced within the organ or tissue in which they are found. Stem cells in the bone marrow can become osteocytes, blood cells, and lymph cells.[4] Cardiac-derived stem cells can become any of the cells that are part of a functioning heart including cardiomyocytes, neurons, or endothelial cells.[5-7] It is debatable whether bone marrow cells can turn into heart cells or any other cell that is not related to blood, bone, or lymph.[4,8]

II. Genetic Engineering

Genetic engineering of stem cells can be useful for increasing cell survival after transplantation, particularly into a hostile environment. Stem cells can be modified to deliver proteins to neighboring cells[9] or reduce graft-host rejection.

Obviously, both embryonic and adult stem cells have great potential for treatments involving cellular repair, replacement, and regeneration. One of the limitations of cell replacement therapy is that most of the grafted cells do not survive when grafted. Even if they are autologous or from a syngenic population, cell transplantation usually results in a loss of cells. Genetic engineering can increase survival of engrafted stem cells when transgenes are inserted into the cell to prevent or reduce apoptosis and inflammatory injury. In genetic modification, a gene cassette is constructed and loaded into a vector for entry into the cell. Once inside the cell, the gene construct can express or overexpress specific genes. The transgene expression can be constant leading to constitutive synthesis of specific proteins, or can be controlled by a gene switch. Constitutive activation of genes is nonphysiological, leading to overproduction of proteins, which downregulates receptors and renders the gene expression ineffective. A gene switch essentially makes the cell "intelligent" because it will then respond to a physiological stimulus, for example, low oxygen, high glucose levels, hormone concentrations, or to drugs or chemical agents.

A key principle in genetic engineering of cells is to mix and match modules of functional domains that are used in nature. Thus we can take a gene module used by yeast and a human virus module to create a chimeric regulator. Wang et al. first described a gene regulatory system for gene transfer by building a gene switch that responds to increases in mifepristone, a progesterone antagonist.[10] They fused a ligand-binding domain of a mutated human progesterone receptor to the transcriptional activator GAL4 DNA-binding domain of yeast and the protein VP16-activated domain of the herpes simplex virus. They demonstrated that this system could be activated by the exogenous administration of mifepristone (RU 486) at low doses to activate transcription of target genes. As described below, we developed a Vigilant Vector[11,12] with a gene switch similar to this concept, but built it to automatically respond to hypoxia, so that no exogenous drug was required to turn the system on or off.

A. Transgenics

A very well established gene modification of embryonic stem cells (ESCs) is in the production of transgenic animals. Transgenic mice with genes knocked out, or genes "knocked in" (where the number of copies of genes is increased),[13] are ubiquitous for gene studies in living animals. They have been very useful for studying the role of specific genes and practical for producing specific human proteins. The method involves harvesting ESCs from the inner cell mass of the blastocyst. Using recombinant DNA, a desired gene is made and inserted in a vector together with promoter sequences to regulate the gene expression. To replace a normal gene or knock one out, two drug-resistant genes are added to the cassette: a neo^R gene, which is resistant to the lethal effects of neomycin, and a thymidine kinase gene (TK), which phosphorylates gancyclovir. Most cells fail to take the vector inside their walls. These cells can be killed by neomycin or its analogs. A few of the remaining cells allow the vector in but the gene is inserted randomly. To avoid this, these cells are killed by gancyclovir. That leaves the cells in which homologous recombination has occurred. The normal gene has been knocked out, and a new, specific gene knocked in. These cells are then injected into a blastocyst, which is implanted in the uterus to produce offspring that can be bred. If the new gene is nonfunctional (i.e., a null allele), the function of the former gene may be revealed through breeding the mice with the knockout gene to homozygosity.

Ideally, the function of the missing gene should be obvious, as if a limb had been cut off. In actuality, several things can happen. The knocked out gene may prevent the embryo from developing (it is embryonically lethal), or the missing gene is fully compensated by other genes, or subtle changes occur in development or in different organs so that the effect is not obvious. Nevertheless, the technique has been hugely influential in revealing the functional effects of proteins especially where antibodies have not been developed. The opposite of knocking in copies of a gene has been used to reveal mechanisms of diseases caused by overexpression of a protein.[13] The transgenic animal approach requires going through embryonic development. This limits the technique when a knocked out gene is embryonically lethal. However, in a method first used by Gu et al.,[14] the Cre/Lox P system is able to induce the same mutation and avoid lethality.

B. Cre/Lox P System

To knock out a target gene in specific cell groups or tissue in adult animals, the Cre/lox P system is a suitable technique. It is based on the viral bacterial phage P1, which produces Cre, a recombinase enzyme. Cre cuts its viral DNA into packages. Cre cuts all the DNA out between two separate lox P sites. The DNA ends, each of which has a half lox P site, are then ligated by the recombinase. Gu et al.[14] used this principle with the strategy of a conventional transgenic mice, in which the Cre transgene plus a promoter was inserted by homologous recombination in a cell-specific manner. This mouse was crossed with a second mouse strain that had a target gene flanked by two lox P sites. In the offspring, the target gene was deleted only in the cells that contained Cre and the lox P "floxed" sequences. The target gene remained functional in all the other cells, and the animals survived development; thus, the function of the targeted gene in specific cells could be studied.

More recent developments have made the technique less laborious to use.[15,16] An example is a study by Sanniyha et al.[16] who made transgenic mice with lox P insertions flanking the gene for angiotensinogen. Angiotensinogen is a substrate for the enzyme renin and is one of the critical components for the synthesis of the peptide angiotensin. Instead of making a separate strain of Cre mice and proceeding with breeding, they simply injected Cre into the floxed mice. This had the advantage of being not only a time saving technique but also a new way to study genes with site-directed, conditional gene ablation in specific cells. As they were working on the brain, they were able to pinpoint anatomically a very small brain structure, the subfornical organ. By injecting Cre into the structure, they showed that angiotensin synthesis could be blocked and proved that it is synthesized in the brain.[17,18]

To inhibit synthesis of proteins by inhibiting gene translation, there are two methods, namely, antisense inhibition and RNA interference.

C. Antisense Inhibition

Antisense is based on the fact that messenger RNA (mRNA) is in the "sense" direction from 5′ to 3′. Antisense is a limited sequence of DNA in the antisense direction 3′–5′ designed from knowing the sequence of a target gene. Antisense oligonucleotides (AS-ODN) are usually built around the initiation codon of a gene (the AUG start site) and are shorter than the full-length gene. This is because the AS-ODN binds to part of the appropriate mRNA sequence and prevents the mRNA from translating the protein it would otherwise produce.[19–21]

For gene modification with antisense within a cell, a viral vector can be fitted with DNA in the antisense direction. We have designed these in the adeno-associated virus (AAV) and shown them to have long-lasting inhibitory effects on designated cell protein synthesis.[22] Antisense inhibition, although widely used in research and approved for clinical treatment,[23] is not perfect. When antisense is put into a cell, it competes with the cell's own mRNA copying machinery. The presence of AS-ODN may actually increase the number of cell-produced mRNA copies, thereby overcoming the endogenously administered AS-ODN. Because of this, antisense as a treatment has not proven to be a killer of cells and therefore not a revolutionary anticancer agent, as it had originally been hoped. However, antisense has played a pivotal role in leading to the next advance in cellular gene inhibition—RNA interference—and more recently, it has reemerged as antagomirs for inhibiting microRNA.

D. siRNA Gene Silencing

Fire and Mello[24] used antisense to study behavioral effects on the primitive worm *Caenorhabditis elegans*. They tested both sense RNA and antisense RNA on the worms, but found no effect of either. However, when they tested a combination of sense and antisense RNA, the worms started to twitch spontaneously. The gene that was holding back the twitching had been silenced. Fire and Mello had discovered gene silencing by double-stranded (ds) RNA, which acted as small interfering RNA (siRNA). RNA interference has become widely recognized as a biological mechanism for the regulation of gene expression and is used for intracellular inhibition. dsRNA is produced in the nucleus. In the cytoplasm, it binds to an enzyme, Dicer. Dicer literally dices up the dsRNA into short strands (15–20 nucleotides).

One of the strands is loaded into a protein complex, the RNA-induced silencing complex (RISC). The RISC now has the single strand of short RNA as a binding site to bind to a complementary sequence on the cell's mRNA. This binding leads to cleavage of mRNA, degrading the message and stopping it from translating a specific protein, and hence it is silenced.

RNA interference (RNAi) is a fundamental cellular process of gene regulation in the cells of animals and plants. Since both animals and plants are subject to diseases induced by viruses, RNAi may have evolved to protect cells from invasion by viruses. The genome of retroviruses is in the double strands of RNA. A retrovirus, lacking cellular mechanisms and DNA, injects its genomic dsRNA into a cell to reproduce itself using the DNA of the invaded cell. RNAi protects the cell by destroying the viral RNA through the RISC mechanism.

siRNA is more powerful than antisense in silencing genes, but it has its difficulties. It is not long lasting, it may silence off-target sites, and it has not been easy to inject systemically as a therapy. We have directly compared siRNA to antisense to inhibit the Beta-1 adrenergic receptor gene.[25] The effect was measured on blood pressure in hypertensive rats and on heart performance, because beta blockers have long been used for hypertension and heart failure treatments. The siRNA and AS-ODN were injected systemically in a lipofectamine vehicle. The result was a significantly better effect on lowering blood pressure and improving heart performance with the siRNA compared to the AS-ODN. Both approaches lasted about 1 week with a single injection.[25]

E. MicroRNA

MicroRNAs (miRs) offer completely new possibilities for gene modification, cell therapy, and drug development. They are involved in almost every biological process regulated by genes, and their absence or mutations could be the cause of many disease states from birth defects to cancer.

Although miRs were discovered over 20 years ago in *C. elegans*[26] and later found in mammals, we are still in the early stages of discovering how many there are, what they do, and how they do what they do. Over 500 miRs have been found in the human genome. A recent review in *Nature Reviews* suggests that miRs regulate one-third of human genes.[27] miRs have become recognized as a new class of gene regulators and are therefore important for gene modification of cells. miRs are small noncoding RNAs that modify gene expression by posttranscriptional inhibition of targeted mRNA. In the nucleus, miR is formed from introns and exons as "primary" or "pri-miRNA." But it is not a messenger RNA—it does not specify or generate a protein. The pri-miRNA, a folded-back structure of 60–70 nucleotides, is processed in the nucleus by the enzymes Drosha and Pasha. Drosha cuts out the stem-loop structure, which is the "pre-miRNA." The pre-miRNA is exported out of the nucleus by exportin and into the cytoplasm where it is diced up by the enzyme Dicer RNase III, as mentioned earlier in the siRNA process. The same effect occurs. Dicer cuts the stem loop into short length (19–25 nucleotides) inverted "mature miRNA." As with siRNA, one strand of the mature miR becomes part of the RISC and targets mRNA by binding to antisense complementary regions and cleaving or degrading the targeted mRNA. Multiple roles for miRNAs in gene regulation have been revealed by gene expression analysis polymerase chain reaction (PCR) and by transgenic mice with knockouts of specific miR. Expression arrays reveal specific miRs in different tissues and cells from invertebrates to humans. Many miRs (miR-1, miR-34, miR-60, miR-87, miR-124a) are highly conserved between vertebrates and invertebrates[28] including the small temporal (st)RNAs discovered in *C. elegans* (e.g., let-7 RNA, lin-4) that are similar to miRs in humans. As these stRNAs are critical for cell differentiation and timing of neural connections, the conservation may indicate functional evolution. A survey of mouse tissues with northern blotting[28] has shown that miR-1 is dominant in the heart (45%). In the liver, miR-122 was 72% of all miRs tested, and in the mouse brain miR-124a was profound.

Although the mechanism of miR action is principally inhibitory on target mRNA, which is essential for normal growth and differentiation in cell and tissue development, miRs can be involved in cancer. They can be depleted or suppressed, allowing oncogenes to be overproduced. Kumar *et al.*[29] recently showed that global suppression of miRs in various cancer cell lines increased cancer cell transformation and enhanced tumorogenesis in mice. To suppress miR, they targeted Drosha and Dicer with siRNA. Noncancerous cells did not become cancerous, but they did not grow. This suggests that increasing miRs could be a new approach to treating cancer either by suppressing oncogenes or by increasing differentiation.

F. Reporter Genes

Manipulation of genes in cells such as stem cells before transplantation can be done at several different levels of sophistication. If one wants to simply label cells with an internal marker so that the cells can be identified after transplantation, then a reporter gene such as green fluorescent protein (gfp), or luciferase (Luc), or beta-galactosidase (Lac Z) gene sequence can be inserted into any of the vectors described above. Each cell marker has its own advantage or disadvantage. Fluorescent labels are visible, but not easily quantified. However, a great advantage is that they are visible using a highly sensitive fluoroscope such that the cells can be located even under the skin in tissues and tumors. Luciferase has the advantage that it is quantifiable using luminometers, dual luciferase assays, or relative luciferase gene expression.

G. Cell-Specific Promoters

At the next level of sophistication, a cell- or tissue-specific promoter is spliced with the selected cell marker transgene so that the transgene can be observed to be expressed in one type of cell. Selecting the promoter raises some problems. A powerful promoter such as cytomegalovirus (CMV) drives a gene but is nonselective for tissue type. A more cell-specific promoter is likely to have weaker power and therefore there will be less gene marker expressed.

Improving promoter power without losing cell specificity is a challenge. Also, fitting a promoter into a cassette for a vector of small loading capacity, such as AAV, may require cutting the promoter into fragments and test-driving for specificity. For example, we used the myosin light chain-2v promoter (MLC-2v) in the heart,[12] which is 1700 bp long. In order to fit this promoter into the AAV, we reduced the MLC-2v to a 250-bp fragment that contained the heart-specific *cis* regulatory elements.[30] To further increase power, a promoter enhancer can be added to the effective promoter fragment. SV40 and Chick beta actin or globin[31] have been tried and found to increase expression by several fold. A feed-forward system can be introduced by making the product of cassette transgenes—the fusion proteins, feedback on an activating sequence to drive the promoter. Thus more and more fusion protein is produced. If this protein is also activating an upstream activating sequence in front of a transgene TATA box, more and more transgene expression will result.

For a therapeutic approach, the gene modification needs to have a gene switch added. A high level of expression powered by CMV or even the lower level of gene expression driven by a cell-specific promoter is constant. This constitutive gene expression could lead to a buildup of protein and unwanted side effects. The design of a transgene construct needs to have a regulator to control the amount of expression.

H. Gene Switches

Several different types of gene switches have been developed. Some require exogenous drugs to be applied to induce expression. These include the "Tet-on Tet-off" system using tetracyline as the switch inducer.[32] Ecdysone,[33] rapamycin,[34] and mifepristone[10] have also been used.

To make a transgene turn on and off to physiological stimulus requires genetic engineering of the cassette to include naturally occurring cellular regulatory elements. The cassette is constructed from modules that can be spliced together in a specific order. To illustrate, we have developed a "Vigilant Vector[TM]" that is switched on by hypoxia in heart cells.[35] To develop the hypoxia switch, there were several possibilities. The natural oxygen-sensitive elements of a cell had been worked out and sequenced.[36] The hypoxia regulatory element contains inducible factors HIF-1α and HIF-1β. When oxygen is low, the HIF-1α combines with the HIF-1β and the fusion product acts as a transcription factor in the nucleus to generate proteins in response to low oxygen, such as vascular endothelial growth factor (VEGF) and erythropoietin. By extracting the oxygen sensor in HIF-1α oxygen dependent domain (ODD) and installing it as the oxygen sensor of a chimeric gene, we could control the genetic response to hypoxia and avoid the production of these and other pro-teins. The ODD module was spliced in an activator system. The DNA-binding domain is the yeast GAL4, and the activating domain is the human p65 derived from human nuclear kappa B protein (Fig. 1). Under normal oxygen levels, the fusion protein of p65/ODD/Gal4 is ubiquitinated and the ubiquitin tail is the signal for transport to and destruction in proteosomes. But as oxygen decreases, a threshold is reached where the fusion protein is not ubiquitinated or destroyed and the GAL4 component of the protein binds to an inserted upstream activating sequence in front of the TATA box that activates gene expression. The lower the oxygen concentration, the greater the number of fusion proteins generated, exponentially increasing gene activity. Combined with a heart-specific promoter (MLC-2v), the whole system acts as a site-specific gene switch for hypoxia. Further, the system allows amplification of gene expression. In practical terms, when the transgene was heme-oxygenase-1, an antioxidant with anti-apoptotic and anti-inflammatory effects, it protected ischemic (mouse) hearts from heart failure.[37]

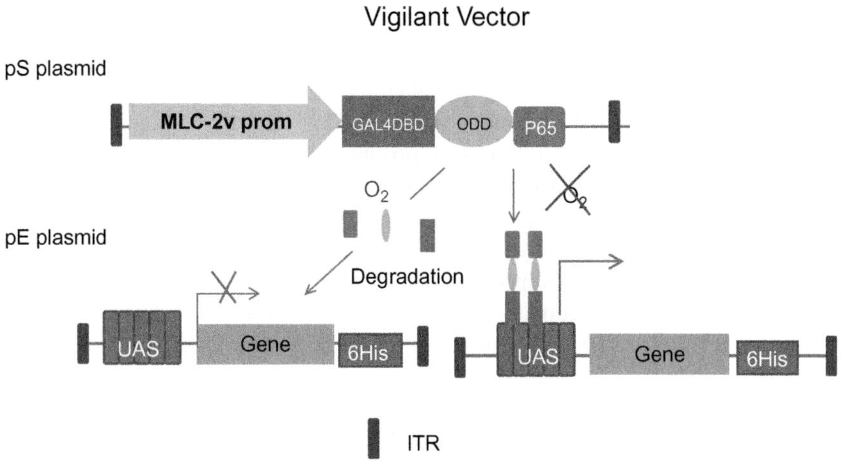

FIG. 1. Diagram of Vigilant Vector™ designed in this version as a cardiac-specific, hypoxia-regulated vector system that can amplify the power of promoters. There are two components delivered together: the sensor plasmid (pS) containing the gene switch for low oxygen (see text) and the MLC-2v promoter; and the effector plasmid (pE), which contains a GAL4 upstream activation sequence (UAS) in front of an adenovirus E1b TATA box and the Gene/6His fused gene. ITR, inverted terminal repeats for rAAV packaging. In normal oxygen, the fusion protein (GAL4ODD-p65AD) is ubiquitinated and destroyed in proteosomes. Under hypoxia, more and more fusion protein is made and not destroyed, so it acts as an amplifying system by binding to the UAS and activating the transgene. (See Color Insert.)

III. The Application of Genetic Modification of Stem Cells

A. Cardiology

1. Increase Graft Cell Survival

Adult stem cells have been proposed as a promising source for heart repair; however, cell-based therapy is confronted with the problem of poor survival in host myocardium.[38] Graft cell survival is limited by various pathological processes such as the inflammatory response, rejection, and ischemia–reperfusion. The survival of engrafted stem cells requires adaptation to the adverse environment in the ischemic myocardium. Different strategies have been developed to increase cell survival after grafting. Pharmacologic preconditioning has been tested successfully in skeletal myoblasts, showing, cytoprotective effects both *in vitro* and *in vivo*.[39] Suzuki *et al.*[40] reported that heat-shock treatment could improve the cell's tolerance to hypoxia–reoxygen insult *in vitro* and enhance its survival when grafted into the heart. Exploiting cell growth and apoptotic regulatory factors to enhance the proliferation of viable stem cells or confer apoptosis resistance to donor cells, by gene modification, is a potential way to improve cell transplant efficiency. Akt is a powerful survival signal in many systems.[41] Akt gene modification of stem cells has been reported by Mangi *et al.*[9] Their work demonstrated that a direct intramuscular injection of 5×10^6 Akt-engineered mesenchymal stromal cells (MSCs) improved the function of infarcted rat hearts. However, the overall application of constitutively active Akt gene may increase the risk of tumorogenesis.[42] HO-1 is the rate-limiting enzyme in the catabolism of heme, followed by production of biliverdin, free iron, and carbon monoxide (CO). All three byproducts exert beneficial actions that protect the cells from oxidative damage and death.[43] Hypoxia-inducible HO-1 plasmid modification of graft mesenchymal stem cells can protect cells from subsequent hypoxia injury *in vitro*, and improve graft cell survival in the ischemic myocardium *in vivo* via anti-inflammatory action and anti-apoptosis.[44] These findings underscore the role of HO-1 in protecting grafted cells from ischemia/inflammation-induced death.

2. Increased Angiogenesis in Ischemic Heart Disease

Myocardial ischemia associated with coronary artery disease is a leading cause of morbidity and mortality in the United States.[45] Although percutaneous transluminal angioplasty (PTCA) and operative coronary revascularization (CABG) procedures are effective for revascularization, there are increasing numbers of patients with extensive atherosclerotic coronary artery disease not amenable to traditional methods of revascularization. Several growth factors have appeared recently as adjuncts to regular revascularization, including VEGF.[46] Although viruses carrying the VEGF gene can maintain a therapeutic angiogenesis, VEGF expression is not under tight control and thus might cause unwanted side effects, such as angioma formation. To develop an approach for safe and long-lasting angiogenesis, we investigated neovascularization in ischemic myocardium via autologous MSC transplantation. Our findings suggest that bone-marrow-derived MSCs play a crucial role in improving regional blood flow in ischemic myocardium, and provide an optimal strategy for therapeutic angiogenesis by secreting a broad spectrum of angiogenic cytokines, including VEGF,[47] HGF,[48] bFGF,[49] and SDF-1α.[49] Increased blood supply from neovascularization would inhibit apoptosis and necrosis of hibernating and stunned myocardium in the border zone. Moreover, autologous MSCs have high proliferative and self-renewal capability, which is critical for maintaining the lasting effects fit for clinical treatment of patients with extensive atherosclerotic coronary disease.[47] Although autologous MSC transplantation can be administrated as "sole therapy" for neovascularization, many laboratories have developed strategies to use MSCs as vehicles for angiogenic gene therapy to enhance the benefits of neovascularization. Table I lists genetically modulated cells carrying exogenous genes encoding for angiogenic factors and MSCs, which have an inherent ability to secrete multiple paracrine factors to achieve superior revascularization. Lei et al.[56] have reviewed improvements in angiogenic outcome via delivery of multiple growth factors with synergic effects.

B. Gene-Modified Stem Cells to Form Surrogate β Cells for Treating Diabetes

A leading cause of type 1 diabetes is the failure of pancreatic islet β cells to survive and produce insulin. Current cell therapy relies mainly on replacement of functional insulin-producing pancreatic β cells via pancreatic islet transplantation. However, the shortage of donor cells as well as the number of donors required (3:1 recipient) limits the application of this treatment. The use of stem cells, a potential renewable source of pancreatic β-like cells, is currently being investigated as an alternative to isolated pancreatic islet transplantation for the treatment of type 1 diabetes mellitus. Stem-cell-derived insulin-producing cells could be a renewable source of insulin-producing cells for cell transplantation. To enhance the maturation process of human ESC (hESC)-derived insulin-producing cells, recent studies have used genetic manipulation methodologies to deliver specific pancreatic transcription factors or developmental control genes to hESCs.[57] They made hESCs "gain-of-function" by constitutively expressing two different transcription factors, Foxa2 and pancreatic duodenum homeobox protein-1(Pdx1). Foxa2 is found in the early endoderm layer[58] and is expressed at a very early stage in pancreas development.[59] Pdx1 is a pancreas-specific transcription factor expressed downstream of Foxa2 and specifically

TABLE I
Gene Modification of Stem Cells to Improve Angiogenesis in the Heart

Heart disease	Cells	Gene modification	Method of transplant	Efficacy	Adverse effects	Follow-up period (week)	References
Mouse MI	MSCs	Adenovirus-hVEGF$_{165}$	MSC i.m. w/cytokine mobilization	Superior therapeutic angiomyogenesis and LV-function recovery	None	4	50
Rat MI	Skeletal myoblasts	Nonviral hSDF-1α	Myoblast i.m.	Enhances angiomyogenesis	None	4	51
Rat MI	MSCs	Adenovirus-Akt + Ang-1	MSC i.m.	Enhanced cell survival, improved angiomyogenesis, and restored global cardiac function	None	4	52
Rat MI	MSCs	lentivirus-hSDF-1α	MSC i.v.	Enhanced angiogenesis effects and improve function	None	5	53
Rat MI	MSCs	Adenovirus-Ang-1	MSC i.m.	Improved angiogenesis and arteriogenesis effects	None	4	54
Pig chronic ischemia	MSCs	Adenovirus-Ang-1	MSC i.m.	Improvement of heart perfusion and function	None	4	55

involved in stem cell differentiation into β-cell progenitors.[60] Pdx1 binds and activates insulin promoter in β cells.[60,61] Their study demonstrated that the constitutive expression of Pdx1 enhances the differentiation of hESCs toward pancreatic endocrine and exocrine cell types. The expression of Pdx1 also increased the expression of several transcription factors that are downstream of Pdx1, such as Ngn3, PAX4, NKX2.2, and ISL1. However, this group also found that the expression of the insulin gene could be demonstrated only when the cells differentiated *in vivo* into teratomas; therefore, additional work is necessary to induce insulin expression by hESCs without forming teratomas.

The hESCs can be used as a source of cells for therapy in diabetes, but one major problem with using hESCs for β-cell transplantation is the immunological incompatibility between the cell donors and the recipients. The levels of MHC-I expression in hESCs increase after *in vitro* differentiation.[62] Therefore, the host immune system will recognize and attack foreign hESCs, leading to rejection of transplanted hESCs. To eliminate the problem of immunoincompatibility and the requirement for the classic immunosuppressive therapy employed for organ transplantation, multipotential stem cells in adult tissues may offer an alternative source as functional insulin-producing cells. Tang *et al.*[63] tested the possibility of reprogramming rat hepatic stem cell-like WB cells into functional insulin-producing cells by overexpression of Pdx1 via lentivirus. Their findings demonstrate that long-term, persistent expression of Pdx1 is effective in converting hepatic stem cells into pancreatic endocrine precursor cells, which upon transplantation into diabetic mice become functional insulin-producing cells and restore euglycemia. Apart from liver adult stem cells, human bone-marrow-derived MSCs (hMSCs) may be a source to produce insulin-producing cells. They are autologous and have rapid renewal capability, low discomfort for the donor, and low risk of graft versus host disease. Li Y *et al.*[64] proved that hMSCs can be induced to differentiate into functional insulin-producing cells by introduction of Pdx1 via recombinant adenoviral vector. Pdx1 gene-modified hMSCs expressed multiple islet-cell genes including neurogenin3 (Ngn3), insulin, GK, Glut2, and glucagon, and produced and released insulin/C-peptide in a weak glucose-regulated manner. Furthermore, Pdx1-modified hMSCs seemed to contribute to the regeneration of pancreatic islets after cell transplantation in streptozotocin (STZ)-induced diabetic mice. Euglycemia can be obtained within 2 weeks and maintained for at least 42 days after the Pdx1-modified hMSC transplantation. Transplanted cells were found in the kidney capsule of the recipient and expressed insulin at 2 weeks after cell transplantation. Therefore, hMSCs can be used as a potential cell source for cell replacement therapy in diabetes.

Ductal progenitor cells in the pancreas were also used for β-cell replacement because they were abundant in the pancreas of these patients. Noguchi *et al.*[65] used adenovirus to mediate PDX-1, Neurogenin3 (Ngn3), NeuroD, or

Pax4 expression in adult mouse and human duct cells, and found that NeuroD was the most effective inducer of insulin expression in primary duct cells. Their work suggested that the overexpression of transcription factors, especially NeuroD, facilitates pancreatic stem/progenitor cell differentiation into insulin-producing cells.

C. Gene-Modified Stem Cells for Stroke

About 700,000 Americans each year suffer a new or recurrent stroke. Strokes kill more than 150,000 people a year, that is about 1 of every 16 deaths. It is the third highest cause of death after diseases of the heart and cancer. Americans paid about $62.7 billion in 2007 for stroke-related medical costs and disability. Bone-marrow-derived stem cells have been demonstrated to cross the blood–brain barrier[66] and to differentiate into neurons and glia.[67] Transplantation of bone marrow stem cells in animal models of cerebral ischemia by either intracerebral or i.v. route has demonstrated therapeutic efficacy in reducing lesion size and improving functional outcome.[68–71] Although bone marrow stem cells have the potential to self-renew, these cells have reduced replicative capacity after about five cell doublings over the course of about 6 weeks in culture.[72] The limitation in life span of these cells is directly related to telomere shortening because of the lack of telomerase activity that is necessary for maintenance of telomeres,[73] and may limit the clinical application of bone marrow stem cells. Overexpression of hTERT (telomerase reverse transcriptase) has been demonstrated to increase or stabilize telomere length, and immortalize human cells.[74,75] The technology of hTERT immortalization could be used to improve stem cell expansion for subsequent therapeutic cell transplantation, especially important for aging patients with stroke. Recently, hTERT-immortalized hMSCs have been used in a rat cerebral ischemia model for brain functional repair.[76] In the experiment, hMSCs were isolated from healthy adult volunteers, and the primary MSCs were immortalized with hTERT-expressing retrovirus. The cell population was expanded in culture within 40 population doublings and intravenously delivered into rats 12 h after induction of transient middle cerebral artery occlusion (MCAO) to study their potential therapeutic benefit. Using histological assay and magnetic resonance spectroscopy, it was found that intravenous infusion of immortalized hMSCs 12 h after transient MCAO in the rat resulted in a reduction in infarction volume. More importantly, behavioral performance was seen to have improved in the hTERT-MSC-treated group using the treadmill test and the Morris water maze test. Therefore, hTERT modification of MSCs appears beneficial to ameliorate functional deficits after stroke and to enhance the efficacy of cell transplants.

MSCs were reported to promote neuronal cell survival and neurogenesis via secretion of a variety of neuroregulatory molecules, such as brain-derived neurotrophic factor (BDNF).[77] To further enhance this paracrine effect, Kurozumi et al.[78,79] transfected telomerized human MSCs with the BDNF gene via a fiber-mutant F/RGD adenovirus vector and investigated whether these cells contributed to improved functional recovery in a rat transient MCAO model. They found that BDNF production by MSC-BDNF cells was 23-fold greater than that seen in uninfected MSC. Rats that received MSC-BDNF showed significantly more functional recovery than did control rats following MCAO. Moreover, magnetic resonance imaging analysis revealed that the rats in the MSC-BDNF group exhibited a more significant recovery from ischemia after 7 and 14 days. The apoptotic cells in the ischemic boundary zone was significantly reduced in animals treated with MSC-BDNF compared to animals in the control group. Their findings suggest that BDNF gene modification of MSC may be used as a novel strategy for the treatment of stroke to promote functional recovery and reduce infarct size in cerebral ischemia.

D. Gene-Modified Stem Cells to Treat Hemophilia

Gene-modified bone marrow stem cell therapy approaches have been used to target the life-threatening genetic bleeding disorder, hemophilia. Hemophilia A is due to a mutation in the Factor VIII gene, and hemophilia B is due to a mutation in the Factor IX gene of the blood clotting cascade; hemophilia A is the more common one. Moayeri et al.[80] used hematopoietic stem cells (HSCs) to express coagulation factor VIII (FVIII) by an oncoretroviral vector. Transduced HSCs were transplanted into immunocompetent hemophilia A mice. Therapeutic levels of FVIII were detected in the serum of the transplant recipient for over 6 months. More importantly, there was only minor anti-FVIII inhibitor antibody production induced following transplantation of gene-modified HSCs. In a related study, Gangadharan et al.[81] compared the therapeutic effect of achieving sustained therapeutic levels of FVIII between gene-modified MSCs and HSCs. To test this, they used the retroviral-mediated porcine FVIII vector to genetically modify bone-marrow-derived MSCs and HSCs, and transplanted the cells into genetically immunocompetent hemophilia A mice. They found that the FVIII activity levels dropped rapidly and returned to baseline in the MSC group as a result of the formation of anti-porcine FVIII neutralizing antibodies; however, FVIII levels stayed high in mice treated with HSCs. They found that FVIII expression was sustained beyond 10 months because of immunologic tolerance. This investigation demonstrates that HSCs, but not MSCs, offer a sufficient and durable strategy for delivery of curative FVIII for treating hemophilia A.[81]

REFERENCES

1. Leor J, Aboulafia-Etzion S, Dar A, Shapiro L, Barbash IM, Battler A, et al. Bioengineered cardiac grafts: a new approach to repair the infarcted myocardium? *Circulation* 2000;**102**:III56–61.
2. Reinecke H, Murry CE. Taking the death toll after cardiomyocyte grafting: a reminder of the importance of quantitative biology. *J Mol Cell Cardiol* 2002;**34**:251–3.
3. Müller-Ehmsen J, Whittaker P, Kloner RA, Dow JS, Sakoda T, Long TI, et al. Survival and development of neonatal rat cardiomyocytes transplanted into adult myocardium. *J Mol Cell Cardiol* 2002;**34**:107–16.
4. Balsam LB, Wagers AJ, Christensen JL, Kofidis T, Weissman IL, Robbins RC. Haematopoietic stem cells adopt mature haematopoietic fates in ischaemic myocardium. *Nature* 2004;**428**:668–73.
5. Tang YL, Shen L, Qian K, Phillips MI. A novel two-step procedure to expand cardiac Sca-1+ cells clonally. *Biochem Biophys Res Commun* 2007;**359**:877–83.
6. Messina E, De Angelis L, Frati G, Morrone S, Chimenti S, Fiodaliso F, et al. Isolation and expansion of adult cardiac stem cells from human and murine heart. *Circ Res* 2004;**95**:852–4.
7. Moretti A, Caron L, Nakano A, Lam JT, Bernshausen A, Chen Y, et al. Multipotent embryonic isl1+ progenitor cells lead to cardiac, smooth muscle, and endothelial cell diversification. *Cell* 2006;**127**:1151–65.
8. Murry CE, Soonpaa MH, Reinecke H, Nakajima H, Nakajima HO, Rubart M, et al. Haematopoietic stem cells do not transdifferentiate into cardiac myocytes in myocardial infarcts. *Nature* 2004;**428**:664–8.
9. Mangi AA, Noiseux N, Kong D, He H, Rezvani M, Ingwall JS, et al. Mesenchymal stem cells modified with Akt prevent remodeling and restore performance of infarcted hearts. *Nat Med* 2003;**9**:1195–201.
10. Wang Y, O'Malley Jr. BW, Tsai SY, O'Malley BW. A regulatory system for use in gene transfer. *Proc Natl Acad Sci USA* 1994;**91**:8180–4.
11. Tang Y, Jackson M, Qian K, Phillips MI. Hypoxia inducible double plasmid system for myocardial ischemia gene therapy. *Hypertension* 2002;**39**:695–8.
12. Phillips MI, Tang Y, Schmidt-Ott K, Qian K, Kagiyama S. Vigilant vector: heart-specific promoter in an adeno-associated virus vector for cardioprotection. *Hypertension* 2002;**39**:651–5.
13. Smithies O. Many little things: one geneticist's view of complex diseases. *Nat Rev Genet* 2005;**6**:419–25.
14. Gu H, Marth JD, Orban PC, Mossmann H, Rajewsky K. Deletion of a DNA polymerase ß gene segment in T cells using cell type-specific gene targeting. *Science* 1994;**265**:103–6.
15. Sakai K, Agassandian K, Morimoto S, Sinnayah P, Cassell MD, Davisson RL, et al. Local production of angiotensin ll in the subfornical organ causes elevated drinking. *J Clin Invest* 2007;**117**:1088–95.
16. Sinnayah P, Lindley TE, Staber PD, Davidson BL, Cassell MD, Davisson RL. Targeted viral delivery of Cre recombinase induces conditional gene deletion in cardiovascular circuits of the mouse brain. *Physiol Genomics* 2004;**18**:25–32.
17. Phillips MI, Sumners C. Angiotensin ll in central nervous system physiology. *Regul Pept* 1998;**78**:1–11.
18. Phillips MI. A Cre-loxP solution for defining the brain renin-angiotensin system. Focus on "Targeted viral delivery of Cre recombinase induces conditional gene deletion in cardiovascular circuits of the mouse brain" *Physiol Genomics* 2004;**18**:1–3.
19. Zamecnik PC, Stephenson ML. Inhibition of Rous sarcoma virus replication and cell transformation by a specific oligodeoxynucleotide. *Proc Natl Acad Sci USA* 1978;**75**:280–4.

20. Wahlestedt C, Pich EM, Koob GF, Yee F, Heilig M. Modulation of anxiety and neuropeptide Y-Y1 receptors by antisense oligodeoxynucleotides. *Science* 1993;**259**:528–31.
21. Gyurko R, Wielbo D, Phillips MI. Antisense inhibition of AT1 receptor mRNA and angiotensinogen mRNA in the brain of spontaneously hypertensive rats reduces hypertension of neurogenic origin. *Regul Pept* 1993;**49**:167–74.
22. Kimura B, Mohuczy D, Tang X, Phillips MI. Attenuation of hypertension and heart hypertrophy by adeno-associated virus delivering angiotensinogen antisense. *Hypertension* 2001;**37**:376–80.
23. Crooke ST. Progress in antisense technology. *Annu Rev Med* 2004;**55**:61–95.
24. Fire A, Xu S, Montgomery MK, Kostas SA, Driver SE, Mello CC. Potent and specific genetic interference by double-stranded RNA in Caenorhabditis elegans. *Nature* 1998;**391**:806–11.
25. Arnold AS, Tang YL, Qian K, Shen L, Valencia V, Phillips MI, et al. Specific beta1-adrenergic receptor silencing with small interfering RNA lowers high blood pressure and improves cardiac function in myocardial ischemia. *J Hypertens* 2007;**25**:197–205.
26. Lee RC, Feinbaum RL, Ambros V. The C. elegans heterochronic gene lin-4 encodes small RNAs with antisense complementarity to lin-14. *Cell* 1993;**75**:843–54.
27. Esquela-Kerscher A, Slack FJ. Oncomirs—microRNAs with a role in cancer. *Nat Rev Cancer* 2006;**6**:259–69.
28. Lagos-Quintana M, Rauhut R, Yalcin A, Meyer J, Lendeckel W, Tuschi T. Identification of tissue-specific microRNAs from mouse. *Curr Biol* 2002;**12**:735–9.
29. Kumar MS, Lu J, Mercer KL, Golub TR, Jacks T. Impaired microRNA processing enhances cellular transformation and tumorigenesis. *Nat Genet* 2007;**39**:673–7.
30. Henderson SA, Spencer M, Sen A, Kumar C, Siddiqui MA, Chien KR. Structure, organization, and expression of the rat cardiac myosin light chain-2 gene. Identification of a 250-base pair fragment which confers cardiac-specific expression. *J Biol Chem* 1989;**264**:18142–8.
31. Rincón-Arano H, Valadez-Graham V, Guerrero G, Escamilla-Del-Arenal M, Recillas-Targa F. YY1 and GATA-1 interaction modulate the chicken 3′-side alpha-globin enhancer activity. *J Mol Biol* 2005;**349**:961–75.
32. Gossen M, Freundlieb S, Bender G, Muller G, Hillen W, Bujard H. Transcriptional activation by tetracyclines in mammalian cells. *Science* 1995;**268**:1766–9.
33. No D, Yao TP, Evans RM. Ecdysone-inducible gene expression in mammalian cells and transgenic mice. *Proc Natl Acad Sci USA* 1996;**93**:3346–51.
34. Pollock R, Issner R, Zoller K, Natesan S, Rivera VM, Clackson T. Delivery of a stringent dimerizer-regulated gene expression system in a single retroviral vector. *Proc Natl Acad Sci USA* 2000;**97**:13221–6.
35. Tang Y, Schmitt-Ott K, Qian K, Kagiyama S, Phillips MI. Vigilant vectors: adeno-associated virus with a biosensor to switch on amplified therapeutic genes in specific tissues in life-threatening diseases. *Methods* 2002;**28**:259–66.
36. Semenza GL. O2-regulated gene expression: transcriptional control of cardiorespiratory physiology by HIF-1. *J Appl Physiol* 2004;**96**:1173–7.
37. Tang YL, Tang Y, Zhang YC, Agarwal A, Kasahara H, Qian K, et al. A hypoxia-inducible vigilant vector system for activating therapeutic genes in ischemia. *Gene Ther* 2005;**12**:1163–70.
38. Zhang M, Methot D, Poppa V, Fujio Y, Walsh K, Murry CE. Cardiomyocyte grafting for cardiac repair: graft cell death and anti-death strategies. *J Mol Cell Cardiol* 2001;**33**:907–21.
39. Niagara MI, Haider HK, Jiang S, Ashraf M. Pharmacologically preconditioned skeletal myoblasts are resistant to oxidative stress and promote angiomyogenesis via release of paracrine factors in the infarcted heart. *Circ Res* 2007;**100**:545–55.
40. Suzuki K, Smolenski RT, Jayakumar J, Murtuza B, Brand NJ, Yacoub MH. Heat shock treatment enhances graft cell survival in skeletal myoblast transplantation to the heart. *Circulation* 2000;**102**:III216–21.

41. Datta SR, Brunet A, Greenberg ME. Cellular survival: a play in three Akts. *Genes Dev* 1999;**13**:2905–27.
42. Meuillet EJ, Mahadevan D, Vankayalapati H, Berggren M, Williams R, Coon A, et al. Specific inhibition of the Akt1 pleckstrin homology domain by D-3-deoxy-phosphatidyl-myo-inositol analogues. *Mol Cancer Ther* 2003;**2**:389–99.
43. Otterbein LE, Choi AM. Heme oxygenase: colors of defense against cellular stress. *Am J Physiol Lung Cell Mol Physiol* 2000;**279**:L1029–37.
44. Tang YL, Tang Y, Zhang YC, Qian K, Shen L, Phillips MI. Improved graft mesenchymal stem cell survival in ischemic heart with a hypoxia-regulated heme oxygenase-1 vector. *J Am Coll Cardiol* 2005;**46**:1339–50.
45. Lenfant C. NHLBI at 50: reflections on a half-century of research on the heart, lungs, and blood. National Heart, Lung, and Blood Institute. Interview by Charles Marwick. *JAMA* 1998;**280**:2062–4.
46. Koransky ML, Robbins RC, Blau HM. VEGF gene delivery for treatment of ischemic cardiovascular disease. *Trends Cardiovasc Med* 2002;**12**:108–14.
47. Tang YL, Zhao Q, Zhang YC, Cheng L, Liu M, Shi J, et al. Autologous mesenchymal stem cell transplantation induce VEGF and neovascularization in ischemic myocardium. *Regul Pept* 2004;**117**:3–10.
48. Lange C, Bassler P, Lioznov MV, Bruns H, Kluth D, Zander AR, et al. Hepatocytic gene expression in cultured rat mesenchymal stem cells. *Transplant Proc* 2005;**37**:276–9.
49. Tang YL, Zhao Q, Qin X, Shen L, Cheng L, Ge J, et al. Paracrine action enhances the effects of autologous mesenchymal stem cell transplantation on vascular regeneration in rat model of myocardial infarction. *Ann Thorac Surg* 2005;**80**:229–36.
50. Wang Y, Haider HK, Ahmad N, Xu M, Ge R, Ashraf M. Combining pharmacological mobilization with intramyocardial delivery of bone marrow cells over-expressing VEGF is more effective for cardiac repair. *J Mol Cell Cardiol* 2006;**40**:736–45.
51. Elmadbouh I, Haider HK, Jiang S, Idris NM, Lu G, Ashraf M. Ex vivo delivered stromal cell-derived factor-1alpha promotes stem cell homing and induces angiomyogenesis in the infarcted myocardium. *J Mol Cell Cardiol* 2007;**42**:792–803.
52. Jiang S, Haider HK, Idris NM, Salim A, Ashraf M. Supportive interaction between cell survival signaling and angiocompetent factors enhances donor cell survival and promotes angiomyogenesis for cardiac repair. *Circ Res* 2006;**99**:776–84.
53. Zhang M, Mal N, Kiedrowski M, Chacko M, Askari AT, Popovic ZB, et al. SDF-1 expression by mesenchymal stem cells results in trophic support of cardiac myocytes after myocardial infarction. *FASEB J* 2007;**21**:3197–207.
54. Sun L, Cui M, Wang Z, Feng X, Mao J, Chen P, et al. Mesenchymal stem cells modified with angiopoietin-1 improve remodeling in a rat model of acute myocardial infarction. *Biochem Biophys Res Commun* 2007;**357**:779–84.
55. Huang SD, Lu FL, Xu XY, Liu XH, Zhao XX, Zhao BZ, et al. Transplantation of angiogenin-overexpressing mesenchymal stem cells synergistically augments cardiac function in a porcine model of chronic ischemia. *J Thorac Cardiovasc Surg* 2006;**132**:1329–38.
56. Lei Y, Haider HK, Shujia J, Sim ES. Therapeutic angiogenesis. Devising new strategies based on past experiences. *Basic Res Cardiol* 2004;**99**:121–32.
57. Lavon N, Yanuka O, Benvenisty N. The effect of overexpression of Pdx1 and Foxa2 on the differentiation of human embryonic stem cells into pancreatic cells. *Stem Cells* 2006;**24**:1923–30.
58. Ang SL, Wierda A, Wong D, Stevens KA, Cascio S, Rossant J, et al. The formation and maintenance of the definitive endoderm lineage in the mouse: invovement of HNF3/forkhead proteins. *Development* 1993;**119**:1301–15.

59. Chakrabarti SK, Mirmira RG. Transcription factors direct the development and function of pancreatic beta cells. *Trends Endocrinol Metab* 2003;**14**:78–84.
60. Ahlgren U, Jonsson J, Jonsson L, Simu K, Edlund H. beta-cell-specific inactivation of the mouse Ipf1/Pdx1 gene results in loss of the beta-cell phenotype and maturity onset diabetes. *Genes Dev* 1998;**12**:1763–8.
61. Wang J, Elghazi L, Parker SE, Kizilocak H, Asano M, Sussel L, et al. The concerted activities of Pax4 and Nkx2.2 are essential to initiate pancreatic beta-cell differentiation. *Dev Biol* 2004;**266**:178–89.
62. Drukker M, Katz G, Urbach A, Schuldiner M, Markel G, Itskovitz-Eldor J, et al. Characterization of the expression of MHC proteins in human embryonic stem cells. *Proc Natl Acad Sci USA* 2002;**99**:9864–9.
63. Tang DQ, Lu S, Sun YP, Rodrigues E, Chou W, Yang C, et al. Reprogramming liver-stem WB cells into functional insulin-producing cells by persistent expression of Pdx1- and Pdx1-VP16 mediated by lentiviral vectors. *Lab Invest* 2006;**86**:83–93.
64. Li Y, Zhang R, Qiao H, Zhang H, Wang Y, Yuan H, et al. Generation of insulin-producing cells from PDX-1 gene-modified human mesenchymal stem cells. *J Cell Physiol* 2007;**211**:36–44.
65. Noguchi H, Xu G, Matsumoto S, Kaneto H, Kobayashi N, Bonner-Weir S, et al. Induction of pancreatic stem/progenitor cells into insulin-producing cells by adenoviral-mediated gene transfer technology. *Cell Transplant* 2006;**15**:929–38.
66. Eglitis MA, Mezey E. Hematopoietic cells differentiate into both microglia and macroglia in the brains of adult mice. *Proc Natl Acad Sci USA* 1997;**94**:4080–5.
67. Brazelton TR, Rossi FM, Keshet GI, Blau HM. From marrow to brain: expression of neuronal phenotypes in adult mice. *Science* 2000;**290**:1775–9.
68. Chen J, Li Y, Wang L, Lu M, Zhang X, Chopp M. Therapeutic benefit of intracerebral transplantation of bone marrow stromal cells after cerebral ischemia in rats. *J Neurol Sci* 2001;**189**:49–57.
69. Li Y, Chen J, Chen XG, Wang L, Gautam SC, Xu YX, et al. Human marrow stromal cell therapy for stroke in rat: neurotrophins and functional recovery. *Neurology* 2002;**59**:514–23.
70. Iihoshi S, Honmou O, Houkin K, Hashi K, Kocsis JD. A therapeutic window for intravenous administration of autologous bone marrow after cerebral ischemia in adult rats. *Brain Res* 2004;**1007**:1–9.
71. Baker AH, Sica V, Work LM, Williams-Ignarro S, de Nigris F, Lerman LO, et al. Brain protection using autologous bone marrow cell, metalloproteinase inhibitors, and metabolic treatment in cerebral ischemia. *Proc Natl Acad Sci USA* 2007;**104**:3597–602.
72. Kobune M, Kawano Y, Ito Y, Chiba H, Nakamura K, Tsuda H, et al. Telomerized human multipotent mesenchymal cells can differentiate into hematopoietic and cobblestone area-supporting cells. *Exp Hematol* 2003;**31**:715–22.
73. Harley CB. Telomere loss: mitotic clock or genetic time bomb? *Mutat Res* 1991;**256**:271–82.
74. Jiang XR, Jimenez G, Chang E, Frolkis M, Kusler B, Sage M, et al. Telomerase expression in human somatic cells does not induce changes associated with a transformed phenotype. *Nat Genet* 1999;**21**:111–4.
75. Bodnar AG, Ouellette M, Frolkis M, Holt SE, Chiu CP, Morin GB, et al. Extension of life-span by introduction of telomerase into normal human cells. *Science* 1998;**279**:349–52.
76. Honma T, Honmou O, Iihoshi S, Harada K, Houkin K, Hamada H, et al. Intravenous infusion of immortalized human mesenchymal stem cells protects against injury in a cerebral ischemia model in adult rat. *Exp Neurol* 2006;**199**:56–66.
77. Crigler L, Robey RC, Asawachaicharn A, Gaupp D, Phinney DG. Human mesenchymal stem cell subpopulations express a variety of neuro-regulatory molecules and promote neuronal cell survival and neuritogenesis. *Exp Neurol* 2006;**198**:54–64.

78. Kurozumi K, Nakamura K, Tamiya T, Kawano Y, Kobune M, Hirai S, et al. BDNF gene-modified mesenchymal stem cells promote functional recovery and reduce infarct size in the rat middle cerebral artery occlusion model. *Mol Ther* 2004;**9**:189–97.
79. Kurozumi K, Nakamura K, Tamiya T, Kawano Y, Ishii K, Kobune M, et al. Mesenchymal stem cells that produce neurotrophic factors reduce ischemic damage in the rat middle cerebral artery occlusion model. *Mol Ther* 2005;**11**:96–104.
80. Moayeri M, Hawley TS, Hawley RG. Correction of murine hemophilia A by hematopoietic stem cell gene therapy. *Mol Ther* 2005;**12**:1034–42.
81. Gangadharan B, Parker ET, Ide LM, Spencer HT, Doering CB. High-level expression of porcine factor VIII from genetically modified bone marrow-derived stem cells. *Blood* 2006;**107**:3859–64.

Role of Heat Shock Proteins in Stem Cell Behavior

GUO-CHANG FAN

Department of Pharmacology and Cell Biophysics, University of Cincinnati College of Medicine, Cincinnati, Ohio, USA

Stress response is well appreciated to induce the expression of heat shock proteins (Hsps) in the cell. Numerous studies have demonstrated that Hsps function as molecular chaperones in the stabilization of intracellular proteins, repairing damaged proteins, and assisting in protein translocation. Various kinds of stem cells (embryonic stem cells, adult stem cells, or induced pluripotent stem cells) have to maintain their stemness and, under certain circumstances, undergo stress. Therefore, Hsps should have an important influence on stem cells. Actually, numerous studies have indicated that some Hsps physically interact with a number of transcription factors as well as intrinsic and extrinsic signaling pathways. Importantly, alterations in Hsp expression have been demonstrated to affect stem cell behavior including self-renewal, differentiation, sensitivity to environmental stress, and aging. This chapter summarizes recent findings related to (1) the roles of Hsps in maintenance of stem cell dormancy, proliferation, and differentiation; (2) the expression signature of Hsps in embryonic/adult stem cells and differentiated stem cells; (3) the protective roles of Hsps in transplanted stem cells; and (4) the possible roles of Hsps in stem cell aging.

Progress in Molecular Biology
and Translational Science, Vol. 111
http://dx.doi.org/10.1016/B978-0-12-398459-3.00014-9

I. Introduction

Stem cells (SCs) are pluripotent cells that can self-renew and differentiate into various cell lineages.[1-5] In mammals, there are two major types of SCs: (1) embryonic stem cells (ESCs), which exist in the inner cell mass of blastocytes and can differentiate into all the specialized cells of organs and (2) adult SCs, identified in most of the tissues, which are believed to act as a repair system to replenish worn out or damaged tissues.[3-7] Indeed, SCs originating from bone marrow, adipose tissue, and blood have been successfully applied to the treatment of bone/blood-related cancers and cardiovascular disease.[8-16] However, the quantity of pluripotent SCs in adult tissues is very small. In addition, a large number of transplanted SCs die within hours/days, which eventually limits the efficacy of cellular therapy.[17-19] Therefore, there is an urgent need to expand our knowledge of SC behavior including SC renewal and differentiation, their survival and stress response, as well as their aging.

Heat shock response (HSR) was first described in 1962 by Ritossa[20] who observed that heat stress caused chromosomal puffs in the salivary glands of the *Drosophila* larva. Since then, HSR (also referred to as stress response) and a vast array of heat shock proteins (Hsps) have been widely identified in various organisms ranging from prokaryotic *Escherichia coli* to eukaryotic mammalians. Actually, the HSR is an evolutionarily conserved and homeostatic genetic response to a multitude of physiological, pathological, chemical, and environmental stresses.[21] This stress response is initiated by activation of the heat shock transcription factors (HSFs), leading to the enhanced transcription and translation of Hsps. To date, several members of the HSF family (HSFs 1–5) have been reported in vertebrates.[22-25] HSF1 can be activated by heat shock (HS), metals, and ethanol, and, consequently, by upregulation of cellular Hsps (i.e., Hsp40, Hsp70, and Hsp90). In contrast, HSF2 is activated only during differentiation and development. HSF3, a heat-responsive transcription factor, is expressed only in avians, whereas HSF4 exhibits tissue-specific expression (i.e., human heart, brain, skeletal muscle, and pancreas) and functions to repress the expression of Hsps (i.e., Hsp27, Hsp70, and Hsp90).[23-25] HSF5 is the most recently discovered HSF in human tissue[26] and its function and characteristics remain unknown.

Based on the molecular weight and/or the stimuli, Hsps are usually categorized into three major groups.[27-29] The first group consists of the high-molecular weight Hsps, including members of the Hsp110, Hsp90, Hsp70, and Hsp60 family. The second group includes the Hsps induced under conditions of glucose deprivation and are referred to as the "minor Hsps," including glucose-regulated proteins (GRPs) 34, 47, 56, 75, 78, 94, and 174. The third group consists of the small Hsps (sHsps) and includes at least 10 members (HspB1–B10) whose molecular weights range from 12 to 30 kDa. It is well accepted that Hsps function as molecular chaperones in the modulation of cellular protein conformational state and protein translocation (i.e., shuttle proteins from one compartment to another inside the cell, and transport old proteins to "garbage disposals" inside the cell).[21] While numerous intrinsic and extrinsic signals are involved in the tight regulation of SC self-renewal, differentiation, survival, and aging, increasing evidence indicates that Hsps play an essential role in the regulation of these behaviors of SCs.[30–35] This chapter summarizes recent findings related to (1) the expression profiles of Hsps in embryonic and adult SCs as well as differentiated SCs; (2) the possible role of Hsps in the maintenance of SC dormancy, proliferation, and differentiation; (3) the protective effects of Hsps in transplanted SCs; and (4) the possible role of Hsps in SC aging.

II. Hsps in the Modulation of SC Self-Renewal

SC self-renewal is a complex regulatory process that is dependent on multiple factors including the transcription factors Nanog,[36,37] Oct4,[38] Sox2,[39] and STAT3.[40,41] While STAT3 appears to be critical for the self-renewal of the mouse, but not human, ESCs,[40,41] the expression of Nanog, a key regulatory protein in the self-renewal of both human and mouse ESCs,[36,37,42] is reported to be controlled by STAT3. Hence, these transcription factors may work in concert to regulate SC renewal. Interestingly, certain Hsps are highly expressed in ESCs, interacting with these transcription factors, and are essential for normal cell development and functioning.[43–46] For instance, Hsp90 constitutes one of the most abundant cytosolic proteins commonly present in two major forms: the stress-induced Hsp90α[43] and the constitutively expressed Hsp90β.[44] Hsp90 functions to exclusively mediate the folding and maturation of transcription regulators and signal transducers; interactions have been mapped with approximately 300 client proteins. Recent studies further indicate that Hsp90 interacts with STAT3/Hop, and that the Hsp90/Hop complex is involved in the activation of STAT3 in mouse ESCs, suggesting that Hsp90 may modulate SC renewal. Indeed, Hsp90β knockout mouse embryos fail to develop past E9/9.5.[47] Moreover, Baharvand et al.[48] conducted an analysis for shared and unique chaperone expression profiles in ESC, mesenchymal stem cell (MSC), and neural stem cell (NSC). They found that all types of SCs shared high expression levels of Hsp70 protein 5 (HspA5), Hsp70 protein 8 (HspA8), and Hop (Stip1). ESCs showed a unique chaperone expression signature of Hsp70 protein 4 (HspA4), Hsp27 (HspB1), and Hsp90β (HspCb). These data indicate that high expression levels of chaperone and co-chaperones in SCs may exert a buffering effect against external and internal stressors, thereby maintaining their stemness.

On the other hand, cancer stem cells (CSCs) are characterized by enhanced expression of both oncogenes and Hsp genes.[49–51] Importantly, CSCs exhibit self-renewal capacity to promote tumor regeneration and metastasis, and contribute to radioresistance and chemoresistance.[52,53] Therefore, deciphering Hsp genes responsible for the maintenance of self-renewal and tumorigenicity in the CSCs is a crucial step in the development of new antitumor drugs. Wu et al.[54] recently identified that the expression of GRP78 (also referred to as HspA5), a central mediator of endoplasmic reticulum homeostasis, was significantly increased in isolated head and neck cancer stem cells (HN-CSCs). These HN-CSCs display cancer-initiating properties *in vitro* and *in vivo*. Downregulation of GRP78 reduced self-renewal properties and inhibited tumorigenicity of HN-CSCs, whereas overexpression of GRP78 in HN-CSCs enhanced *in vitro* malignant potentials. Indeed, numerous studies have demonstrated that GRP78 is required for the survival of ESCs and is also highly expressed in hematopoietic stem cells (HSCs).[55] Furthermore, Wu et al.[54] found that knockdown of GRP78 increased the expression of PTEN, Bax, and Caspase-3 but reduced the expression of p-MAPK in HN-CSCs. Additionally, they observed that downregulation of GRP78 reduced the expression of Nanog in HN-CSCs. Overall, these studies suggest that GRP78 may significantly contribute to signaling and transcriptional regulatory networks in CSC renewal.

III. Expression Profiles of Hsp in Differentiated SCs

ESCs have the capacity to self-renew in an undifferentiated state but may also differentiate into cell types representing the three embryonic germ lineages (ectoderm, mesoderm, and endoderm), thereby demonstrating their pluripotent potential.[1,5] Progress has been made toward the molecular understanding of the process of SC differentiation, using wide-scale transcriptional and translational profiling approaches. Especially, proteomics has emerged as a robust method for performing large-scale studies of proteins to complement high-throughput gene expression analyses at the RNA level. Accordingly, there have been an increasing number of publications comparing protein profiles in differentiated and undifferentiated SCs.[56–61] Saretzki et al.[58] reported that four Hsps, namely, HspB1 (Hsp27), HspA1a (Hsp70a1), HspA1b (Hsp70b1), and HspA9a (Hsp70a9, mortalin), were highly expressed in mESCs, which perhaps contributes to their remarkable resistance to potential genotoxic stress. However, these four Hsps were significantly decreased upon the differentiation of mESCs. Likewise, Baharvand et al.[59] also observed that Hsp70, Hsp60, and co-chaperone Hop were downregulated in differentiated mESC cell lines Royan B1 and D3. Similarly, during the directed differentiation of mouse ESCs into neurogenic embryoid bodies (NEBs), a significant reduction of Hsp27 and mortalin (mtHsp70/mot-2/GRP75) was observed.[60] Interestingly, during differentiation of human ESCs, the expression of *HSPA1b* decreased significantly whereas Hsp27 did not show any significant alteration.[61] Together, these studies suggest that ESC differentiation may involve the restriction of protein expression rather than the addition of newly expressed genes, and that alterations in certain Hsps could serve as differentiation markers of ESCs. Nonetheless, it is important to note that downregulation of Hsp may occur before detectable expression of traditional differentiation markers, for example, Sox1. Furthermore, in human adipose-derived adult stem cells, differentiation enhances the expression of Hsp20 (HspB6), Hsp27 (HspB1), Hsp60, and αB-crystallin (HspB5).[62] In contrast, the relative levels of Hsp47, Hsp70, Hsp90, and FK506-binding protein showed no change following adipogenesis.[62] These results suggest that alterations of Hsp expression during adult SC differentiation may be different from ESC differentiation and may be organ-specific, as reviewed below.

IV. Roles of Hsps in Tissue Genesis

A. Hsps in Hematopoietic Differentiation

Hsps have also been recognized to be involved in differentiation events of hematopoietic progenitors toward erythroid,[63] myeloid,[64] and lymphoid[65] lineages through a protective mechanism of peculiar transcription factors which are able to direct differentiation events in immature HSCs. For example, Ribeil *et al.*[63] have demonstrated the role of Hsp70 in the prevention of the inactivation of transcription factor GATA-1 modulated by a caspase-mediated proteolysis, indicating that Hsp70 might indirectly trigger erythroid differentiation. HSF2 is rapidly downregulated during megakaryocytic differentiation, but activated during hemin-induced differentiation of human erythroleukemia cells. HSPA8, referred to as Hsc70, has been demonstrated to play a role in cytokine-mediated HSC survival and to negatively influence the stability of proapoptotic Bim mRNA, thus preventing apoptosis in hematopoiesis and leukemogenesis.[66] Hsp27 and Hsp60 have been associated with myeloid commitment by functioning as cognate triggers for crucial monocyte–macrophage receptors, which induce activation and proliferation of these cells.[64,67] Loss of HSPA9b (mortalin) recapitulates the ineffective hematopoiesis of myelodysplastic syndrome in zebrafish.[68] Accordingly, Hsps could also be involved in differentiation events of umbilical cord blood-derived HSCs.[69] Together, these studies suggest that Hsp might represent a key target either for future investigations or for approaches addressing mechanisms underlying the induction of SC proliferation and modulation of their differentiation events.

B. Essential Role of αB-crystallin in Muscle Differentiation

Myogenic differentiation is under the strict control of myogenic transcription factors such as MyoD, Myf5, myogenin, and MRF4 in coordination with a second class of transcription factors including MEF2A-2D.[70–72] Mice lacking MyoD develop normally but are severely impaired in their ability to regenerate muscle after tissue injury.[73] MyoD is found to be a master regulator of muscle differentiation and is involved in the determination of the muscle cell lineage.[74] Importantly, the promoter of sHsps is predicted to contain at least one MyoD/myogenin binding site. Sugiyama et al.[75] reported that sHsps form two different complexes during muscle differentiation, implying their importance in muscle development. Of these sHsps, αB-crystallin (HspB5) seems to play an essential role in muscle differentiation and its maintenance. During muscle differentiation, the level of αB-crystallin goes up 10-fold, and mice lacking αB-crystallin have lower muscle mass.[76,77] A point mutation in αB-crystallin, namely, R120G, has been found to be associated with desmin-related myopathies, suggesting its importance in muscle tissue homeostasis.[78,79] Recently, Singh et al.[80] demonstrated that overexpression of αB-crystallin in C2C12 cells severely affected myotube formation. They further revealed that αB-crystallin modulates MyoD activity by the combined effect of its reduced synthesis and increased degradation, thus retarding muscle differentiation. αB-crystallin can also activate the NF-κB signaling pathway, which may cause a reduction of MyoD mRNA and augmentation of the cyclin D1, a growth stimulatory molecule.[81] In addition, overexpression of αB-crystallin delayed p21 expression and prevented the activation of caspase-3, which might contribute to the delay in the muscle differentiation process.[80] Taken together, these data suggest that αB-crystallin is essential for muscle differentiation because increased proliferation and/or degradation of MyoD by αB-crystallin are antagonistic to myogenesis.

C. Altered HSR in the Differentiated Neural Cells

The HSR (stress response) is a primary, evolutionarily conserved, and homeostatic genetic response to many stressors. Activation of HSR causes the induction of Hsps that function as molecular chaperones to facilitate the folding and refolding of non-native proteins and other proteins essential for recovery from cell damage.[21,82–84] Numerous studies have shown that HSR and the ability to induce Hsp expression are attenuated in various brain and spinal cord neurons in vivo and in vitro.[84,85] A recent study by Yang et al.[86] reported that differentiation of neural progenitor cells was associated with an attenuated HSR. In consistence with this, Battersby et al.[60] performed proteomic analysis in mESC-derived NEBs. They observed that, following mESC differentiation, expression of Hsp25 became excluded from neural precursors as well as other differentiating cells. The reduced HSR in the differentiated neural cells is likely to contribute to their vulnerability to stress-induced pathologies and death. As shown by Yang *et al.*[86], the differentiated NG108-15 cells (mouse neuroblastoma–glioma hybrid cell line used as the prototype neural progenitor cells) were more sensitive to the cytotoxic effect of an oxidizer, namely, sodium arsenite, than the undifferentiated cells. Furthermore, they observed that preconditioning with HS or forced expression of Hsp70 conferred cytoprotection when the differentiated cells were challenged. These data provide evidence of an attenuated HSR as part of the neural differentiation program. Hence, development of a treatment regimen or a pharmacological agent to rectify the defective HSR in the differentiated neuron would be a promising approach to mitigate the dire consequences of protein misfolding and boost neuron survival under stress.

On the other hand, NSCs, present in the developing mammalian CNS, are immature cells that are capable of self-renewal and can differentiate into neurons, astrocytes, or oligodendrocytes. NSCs play a pivotal role in the development of the CNS.[87] However, it has been reported that the exposure of fetuses or neonates to alcohol during the critical periods of brain development can result in fetal alcohol syndrome (FAS).[88] While numerous studies have been conducted to investigate the effects of ethanol on the brain tissue, the genes regulated by alcohol and their associated biological pathways remain obscure. Recently, Choi *et al.*[89] performed a genome-wide transcriptional analysis in differentiated NSCs derived from the forebrain of E15 mouse embryos with/out ethanol exposure. Interestingly, they observed that genes of the sHsp family including HspB2, HspB7, and HspB8, as well as Hsp40 (Hsp40 is responsible for Hsp70 recruitment and stimulates Hsp70 ATPase activity), were all upregulated in NSCs during short-term differentiation without ethanol exposure, and were more pronounced in the presence of ethanol, compared to undifferentiated NSCs. In contrast, HspB1 was downregulated during NSC differentiation regardless of ethanol exposure. Furthermore, they examined which HSF genes were expressed during the short-term differentiation of NSCs in the presence of ethanol, and observed that the gene expression patterns of HSF1 and HSF4 were not modulated by differentiation or by the presence of ethanol. Conversely, the gene expression levels of HSF2 and HSF5 increased during differentiation, and were further increased following treatment with ethanol. Interestingly, other studies have reported that HSF2 is not activated in response to classical stress but rather during tissue development.[90,91] Therefore, the results of Choi *et al.*[89] suggest that exposure to ethanol might activate HSF2 in the CNS, resulting in its translocation to the nucleus and the subsequent transcriptional regulation of various sHsp target genes. Overall, these data implicate the possibility that the upregulation of HspB2/7/8 and HSF2/5 in response to ethanol may function to reduce cell damage and may be a good candidate for the diagnosis of FAS.

D. Hsps in Mouse Trophoblast SC Differentiation

Increasing evidence has shown that Hsps are necessary for placental development. For example, targeted deletion of Hsp90 results in embryonic lethality due to placental defects as a result of a failure to form a placental labyrinth layer.[92] HSF1 is the major HSF controlling rapid Hsp induction. Mice mutant for *Hsf1* has a common phenotype that produces a thinning of the spongiotrophoblast layer, leading to embryonic lethality.[93] In the rat, HspB1 was strongly expressed in trophoblast giant cells beginning at Day 11 of gestation.[94] In human pregnancies, HspB1 can be detected in the intermediate trophoblast and syncytiotrophoblast cells during the first two trimesters.[95] In addition, HspB1 is more highly expressed in placenta from patients with severe preeclampsia compared with that from normal-term gestations.[96] A study by Winger *et al.*[97] demonstrated that expression levels of HspB1 significantly increased by the second day of trophoblast stem cell (TSC) differentiation and greater at Days 4 and 6 of differentiation. Meanwhile, levels of phosphorylated HspB1 and MAPKAPK2 protein were also increased during differentiation, and this increase could be inhibited when cells were differentiated in the presence of an inhibitor of MAPK14. These results indicate that the MAPK14–HspB1 axis may be playing an important role during TSC differentiation.

E. HS Enhances Inducible Differentiation of MSC/ESC

MSCs can differentiate into a variety of cell types including osteoblasts, adipocytes, chondrocytes, cardiomyocytes, and endothelial-like cells.[2–7] In order to improve the therapeutic use of SCs, it is important to find ways to enhance the qualitative and quantitative extent of differentiation, and thereby increase the chances of the cells to grow into the right kind of tissue or structure. Given that Hsps have been recognized to maintain cellular homoeostasis during changes in the microenvironment, Norgaard *et al.*[98] examined the effects of HS on the differentiation of human MSCs as measured by the synthesis of an osteoblastic marker, alkaline phosphatase, and cell matrix mineralization. They used a telomerase-immortalized human MSC line designated hMSC-TERT, which was treated with HS 3 h prior to adding the medium, promoting osteoblast differentiation and mineralized matrix formation, respectively. They observed that mild heat stress induced the levels of Hsps and enhanced the responsiveness of SCs to differentiate, and concluded that a 1-h HS of 42.5 °C is the most beneficial pretreatment to hMSC-TERT cells when inducing them to differentiate into osteoblasts. Afzal *et al.*[30] also determined the effects of HS on the neural differentiation of mouse embryonal carcinoma SCs (P19). They induced neural differentiation from pre-HS treated EBs (embryoid bodies) in the presene of 1 mM RA (retinoic acid) and 5% FBS (fetal bovine serum). Interestingly, they observed that a HS 12 h before the initiation of differentiation did not affect the expression of neuroectodermal and neural markers nestin and β-tubulin III, respectively. However, both markers increased remarkably when heat stress was induced after treatment and when EBs were formed. Taken together, these data suggest that Hsps could regulate inducible differentiation of SCs.

V. Protective Effects of Hsps in Transplanted SCs

SC therapy is quickly emerging as an attractive strategy in the treatment of neurodegenerative diseases, stroke, some striated muscle pathologies, and ischemic heart diseases. The promising therapeutic effect(s) of SCs is dependent on their capacity to survive and engraft in the target tissue. However, poor viability of transplanted SCs was observed, which yields only marginal beneficial effects. The Yacoub group showed that 7% of skeletal myoblasts survive 3 days after grafting into the infarcted mouse heart,[99] and Muller-Ehmsen et al.[100] found that 28% of implanted neonatal cardiomyocytes survived 1 week after grafting into normal rat hearts. Similarly, Freyman et al.[101] observed that ~5% of MSCs survived after transplanting into the infarcted porcine heart. Toma et al.[102] reported that less than 0.44% MSCs survive till day 4 after engraftment in an immunodeficient mouse heart model. Clearly, it is imperative to reinforce the transplanted cells to withstand the rigors of the microenvironment of the damaged tissue incurred from ischemia, inflammatory responses, and proapoptotic factors in order to acquire an effective therapeutic modality.

Hsps are well appreciated to provide protection against various stress stimuli such as high temperature, oxidative stress, pressure overload, and hypoxia/ischemia.[21] A number of protective phenotypes have been attributed to Hsps, including ion channel repair, redox balance restoration, inhibition of proinflammatory cytokines, and activation of survival signaling or inhibition of apoptotic pathways.[103,104] Therefore, increased Hsp levels would be expected to improve the survival of transplanted SCs and, consequently, their therapeutic efficacy. Indeed, myriad studies have utilized multiple approaches to enhance Hsp expression in SCs, which significantly augments SC therapeutic effectiveness, as reviewed below.

Over the past decade, many groups have consistently demonstrated that exposure of myoblasts to short-term HS resulted in the upregulation of Hsps including Hsp70/72 and HO-1 (Hsp32) without loss of their myogenic characteristics and improved the survival rate after autotransplantation in intact skeletal muscle.[105–109] For example, Laumonier et al.[109] demonstrated that lentivirus-mediated HO-1 gene transfer enhanced survival of porcine myogenic precursor cells after autologous transplantation. Likewise, exposure of rat MSCs to the chemical inducer 2 mM β-mercaptoethanol for 1–3 h followed by 24-h incubation led to Hsp72 synthesis and exhibited higher resistance to oxidative stress.[110] Preconditioning rat MSCs with recombinant human Hsp90α for 24 h was shown to protect against hypoxia and serum-deprivation-triggered cell death via activation of the Akt and ERK1/2 pathway.[111] Similarly, Chang et al.[112] directly delivered Hsp70 into rat MSCs, using the Hph-1 protein transduction domain ex vivo, and found that Hph-1-Hsp70-MSCs displayed higher viability and antiapoptotic properties upon hypoxic stress, evidenced by the upregulation of Bcl-2 expression, reduction of Bax, phosphorylation of JNK, and activity of caspase-3. In addition, Hph-1-Hsp70-MSCs retained function in the infarcted region of the myocardium and promoted the expression of cardiac-specific markers on MSCs. Transplantation of Hph-1-Hsp70-MSCs into infarcted hearts significantly improved left ventricular (LV) function and limited LV remodeling. Notably, neither cytotoxic effects on MSCs with significantly higher concentrations of Hph-1-Hsp70 nor tumorigenesis of Hph-1-Hsp70-treated MSCs were detected. Together, these data suggest that enhancing the viability and integration potential of MSCs by Hph-1-Hsp70 treatment before transplantation may provide a novel therapeutic opportunity for the treatment of myocardial infarction and end-stage cardiac failure. In consistence with this, Wang et al.[113] modified rat MSCs with the Hsp20 gene (Hsp20-MSCs), and observed that adenovirus-mediated Hsp20 delivery protected MSCs against H_2O_2-triggered cell death via activation of Akt and enhanced release of growth factors (VEGF, FGF-2, and IGF). Importantly, manipulation of MSCs with the Hsp20 gene augmented the paracrine effects of MSCs on infarcted hearts, leading to a significantly increased number of new blood vessels. In addition, in vivo transplantation of Hsp20-MSCs enhanced the survival rate of MSCs in the infarcted heart, which resulted in the attenuation of postinfarction LV remodeling and functional improvement.

Similar protective effects were also observed in Hsp27-overexpressing mouse ESCs, which correlates with resistance to the toxicity of cadmium chloride, mercuric chloride, cis-platinum (II)-diammine dichloride, and sodium arsenite, as well as in Hsp27-transfected hippocampal progenitor cells against glucocorticoid-evoked apoptotic cell death.[114,115]

In summary, increased levels of Hsps in SCs by HS or pharmacological preconditioning, direct delivery, or genetic modification has been validated as an effective approach for improvement of in vivo cell therapy efficacy, because most of the Hsps identified so far (i.e., Hsp70, Hsp20, Hsp27, HO-1) have displayed the features instrumental for antiapoptosis, antioxidative stress, anti-inflammation, enhanced paracrine effects, and neo-angiogenesis (Fig. 1). Therefore, Hsp-engineered SCs could become a promising means for the better survival of clinically transplanted stem cells.

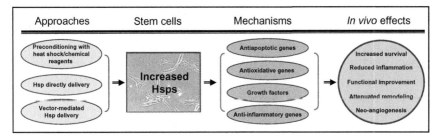

FIG. 1. A scheme for the approaches to increasing Hsp levels in stem cells and the beneficial consequences of *in vivo* transplantation. (For color version of this figure, the reader is referred to the online version of this chapter.

VI. Roles of Hsps in SC Aging

As reviewed above, mild heat stress or preconditioning enhances the defense capacity of SCs to adapt to environmental changes and increases implanted cell survival. However, this is not the case in aged cells/tissues. For example, McArdle *et al.*[116] demonstrated that the maximum force generated by muscles of young mice following a period of damaging lengthening contractions *in vivo* was severely diminished but then slowly recovered to reach pre-exercise values by approximately 28 days following the damaging contractions. In contrast, muscles of old mice maintained a significant deficit in maximum force generation at this time point, and this deficit persisted for at least 2 months. This inability to recover is found to be associated with the attenuated ability of muscles of old mice to generate Hsps in response to stress because muscles of old mice overexpressing Hsp70 recover successfully following damage. Another study by McArdle *et al.*[117] observed that the levels of HSF1 and HSF2 were temporally altered during muscle maturation. The HSF2 content of myotubes was significantly increased during the early stages of skeletal muscle regeneration. In contrast, the HSF1 content of myotubes remained relatively low until late during regeneration. These data suggest that abnormal activation of HSF1 may play a role in the defective regeneration seen in muscles of old mice.

Likewise, Stolzing *et al.*[33] showed that, in MSCs from young (6-week) rats, cold temperature (32 °C) significantly upregulated the expression of Hsp27, Hsp70, and Hsp90, but downregulated the Hsp60 expression. In contrast, aged MSCs from 56-week-old rats did not display or attenuate these alterations in Hsp levels. Indeed, Yu *et al.*[118] recently compared the ability of stress response in MSCs from rhesus monkeys among three age groups: young (<5 years), middle (8–10 years), and old (>12 years). They observed that MSCs exhibited decreased capacities for proliferation and differentiation with aging. While the levels of HSF-2, Hsp27, Hsp47, Hsp60, Hsp90A, and Hsp90B did not alter, there was a twofold decrease in the expression levels of HSF-1 and a fourfold reduction in mRNA levels of Hsp70 in middle-age and old MSCs compared to young MSCs. Together, these results indicate that SCs derived from aged tissue may down-regulate the Hsp expression and impair the capacity of stress response.

VII. Conclusions

A multitude of physiologic stresses and diseases result in cellular protein damage and misfolded protein structure, leading to the alteration of cell behavior/phenotype and dysfunction. The cell has evolved a number of protective means to maintain homeostatsis and promote survival during these periods of environmental stress and disease. One of the most highly conserved mechanisms of cellular protection involves the expression of a class of heat shock or stress proteins (Hsps). Hsps are molecular chaperones whose functions include assistance with native protein folding, maintenance of multiprotein complexes, intraorganellar protein shuttling, and degradation of senescent proteins. In general, various kinds of SCs (ESCs, adult SCs, induced pluripotent SCs) have to maintain their stemness and, under certain circumstances, undergo stress. Therefore, Hsps should be playing an important role in SCs. As reviewed above, ESCs contain abundant Hsps and display resistance to stress. Furthermore, Hsps interact with transcript factors (i.e., STAT3, Nanog, Oct4) and signaling pathways, which may modulate SC differentiation and proliferation. Importantly, SCs originating from aged tissues/organs exhibit impaired stress response and reduced levels of Hsps, which consequently affects SC behavior. Finally, considering the poor survival rate of transplanted SCs, upregulation of cellular Hsp levels should be a promising approach to the improvement of SC therapy.

REFERENCES

1. Denker HW. Potentiality of embryonic stem cells: an ethical problem even with alternative stem cell sources. *J Med Ethics* 2006;**32**(11):665–71.
2. Ratajczak MZ, Zuba-Surma EK, Wysoczynski M, Wan W, Ratajczak J, Wojakowski W, et al. Hunt for pluripotent stem cell—regenerative medicine search for almighty cell. *J Autoimmun* 2008;**30**(3):151–62.
3. Kucia M, Ratajczak MZ. Stem cells as a two edged sword—from regeneration to tumor formation. *J Physiol Pharmacol* 2006;**57**(Suppl. 7):5–16.
4. Yu J, Thomson JA. Pluripotent stem cell lines. *Genes Dev* 2008;**22**(15):1987–97.
5. Condic ML, Rao M. Alternative sources of pluripotent stem cells: ethical and scientific issues revisited. *Stem Cells Dev* 2010;**19**(8):1121–9.
6. Motaln H, Schichor C, Lah TT. Human mesenchymal stem cells and their use in cell-based therapies. *Cancer* 2010;**116**(11):2519–30.

7. Salem HK, Thiemermann C. Mesenchymal stromal cells: current understanding and clinical status. *Stem Cells* 2010;**28**(3):585–96.
8. Trounson A, Thakar RG, Lomax G, Gibbons D. Clinical trials for stem cell therapies. *BMC Med* 2011;**9**:52.
9. Bladé J, Rosiñol L, Cibeira MT, Rovira M, Carreras E. Hematopoietic stem cell transplantation for multiple myeloma beyond 2010. *Blood* 2010;**115**(18):3655–63.
10. Madonna R, Geng YJ, De Caterina R. Adipose tissue-derived stem cells: characterization and potential for cardiovascular repair. *Arterioscler Thromb Vasc Biol* 2009;**29**(11):1723–9.
11. Burt RK, Loh Y, Pearce W, Beohar N, Barr WG, Craig R, et al. Clinical applications of blood-derived and marrow-derived stem cells for nonmalignant diseases. *JAMA* 2008;**299** (8):925–36.
12. Velazquez OC. Angiogenesis and vasculogenesis: inducing the growth of new blood vessels and wound healing by stimulation of bone marrow-derived progenitor cell mobilization and homing. *J Vasc Surg* 2007;**45**(Suppl. A):A39–47.
13. Tendera M, Wojakowski W. Clinical trials using autologous bone marrow and peripheral blood-derived progenitor cells in patients with acute myocardial infarction. *Folia Histochem Cytobiol* 2005;**43**(4):233–5.
14. Steward CG, Jarisch A. Haemopoietic stem cell transplantation for genetic disorders. *Arch Dis Child* 2005;**90**(12):1259–63.
15. Lee N, Thorne T, Losordo DW, Yoon YS. Repair of ischemic heart disease with novel bone marrow-derived multipotent stem cells. *Cell Cycle* 2005;**4**(7):861–4.
16. Dzau VJ, Gnecchi M, Pachori AS, Morello F, Melo LG. Therapeutic potential of endothelial progenitor cells in cardiovascular diseases. *Hypertension* 2005;**46**(1):7–18.
17. Haider HKh, Ashraf M. Bone marrow cell transplantation in clinical perspective. *J Mol Cell Cardiol* 2005;**38**(2):225–35.
18. Pasha Z, Wang Y, Sheikh R, Zhang D, Zhao T, Ashraf M. Preconditioning enhances cell survival and differentiation of stem cells during transplantation in infarcted myocardium. *Cardiovasc Res* 2008;**77**(1):134–42.
19. Haider HKh, Ashraf M. Strategies to promote donor cell survival: combining preconditioning approach with stem cell transplantation. *J Mol Cell Cardiol* 2008;**45**(4):554–66.
20. Ritossa FM. Behavior of RNA and DNA synthesis at the puff level in salivary gland chromosomes of Drosophila. *Exp Cell Res* 1964;**36**:515–23.
21. Benjamin IJ, McMillan DR. Stress (heat shock) proteins: molecular chaperones in cardiovascular biology and disease. *Circ Res* 1998;**83**:117–32.
22. Wu C. Heat shock transcription factors: structure and regulation. *Annu Rev Cell Dev Biol* 1995;**11**:441–69.
23. Pirkkala L, Nykanen P, Sistonen L. Roles of the heat shock transcription factors in regulation of the heat shock response and beyond. *FASEB J* 2001;**15**:1118–31.
24. Yamamoto N, Takemori Y, Sakurai M, Sugiyama K. Differential recognition of heat shock elements by members of the heat shock transcription factor family. *FEBS J* 2009;**276**:1962–74.
25. Nakai A, Tanabe M, Kawazoe Y, Inazawa J, Morimoto RI, Nagata K. HSF4, a new member of the human heat shock factor family which lacks properties of a transcriptional activator. *Mol Cell Biol* 1997;**17**:469–81.
26. Ota T, Suzuki Y, Nishikawa T, Otsuki T, Sugiyama T, Irie R, et al. Complete sequencing and characterization of 21, 243 full-length human cDNAs. *Nat Genet* 2004;**36**:40–5.
27. Sciandra JJ, Subjeck JR. The effects of glucose on protein synthesis and thermosensitivity in Chinese hamster ovary cells. *J Biol Chem* 1983;**258**:12091–3.
28. Kappé G, Franck E, Verschuure P, Boelens WC, Leunissen JA, de Jong WW. The human genome encodes 10 alpha-crystallin-related small heat shock proteins: HspB1-10. *Cell Stress Chaperones* 2003;**8**:53–61.

29. Fan GC, Kranias EG. Small heat shock protein 20 (HspB6) in cardiac hypertrophy and failure. *J Mol Cell Cardiol* 2011;**51**:574–7.

30. Afzal E, Ebrahimi M, Najafi SM, Daryadel A, Baharvand H. Potential role of heat shock proteins in neural differentiation of murine embryonal carcinoma stem cells (P19). *Cell Biol Int* 2011;**35**(7):713–20.

31. Chen TH, Kambal A, Krysiak K, Walshauser MA, Raju G, Tibbitts JF, et al. Knockdown of Hspa9, a del(5q31.2) gene, results in a decrease in hematopoietic progenitors in mice. *Blood* 2011;**117**(5):1530–9.

32. Turturici G, Geraci F, Candela ME, Cossu G, Giudice G, Sconzo G. Hsp70 is required for optimal cell proliferation in mouse A6 mesoangioblast stem cells. *Biochem J* 2009;**421** (2):193–200.

33. Stolzing A, Sethe S, Scutt AM. Stressed stem cells: temperature response in aged mesenchymal stem cells. *Stem Cells Dev* 2006;**15**(4):478–87.

34. Patterson ST, Li J, Kang JA, Wickrema A, Williams DB, Reithmeier RA. Loss of specific chaperones involved in membrane glycoprotein biosynthesis during the maturation of human erythroid progenitor cells. *J Biol Chem* 2009;**284**(21):14547–57.

35. Sauvageot CM, Weatherbee JL, Kesari S, Winters SE, Barnes J, Dellagatta J, et al. Efficacy of the HSP90 inhibitor 17-AAG in human glioma cell lines and tumorigenic glioma stem cells. *Neuro Oncol* 2009;**11**(2):109–21.

36. Mitsui K, Tokuzawa Y, Itoh H, Segawa K, Murakami M, et al. The homeoprotein nanog is required for maintenance of pluripotency in mouse epiblast and ES cells. *Cell* 2003;**113**:631–42.

37. Chambers I, Colby D, Robertson M, Nichols J, Lee S, et al. Functional expression cloning of Nanog, a pluripotency sustaining factor in embryonic stem cells. *Cell* 2003;**113**:643–55.

38. Nichols J, Zevnik B, Anastassiadis K, Niwa H, Klewe-Nebenius D, et al. Formation of pluripotent stem cells in the mammalian embryo depends on the POU transcription factor Oct4. *Cell* 1998;**95**:379–91.

39. Botquin V, Hess H, Fuhrmann G, Anastassiadis C, Gross MK, et al. New POU dimmer configuration mediates antagonistic control of an osteopontin preimplantation enhancer by Oct-4 and Sox-2. *Genes Dev* 1998;**12**:2073–90.

40. Niwa H, Burdon T, Chambers I, Smith A. Self-renewal of pluripotent embryonic stem cells is mediated via activation of STAT3. *Genes Dev* 1998;**12**:2048–60.

41. Raz R, Lee CK, Cannizaro LA, D'eustachio P, Levy DE. Essential role of STAT3 for embryonic stem cell pluripotency. *Proc Natl Acad Sci USA* 1999;**96**:2846–51.

42. Hyslop L, Stojkovic M, Armstrong L, Walter T, Stojkovic P, et al. Downregulation of NANOG induces differentiation of human embryonic stem cells to extraembryonic lineages. *Stem Cells* 2005;**23**:1035–43.

43. Sreedhar AS, Kalmar E, Csermely P, Shen YF. Hsp90 isoforms: functions, expression and clinical importance. *FEBS Lett* 2004;**562**:11–5.

44. Lai BT, Chin NW, Stanek AE, Keh W, Lanks KW. Quantitation and intracellular localization of the 85 K heat shock protein by using monoclonal and polyclonal antibodies. *Mol Cell Biol* 1984;**4**:2802–10.

45. Richter K, Buchner J. Hsp90: chaperoning signal transduction. *J Cell Physiol* 2001;**188**:281–90.

46. Setati MM, Prinsloo E, Longshaw VM, Murray PA, Edgar DH, Blatch GL. Leukemia inhibitory factor promotes Hsp90 association with STAT3 in mouse embryonic stem cells. *IUBMB Life* 2010;**62**(1):61–6.

47. Voss AK, Thomas T, Gruss P. Mice lacking Hsp90β fail to develop a placental labyrinth. *Development* 2000;**127**:1–11.

48. Baharvand H, Fathi A, Van Hoof D, Salekdeh GH. Concise review: trends in stem cell proteomics. *Stem Cells* 2007;**25**:1888–903.

49. Bensaude O, Morange M. Spontaneous high expression of heat-shock proteins in mouse embryonal carcinoma cells and ectoderm from day 8 mouse embryo. *EMBO J* 1983;**2** (2):173–7.
50. Creagh EM, Sheehan D, Cotter TG. Heat shock proteins—modulators of apoptosis in tumour cells. *Leukemia* 2000;**14**(7):1161–73.
51. Setroikromo R, Wierenga PK, van Waarde MA, Brunsting JF, Vellenga E, Kampinga HH. Heat shock proteins and Bcl-2 expression and function in relation to the differential hyperthermic sensitivity between leukemic and normal hematopoietic cells. *Cell Stress Chaperones* 2007;**12**(4):320–30.
52. Lathia JD, Venere M, Rao MS, Rich JN. Seeing is believing: are cancer stem cells the Loch Ness monster of tumor biology? *Stem Cell Rev* 2011;**7**(2):227–37.
53. Gil J, Stembalska A, Pesz KA, Sasiadek MM. Cancer stem cells: the theory and perspectives in cancer therapy. *J Appl Genet* 2008;**49**(2):193–9.
54. Wu MJ, Jan CI, Tsay YG, Yu YH, Huang CY, Lin SC, et al. Elimination of head and neck cancer initiating cells through targeting glucose regulated protein78 signaling. *Mol Cancer* 2010;**9**:283.
55. Luo S, Mao C, Lee B, Lee AS. GRP78/BiP is required for cell proliferation and protecting the inner cell mass from apoptosis during early mouse embryonic development. *Mol Cell Biol* 2006;**26**:5688–97.
56. Guo X, Ying W, Wan J, Hu Z, Qian X, Zhang H, et al. Proteomic characterization of early-stage differentiation of mouse embryonic stem cells into neural cells induced by all-trans retinoic acid in vitro. *Electrophoresis* 2001;**22**(14):3067–75.
57. Wang D, Gao L. Proteomic analysis of neural differentiation of mouse embryonic stem cells. *Proteomics* 2005;**5**(17):4414–26.
58. Saretzki G, Armstrong L, Leake A, Lako M, von Zglinicki T. Stress defense in murine embryonic stem cells is superior to that of various differentiated murine cells. *Stem Cells* 2004;**22**(6):962–71.
59. Baharvand H, Fathi A, Gourabi H, Mollamohammadi S, Salekdeh GH. Identification of mouse embryonic stem cell-associated proteins. *J Proteome Res* 2008;**7**:412–23.
60. Battersby A, Jones RD, Lilley KS, McFarlane RJ, Braig HR, et al. Comparative proteomic analysis reveals differential expression of Hsp25 following the directed differentiation of mouse embryonic stem cells. *Biochim Biophys Acta* 2007;**1773**:147–56.
61. Saretzki G, Walter T, Atkinson S, Passos JF, Bareth B, Keith WN, et al. Downregulation of multiple stress defense mechanisms during differentiation of human embryonic stem cells. *Stem Cells* 2008;**26**(2):455–64.
62. DeLany JP, Floyd ZE, Zvonic S, Smith A, Gravois A, Reiners E, et al. Proteomic analysis of primary cultures of human adipose-derived stem cells: modulation by Adipogenesis. *Mol Cell Proteomics* 2005;**4**(6):731–40.
63. Ribeil JA, Zermati Y, Vandekerckhove J, et al. Hsp70 regulates erythropoiesis by preventing caspase-3-mediated cleavage of GATA-1. *Nature* 2006;**445**:102–5.
64. Kol A, Lichtman AH, Finberg RW, Libby P, Kurt-Jones EA. Cutting edge: heat shock protein (HSP) 60 activates the innate immune response: CD15 is an essential receptor for HSP60 activation of mononuclear cells. *J Immunol* 2000;**164**:13–7.
65. Ripley BJ, Stephanou A, Isenberg DA, Latchman DS. Interleukin-10 activates heat-shock protein 90beta gene expression. *Immunology* 1999;**97**:226–31.
66. Matsui H, Asou H, Inaba T. Cytokines direct the regulation of Bim mRNA stability by heat-shock cognate protein 70. *Mol Cell* 2007;**25**:99–112.
67. Garcia-Bermejo L, Vilaboa NC, Perez C, deBlas E, Calle C, Aller P. Modulation of hsp70 and hsp27 gene expression by the differentiation inducer sodium butyrate in U-937 human promonocytic leukemia cells. *Leuk Res* 1995;**19**:713–8.

68. Craven SE, French D, Ye W, de Sauvage F, Rosenthal A. Loss of Hspa9b in zebrafish recapitulates the ineffective hematopoiesis of the myelodysplastic syndrome. *Blood* 2005;**105**:3528–34.

69. D'Alessandro A, Grazzini G, Giardina B, Zolla L. In silico analyses of proteomic data suggest a role for heat shock proteins in umbilical cord blood hematopoietic stem cells. *Stem Cell Rev* 2010;**6**(4):532–47.

70. Weintraub H. The MyoD family and myogenesis: redundancy, networks, and thresholds. *Cell* 1993;**75**:1241–4.

71. Walsh K, Perlman H. Cell cycle exit upon myogenic differentiation. *Curr Opin Genet Dev* 1997;**7**:597–602.

72. Maione R, Amati P. Interdependence between muscle differentiation and cell-cycle control. *Biochim Biophys Acta, Rev Cancer* 1997;**332**:M19–30.

73. Yablonka-Reuveni Z, Rudnicki MA, Rivera AJ, Primig M, Anderson JE, Natanson P. The transition from proliferation to differentiation is delayed in satellite cells from mice lacking MyoD. *Dev Biol* 1999;**210**:440–55.

74. Weintraub H, Tapscott SJ, Davis RL, Thayer MJ, Adam MA, Lassar AB, et al. Activation of muscle-specific genes in pigment, nerve, fat, liver, and fibroblast cell lines by forced expression of MyoD. *Proc Natl Acad Sci USA* 1989;**86**:5434–8.

75. Sugiyama Y, Suzuki A, Kishikawa M, Akutsu R, Hirose T, Waye MM, et al. Muscle develops a specific form of small heat shock protein complex composed of MKBP/HSPB2 and HSPB3 during myogenic differentiation. *J Biol Chem* 2000;**275**:1095–104.

76. Ito H, Kamei K, Iwamoto I, Inaguma Y, Kato K. Regulation of the levels of small heat-shock proteins during differentiation of C2C12 cells. *Exp Cell Res* 2001;**266**:213–21.

77. Brady JP, Garland DL, Green DE, Tamm ER, Giblin FJ, Wawrousek EF. AlphaB-crystallin in lens development and muscle integrity: a gene knockout approach. *Invest Ophthalmol Vis Sci* 2001;**42**:2924–34.

78. Vicart P, Caron A, Guicheney P, Li Z, Prévost MC, Faure A, et al. A missense mutation in the alphaB-crystallin chaperone gene causes a desmin-related myopathy. *Nat Genet* 1998; **20**:92–5.

79. Sanbe A, Osinska H, Saffitz JE, Glabe CG, Kayed R, Maloyan A, et al. Desmin-related cardiomyopathy in transgenic mice: a cardiac amyloidosis. *Proc Natl Acad Sci USA* 2004; **101**:10132–6.

80. Singh BN, Rao KS, Rao ChM. Ubiquitin-proteasome-mediated degradation and synthesis of MyoD is modulated by alphaB-crystallin, a small heat shock protein, during muscle differentiation. *Biochim Biophys Acta* 2010;**1803**(2):288–99.

81. Adhikari AS, Singh BN, Rao KS, Rao ChM. αB-crystallin, a small heat shock protein, modulates NF-κB activity in a phosphorylation-dependent manner and protects muscle myoblasts from TNF-α induced cytotoxicity. *Biochim Biophys Acta* 2011;**1813** (8):1532–42.

82. Morimoto RI, Tissieres A, Georgopoulos C. *The biology of heat shock proteins and molecular chaperones.* New York: Cold Spring Harbor Laboratory Press; 1994 p. 610.

83. Morimoto RI. Regulation of the heat shock transcriptional response: cross talk between a family of heat shock factors, molecular chaperones, and negative regulators. *Genes Dev* 1998;**12**:378–3796.

84. Morimoto RI. Stress, aging, and neurodegenerative disease. *N Engl J Med* 2006;**355**:2254–5.

85. Forman MS, Trojanowski JQ, Lee VM-Y. Neurodegenerative diseases: a decade of discoveries paves the way for therapeutic breakthroughs. *Nat Med* 2004;**10**:1055–63.

86. Yang J, Oza J, Bridges K, Chen KY, Liu AY. Neural differentiation and the attenuated heat shock response. *Brain Res* 2008;**1203**:39–50.

87. Gage FH. Mammalian neural stem cells. *Science* 2000;**287**:1433–8.

88. Livy DJ, Miller EK, Maier SE, West JR. Fetal alcohol exposure and temporal vulnerability: effects of binge-like alcohol exposure on the developing rat hippocampus. *Neurotoxicol Teratol* 2003;**25**:447–58.
89. Choi MR, Jung KH, Park JH, Das ND, Chung MK, Choi IG, et al. Ethanol-induced small heat shock protein genes in the differentiation of mouse embryonic neural stem cells. *Arch Toxicol* 2011;**85**(4):293–304.
90. Rallu M, Loones MT, Lallemand Y, Morimoto RI, Morange M, Mezger V. Function and regulation of heat shock factor 2 during mouse embryogenesis. *Proc Natl Acad Sci USA* 1997;**94**:2392–7.
91. Mandrekar P, Catalano D, Jeliazkova V, Kodys K. Alcohol exposure regulates heat shock transcription factor binding and heat shock proteins 70 and 90 in monocytes and macrophages: implication for TNF-alpha regulation. *J Leukoc Biol* 2008;**84**:1335–45.
92. Voss AK, Thomas T, Gruss P. Mice lacking HSP90beta fail to develop a placental labyrinth. *Development* 2000;**127**:1–11.
93. Xiao X, Zuo X, Davis AA, McMillan DR, Curry BB, Richardson JA, et al. HSF1 is required for extra-embryonic development, postnatal growth and protection during inflammatory responses in mice. *EMBO J* 1999;**18**:5943–52.
94. Ciocca DR, Stati AO, Fanelli MA, Gaestel M. Expression of heat shock protein 25,000 in rat uterus during pregnancy and pseudopregnancy. *Biol Reprod* 1996;**54**:1326–35.
95. Shah M, Stanek J, Handwerger S. Differential localization of heat shock proteins 90, 70, 60 and 27 in human decidua and placenta during pregnancy. *Histochem J* 1998;**30**:509–18.
96. Geisler JP, Manahan KJ, Geisler HE, Tammela JE, Rose SL, Hiett AK, et al. Heat shock protein 27 in the placentas of women with and without severe preeclampsia. *Clin Exp Obstet Gynecol* 2004;**31**:12–4.
97. Winger QA, Guttormsen J, Gavin H, Bhushan F. Heat shock protein 1 and the mitogen-activated protein kinase 14 pathway are important for mouse trophoblast stem cell differentiation. *Biol Reprod* 2007;**76**(5):884–91.
98. Nørgaard R, Kassem M, Rattan SI. Heat shock-induced enhancement of osteoblastic differentiation of hTERT-immortalized mesenchymal stem cells. *Ann N Y Acad Sci* 2006;**1067**: 443–7.
99. Suzuki K, Murtuza B, Beauchamp JR, Brand NJ, Barton PJ, Varela-Carver A, et al. Role of interleukin-1beta in acute inflammation and graft death after cell transplantation to the heart. *Circulation* 2004;**110**(11 Suppl. 1):II219–24.
100. Muller-Ehmsen J, Whittaker P, Kloner RA, Dow JS, Sakoda T, Long TI, et al. Survival and development of neonatal rat cardiomyocytes transplanted into adult myocardium. *J Mol Cell Cardiol* 2002;**34**(2):107–16.
101. Freyman T, Polin G, Osman H, Crary J, Lu M, Cheng L, et al. A quantitative, randomized study evaluating three methods of mesenchymal stem cell delivery following myocardial infarction. *Eur Heart J* 2006;**27**(9):1114–22.
102. Toma C, Pittenger MF, Cahill KS, Byrne BJ, Kessler PD. Human mesenchymal stem cells differentiate to a cardiomyocyte phenotype in the adult murine heart. *Circulation* 2002;**105**:93–8.
103. Jäättelä M. Heat shock proteins as cellular lifeguards. *Ann Med* 1999;**31**:261–71.
104. Latchman DS. Heat shock proteins and cardiac protection. *Cardiovasc Res* 2001;**51**:637–46.
105. Bouchentouf M, Benabdallah BF, Tremblay JP. Myoblast survival enhancement and transplantation success improvement by heat-shock treatment in mdx mice. *Transplantation* 2004;**77**(9):1349–56.
106. Riederer I, Negroni E, Bigot A, Bencze M, Di Santo J, Aamiri A, et al. Heat shock treatment increases engraftment of transplanted human myoblasts into immunodeficient mice. *Transplant Proc* 2008;**40**(2):624–30.

107. Suzuki K, Smolenski RT, Jayakumar J, Murtuza B, Brand NJ, Yacoub MH. Heat shock treatment enhances graft cell survival in skeletal myoblast transplantation to the heart. *Circulation* 2000;**102**(19 Suppl. 3):III216–21.

108. Maurel A, Azarnoush K, Sabbah L, Vignier N, Maurel A, Azarnoush K, et al. Can cold or heat shock improve skeletal myoblast engraftment in infarcted myocardium? *Transplantation* 2005;**80**(5):660–5.

109. Laumonier T, Yang S, Konig S, Chauveau C, Anegon I, Hoffmeyer P, et al. Lentivirus mediated HO-1 gene transfer enhances myogenic precursor cell survival after autologous transplantation in pig. *Mol Ther* 2008;**16**(2):404–10.

110. Cízková D, Rosocha J, Vanický I, Radonák J, Gálik J, Cízek M. Induction of mesenchymal stem cells leads to HSP72 synthesis and higher resistance to oxidative stress. *Neurochem Res* 2006;**31**(8):1011–20.

111. Gao F, Hu XY, Xie XJ, Xu QY, Wang YP, Liu XB, et al. Heat shock protein 90 protects rat mesenchymal stem cells against hypoxia and serum deprivation-induced apoptosis via the PI3K/Akt and ERK1/2 pathways. *J Zhejiang Univ Sci B* 2010;**11**(8):608–17.

112. Chang W, Song BW, Lim S, Song H, Shim CY, Cha MJ, et al. Mesenchymal stem cells pretreated with delivered Hph-1-Hsp70 protein are protected from hypoxia-mediated cell death and rescue heart functions from myocardial injury. *Stem Cells* 2009;**27**(9):2283–92.

113. Wang X, Zhao T, Huang W, Wang T, Qian J, Xu M, et al. Hsp20-engineered mesenchymal stem cells are resistant to oxidative stress via enhanced activation of Akt and increased secretion of growth factors. *Stem Cells* 2009;**27**(12):3021–31.

114. Wu W, Welsh MJ. Expression of the 25-kDa heat-shock protein (HSP27) correlates with resistance to the toxicity of cadmium chloride, mercuric chloride, cis-platinum(II)-diammine dichloride, or sodium arsenite in mouse embryonic stem cells transfected with sense or antisense HSP27 cDNA. *Toxicol Appl Pharmacol* 1996;**141**(1):330–9.

115. Son GH, Geum D, Chung S, Park E, Lee KH, Choi S, et al. A protective role of 27-kDa heat shock protein in glucocorticoid-evoked apoptotic cell death of hippocampal progenitor cells. *Biochem Biophys Res Commun* 2005;**338**(4):1751–8.

116. McArdle A, Dillmann WH, Mestril R, Faulkner JA, Jackson MJ. Overexpression of HSP70 in mouse skeletal muscle protects against muscle damage and age-related muscle dysfunction. *FASEB J* 2004;**18**(2):355–7.

117. McArdle A, Broome CS, Kayani AC, Tully MD, Close GL, Vasilaki A, et al. HSF expression in skeletal muscle during myogenesis: implications for failed regeneration in old mice. *Exp Gerontol* 2006;**41**(5):497–500.

118. Yu JM, Wu X, Gimble JM, Guan X, Freitas MA, Bunnell BA. Age-related changes in mesenchymal stem cells derived from rhesus macaque bone marrow. *Aging Cell* 2011;**10**(1):66–79.

Preconditioning Approach in Stem Cell Therapy for the Treatment of Infarcted Heart

KHAWAJA HUSNAIN HAIDER AND
MUHAMMAD ASHRAF

Department of Pathology and Laboratory Medicine, University of Cincinnati, Cincinnati, Ohio, USA

Nearly two decades of research in regenerative medicine have been focused on the development of stem cells as a therapeutic option for treatment of the ischemic heart. Given the ability of stem cells to regenerate the damaged tissue, stem-cell-based therapy is an ideal approach for cardiovascular disorders. Preclinical studies in experimental animal models and clinical trials to determine the safety and efficacy of stem cell therapy have produced encouraging results that promise angiomyogenic repair of the ischemically damaged heart. Despite these promising results, stem cell therapy is still confronted with issues ranging from uncertainty about the as-yet-undetermined "ideal" donor cell type to the nonoptimized cell delivery strategies to harness optimal clinical benefits. Moreover, these lacunae have significantly hampered the progress of the heart cell therapy approach from bench to bedside for routine clinical applications. Massive death of donor cells in the infarcted myocardium during acute phase postengraftment is one of the areas of prime concern, which immensely lowers the efficacy of the procedure. An overview of the published data relevant to stem cell therapy is provided here and the various strategies that have been adopted to develop and optimize the protocols to enhance

Progress in Molecular Biology
and Translational Science, Vol. 111
http://dx.doi.org/10.1016/B978-0-12-398459-3.00015-0

323

donor stem cell survival posttransplantation are discussed, with special focus on the preconditioning approach.

Abbreviations: [Control]Sca-1$^+$, control nontransduced native Sca-1$^+$ cells; Cx43, connexin-43; ES cells, embryonic stem cells; HGF, hepatocyte growth factor; HIF-1α, hypoxia-inducible factor-1α; IGF-1, insulin-like growth factor; iPS cells, induced pluripotent stem cells; LDH, lactate dehydrogenase; [Native]SM, native nontransfected skeletal myoblast; [Netrin]Sca-1$^+$, netrin-1 transduced Sca-1$^+$ cells; polyethyleneimine; PKC, protein kinase C; SDF-1α, stromal cell-derived factor-1α; [Trans]SM, skeletal myoblast transfected with four growth factors; TUNEL, terminal deoxynucleotidyl transferase dUTP nick end labeling; VEGF, vascular endothelial growth factor

I. Introduction

Preconditioning of the heart refers to an adaptive mechanism brought on by brief periods of reversible ischemia, which increase the resistance of the heart to subsequent episodes of lethal ischemia.[1] Besides the strategies of early preconditioning (which involves activation of various membrane receptors and is short lived),[2] late preconditioning (also called the "second window of protection," characterized by gene expression changes for increased synthesis of cardioprotective stress proteins),[3–6] and postconditioning (which involves brief episodes of ischemia/reperfusion after a lethal ischemic insult),[7–9] a recent advancement in this area is remote preconditioning (which entails that preconditioning of a remote organ renders beneficial effects to the heart).[10,11] The effect of preconditioning in terms of adaptive response consistently results in reduced infarct size,[7,12,13] but its effects on preservation of global cardiac function as well as antiarrhythmic protection are inconsistent between various experimental animal species and the experimental conditions.[14,15] Regardless of the signaling pathway and the underlying mechanism involved therein, which range from activation of classical survival signaling[9,16,17] to protection of subcellular organelles[18,19] and the recruitment of stem cells,[16,20] the major mechanisms that allow cardioprotection encompass slowed adenosine triphosphate (ATP) usage and limitation of acidosis during a protracted period of ischemia. There is also an associated acute-phase activation of phosphatidyl inositol 3-kinase (PI3K)/Akt.[1,21,22] Some of the major players studied for their relationship with preconditioning are summarized in Table I. This chapter elaborates the pioneering efforts of our research group in the extrapolation of the concept of preconditioning to cellular and subcellular levels. The novel approach of stem cell preconditioning fortifies the cellular defenses and brings the cells to a "state of readiness," which helps them to withstand the rigors of any subsequent ischemic insult.

TABLE I

SOME OF THE KEY PLAYERS INVOLVED IN ISCHEMIC PRECONDITIONING OF THE HEART

- Adenosine—increases delivery by dilation of resistance vessels and decrease inotropy to decrease myocardial oxygen demand.
- Opioid receptors—widely studied in relation to preconditioning and more research is needed to fully understand their uses.
- Bradykinin—believed to reduce the infarct size.
- Prostaglandins, Norepinephrine, angiotensin, and endothelin—these may have a role in ischemic preconditioning, but nothing has been verified yet and is still being debated.
- Nictric oxide—unclear role in ischemic preconditioning. Blocking nitric oxide synthase decreases the effect of ischemic preconditioning in rats, but it is not a trigger or mediator in the early stages of preconditioning.
- Free radicals—activate G-proteins, protein kinases, and ATP-dependent potassium channels.
- Calcium—unclear role, but calcium 1-type channel blockade prevents ischemic preconditioning in the human myocardium.
- Various end effectors—these are unclear and various hypotheses have been developed to explain the cardiac protection that results.
- Release of cardioprotective humoral factors from the preconditioned heart

II. Stem Cell Therapy and the Heart

A. Stem Cell Types Used for Transplantation

Stem cells are characterized by their ability of unlimited self-renewal without any change in their differentiation status. Besides cologenicity, they have the capability of differentiation to adopt different phenotypes. In the mammalian body, the pool of stem cells is classified as derivatives of the adult tissues (i.e., adult stem cells) and those derived from the inner mass of the blastocyst (i.e., embryonic stem cells or ES cells). Based on their differentiation capacity, stem cells have been classified as totipotent (cells that can differentiate into embryonic and extraembryonic cell types), pluripotent (i.e., ES cells and induced pluripotent stem cells, iPS cells), multipotent (i.e., bone marrow stem cells), and unipotent (i.e., myoblasts) on the basis of their potential to cross lineage restriction. Table II lists some of the important cell types that have been extensively studied for their vasculogenic and myogenic potential for myocardial repair and regeneration.[23] Of all these cell types, ES cells possess differentiation characteristics required of an "ideal" donor cell type because of their ability to differentiate and adopt phenotypes of almost all cell types required for repair and regeneration of the injured heart.[24] Moreover, ES cells release bioactive molecules as part of their paracrine activity, which contributes toward myocardial protection and repair process.[25–27] However, the ethical issues regarding their availability, tumorigenicity, and the

TABLE II

Cell Types Used in the Experimental Animal Models and Clinical Studies for Myocardial Repair and Regeneration

Adipose tissue-derived stem cells
Bone-marrow-derived stem cells
Cardiomyocytes
Cardiac stem/progenitor cells
Embryonic stem cells
Endothelial progenitor cells
Induced pluripotent stem cells
Skeletal myoblasts

immunological considerations after transplantation have prevented their progress toward clinical applications.[28,29] More recently, ES cell-derived progenitor cells and their derivative cardiomyocytes are being assessed for myocardial reparability and regeneration.[30–32] The development and optimization of reprogramming protocols to restore pluripotent status in somatic cells have reinvigorated researchers' interest in pluripotent stem cells.[33–37] The availability of pluripotent stem cells without attached moral or ethical issues, but with differentiation characteristics similar to those of ES cells, makes iPS cells a superior choice in regenerative medicine.[38,39] Their autologous character and disease-specific availability avoid ethical and immunological issues and alleviate the need for immunosuppression.[40,41] Despite these advantages, tumorigenicity of iPS cells still remains a major issue impeding their progress toward clinical use.[28,42] Therefore, strategies are being developed to curtail the tumorigenicity of iPS cells, which include optimization of reprogramming protocols,[43–48] partial differentiation of iPS cells into cardiac-like cells before transplantation[49], and transplantation of progenitor cell populations from iPS cells.[50] Another important breakthrough in stem cell research during the last decade is the novel finding from Anversa's group that the heart possesses a pool of resident stem/progenitor cells.[51] The resident cardiac stem/progenitor cells are capable of cardiogenesis because of their vasculogenic as well as cardiomyogenic potential.[52] However, lack of specific surface markers makes their isolation protocol difficult. Besides, there are still many questions regarding their availability for transplantation therapy. In the same context, skeletal myoblasts[53–58] and bone-marrow-derived stem cells[59–64] are the most extensively studied and well characterized cell types both *in vitro* and for their posttransplantation fate in experimental animal models. Skeletal myoblasts and bone marrow stem cells share many advantages, including their ease of availability without ethical issues, easy *in vitro* expansion to achieve the large numbers required for transplantation, differentiation to adopt angio/myogenic phenotypes, and autologous availability that eliminates the use of immunosuppressive therapy.[65,66] However, failure of skeletal myoblasts to form gap junctions and functionally integrate with the host myocytes,[67] as well as the limited cardiac differentiation of bone marrow stem cells and their adoption of undesired phenotypes, has raised concerns about their routine clinical application.[68,69] Despite these limitations, both these cell types have been studied for cardiac reparability and regenerative potential in humans with encouraging results.[70–75] In addition to these stem/progenitor cell types, adipose-tissue-derived stem cells,[76–79] umbilical cord stem cells[80–82], and very small embryonic-like stem cells[83–85] are being intensively characterized and studied, both *in vitro* and in experimental animal models. However, it will be quite some time before these cells will make it to clinical usage.

B. Stem Cell Therapy Approach

The heart has long been considered as a terminally differentiated organ and, therefore, in the event of injury, dead cardiomyocytes are replaced with noncontractile scar tissue. The therapeutic strategy of cellular cardiomyoplasty was developed to support the inefficient intrinsic repair mechanism in the heart, which fails to compensate for the extensive cardiomyocyte loss that ensues especially subsequent to massive myocardial infarction.[86] The strategy was based on transplantation of the cells with inherent capacity to adopt the cardiac phenotype in response to appropriate cues from the myocardial microenvironment and get functionally integrated with the host cardiomyocytes. Since 1992, when the first successful experimental animal study was reported using satellite cells in a dog model of a cryo-injured heart, the field of regenerative medicine has progressed at a rapid pace.[87] Although the study results did not provide any information about the functional outcome from intramyocardial cell transplantation, successful integration of the transplanted cells was observed at the site of the cell graft. Since the publication of these data, heart cell therapy has moved forward from experimental animal models[49,63,88–91] to clinical studies[75,92–95] during the last two decades. Despite the encouraging results and the big hype thus generated about the therapeutic potential of cell transplantation therapy, progress in the field from bench to bedside still suffers from various shortcomings and nonoptimal cell transplantation protocols (Table III).

III. Stem Cell Survival: Major Determinant of Efficacy of Stem Cell Therapy

As described earlier, stem cell therapy for the heart suffers from many limitations which significantly reduce the efficacy of the procedure and give poor prognosis. One of the prime issues faced in cell transplantation therapy is

TABLE III

SOME OF THE MAJOR FACTORS AFFECTING STEM CELL THERAPY FOR MYOCARDIAL REPAIR

- Choice of the donor cell type (adult stem cells, embryonic stem cells, iPS cells, etc.)
- Source of the donor cells (autologous, allergenic, xenogeny)
- Cell culture conditions
- *In vitro* manipulations
- The optimal number of donor cells for transplantation
- Route of administration
- Limited differentiation capacity of the donor cells
- Limited functional integration capacity of the donor cells
- Time of cell transplantation after onset of ischemic insult
- Massive death of the transplanted cells in the infarcted heart

the extensive death of donor cells that ensues during the acute phase after cell transplantation and significantly hampers the efficacy of the cell therapy approach.[96,97] Studies on the dynamics of donor cell death revealed that the first 24–48 h after transplantation are crucial because a large percentage of the donor cells die during this period.[98] The severity of the problem may be judged from the reports that less than 1% of the injected cells (~0.44%) survive to day 4 after cell transplantation.[99] Similar observations have also been made during clinical studies.[92] Associated with the problem of poor rate of donor cell survival is the variability of published data, which may be attributed to the lack of one single optimum method to determine the survival fate of the donor cells. Some of the commonly employed approaches to study the fate of the transplanted cells include the use of fluorescence markers such as cell tracker dyes,[89] genetic labels such as *lac-z*, green fluorescence protein,[99–101] and biological markers such as Y-chromosome studies (by PCR and *in situ* hybridization),[98,102] which involve sex-mismatched transplantation of male donor cells into female recipient hearts, the latter being the most reliable markers for detection of the transplanted cells.[103]

The cause of the massive donor cell death after transplantation is multifactorial. The determinant factors involved range from donor-cell-relevant factors to the cell processing conditions *in vitro* before transplantation and the host-related tissue inflammatory and immunological response mediators at the site of the cell graft.[104–106] The host myocardium after an ischemic insult offers a hostile microenvironment to the donor cells and becomes even more unfavorable because of the infiltration of inflammatory cells. Some earlier studies have shown cell necrosis as the major mechanism affecting cell survival during the early hours after transplantation.[107] However, mounting evidence supports apoptosis as predominantly responsible for the poor survival of the transplanted cells.[108] Given the complexity of the events and the multifactorial nature of the problem at hand, various strategies have been adopted to enhance donor cell survival. These strategies involve anti-inflammatory and immuno-suppressive therapies,[101,109,110] pretreatment of the donor cells[111,112] or the host tissue with growth factors,[113] pharmacological treatment of donor cells to induce cytoprotective proteins involved in survival signaling,[114] and genetic modification of donor cells with growth factor genes[115–117] as well as with genes encoding for important mediators of cell survival signaling.[118–120] The follow-ing sections of this chapter highlight some of the advances made in the development of novel strategies and the optimization of protocols to enhance the resistance of donor cells to ischemic stress.

IV. Preconditioning: A Strategy to "Prime" the Cells for Improved Survival Under Stress

Since the inception of the concept that multiple, brief intermittent episodes of ischemia and reperfusion (called ischemic preconditioning) can protect the myocardium during subsequent exposure to sustained ischemic insult, a pleth-ora of publications from various research groups have shown the feasibility of this approach in cardiac protection against ischemic injury.[22] Although the exact mechanism of protection remains largely unknown, it is hypothesized that preservation of high-energy phosphates is ensured in the preconditioned heart because of the slow energy consumption during ischemic episodes. These molecular changes and altered energy requirements also allow the cells to acclimatize to the ischemic microenvironment. While elucidating the under-lying mechanism, subsequent studies have shown that preconditioning results from the opening of ATP-sensitive potassium (K_{ATP}) channels in response to multiple cycles of ischemia–reperfusion.[12,121–123] Pretreatment with selective K_{ATP} channel antagonists such as glibenclamide abrogates the effects of pre-conditioning.[121] These study results indicate that K_{ATP} channels have a signif-icant role in the endogenous myocardial protective mechanism. Further studies elucidating the molecular mechanism of preconditioning provide clear evidence that stimulation of G-protein coupled receptors with agonists such as adenosine, bardykinins, and opioids occurs,[124–129] with downstream activation of multiple kinases[130] including the protein kinase C (PKC) signaling pathway.[131,132] These molecular events lead to the subsequent opening of mitochondrial ATP-sensitive mitoK$_{ATP}$ channels, which act as mediators or end-effectors.

A. Hypoxic Preconditioning of Stem Cells

Hypoxia-inducible factor-1α (HIF-1α) constitutes the oxygen sensing mechanism of cells[133,134] and has been extensively researched for its role as a critical regulator of cellular functions including metabolic activity, proliferation, differentiation, paracrine behavior, emigrational activity, cell-to-cell communication, and survival.[135–142] Hence, HIF-1α is critical for adaptation of the cells to the low oxygen presence in their microenvironment by direct or indirect regulation of hundreds of genes.[143,144] Therefore, activation of HIF-1α is being sought via hypoxic/anoxic preconditioning of stem cells to enhance their resistance to any subsequent exposure to lethal ischemia.[145,146] In a recent report, 24-h treatment of mesenchymal stem cells activated HIF-1α and Akt and reduced cytochrome c release, with simultaneous reduction in apoptosis of the preconditioned cells under severe hypoxic conditions.[147] Similar prosurvival effects of hypoxic preconditioning have also been reported by other groups and, therefore, preculturing of stem cells under hypoxic conditions is fast becoming an established method to prime the cells for improved survival under ischemic stress *in vitro* as well as posttransplantation in experimental animal models.[146,148–150] A brief culturing of cells under hypoxic conditions not only conditions the cells but also accentuates their regenerative potential.[151] Progenitor cells preconditioned by 6 h of hypoxia treatment showed induction of the CXCR4 receptor and enhanced mobilization to the infarcted myocardium for participation in the repair process after intravenous injection.[152]

Given the importance of HIF-1 signaling in physiological and pathophysiological responses of the biological system,[144,153] strategies have been developed where HIF-1α stability and activity are achieved under nonhypoxic conditions.[154–156] During one of our recent studies, we showed that simultaneous overexpression of the angio-competent molecule angiopoietin-1 and the prosurvival molecule Akt (using adenoviral vectors encoding the respective transgene) increased nonhypoxic stabilization and activity of HIF-1α in bone-marrow-derived mesenchymal stem cells (our unpublished data). Moreover, we also observed higher levels of expression of heme oxygenase-1, endothelial specific markers including Flt1, Flk1, Tie2, VCAM-1, and von Willebrand Factor-VIII (vWFactor VIII), and vascular endothelial growth factor (VEGF) expression in the genetically modified mesenchymal stem cells. These molecular changes occurred in an HIF-1α-dependent fashion and got accentuated under anoxic conditions. The vascular commitment of these cells was confirmed by fluorescence immunostaining, Western blotting, and flow cytometry for early and late markers specific for endothelial commitment. These cells were committed for endothelial phenotype and exhibited higher emigrational activity and angiogenesis *in vitro* and post engraftment in a rat heart model of acute myocardial infarction. These cells continued to secrete higher levels of VEGF, which also promoted mobilization of intrinsic Flk1[+] and Flt1[+] cells into the infarcted heart.

Given the importance and magnitude of HIF-1α's involvement in prosurvival signaling and the very short biological half-life of the HIF-1α protein,[157] treatment strategies based on HIF-1α transgene delivery have been developed that involve genetic modification of stem cells before transplantation as well as coadministration of HIF-1α transgene with stem cell transplantation.[158–160] Stem cells overexpressing HIF-1α showed improved survival and angiogenic response posttransplantation. On the same note, mutagenesis of HIF-1α produced by substitution of alanine (Ala) for proline (Pro) at position 564 and asparagine (Asp) at position 803 was performed to reduce its degradation and enhance its biological half-life. The strategy resulted in high stability and activity of HIF-1α in bone marrow cells to promote their cardiomyogenic differentiation upon coculture with cardiomyocytes.[157]

B. Ischemic Preconditioning and Role of HIF-1α-Dependent MicroRNAs in Cytoprotection

Following the basic principle that "what doesn't kill you, makes you stronger,"[161] and given that ischemic preconditioning imparts the most powerful cytoprotective stimulus, we pioneered the concept of stem cell preconditioning to support their survival under ischemic stress *in vitro* and posttransplantation in the infarcted heart of an experimental animal model. Using skeletal myoblasts as a cellular model, we optimized a protocol to treat cells with repeated 10-min cycles of anoxia interrupted by 30 min of reoxygenation at 37 °C. The preconditioned skeletal myoblasts were later subjected to lethal anoxia for 8 h, using nonpreconditioned native skeletal myoblasts as controls (our unpublished data). Interestingly, the preconditioned skeletal myoblasts showed significantly higher survival under lethal anoxia compared to the control cells as determined by terminal deoxynucleotidyl transferase dUTP nick end labeling (TUNEL) and annexin-V staining followed by flow cytometry. We also assessed the versatility of our approach in cells from different sources including bone-marrow-derived mesenchymal stem cells.[162] In one of our recently concluded studies, we observed that ischemic preconditioning of mesenchymal stem cells by exposure to 30 min of two intermittent anoxia–reoxygenation cycles primed the cells via phosphorylation of Akt (ser-473) and Erk1/2 (Thr-202/Tyr-204).[162] More importantly, we also observed downstream activation of the HIF-1α, which was evident from HIF-1α accumulation in the nucleus in the preconditioned cells. As discussed earlier, HIF-1α is central to the oxygen sensing mechanism of the cells and helps in the maintenance of oxygen homoeostasis in the biological system.[162] Our results showed that abrogation of HIF-1α with specific RNA interference abolished the effects of preconditioning and resulted in significant cell death upon subsequent exposure of the cells to lethal anoxia. Given that microRNAs (miRs) are critical regulators of physiological cellular processes ranging from cell proliferation to differentiation and survival in the unfavorable microenvironment,[163] we analyzed the preconditioned cells for changes in their miR profile in response to ischemic preconditioning. One of the miRs that showed altered expression in response to preconditioning was miR-210, also known as hypoxia-dependent miR or hypoxamir-210,[164–166] as its expression changes have a pivotal role in the determination of the cell's response to the oxygen level in the microenvironment.[167,168] We observed a multifold increase in miR-210 expression in the preconditioned stem cells as compared to the baseline expression observed in the nonpreconditioned cells. Pretreatment of the cells with miR-210 antagomir abolished the protective effects of preconditioning, whereas transgenic overexpression of miR-210 using an miR-210 mimic simulated the protective effects of preconditioning. Real-time PCR arrays for rat apoptotic genes, computational target gene analyses, and luciferase reporter assays identified the FLASH/caspase-8-associated protein-2 (CASP8AP2) as a putative target gene of miR-210 in the preconditioned cells. Induction of FLASH/CASP8AP2 in preconditioned stem cells pretreated with anti-miR-210 resulted in increased cell apoptosis. In summation, these data provided clear demonstration that the cytoprotective effects afforded by ischemic preconditioning were mediated via miR-210 induction which suppressed FLASH/CASP8AP2 as its putative target. Extrapolation of these data to a rat model of acute coronary artery ligation showed significantly improved survival of the preconditioned stem cells compared to their nonpreconditioned counterparts.

In one of our ongoing studies, we have observed that transplanted stem cells, either preconditioned or transfected with an miR-210 mimic, transferred mir-210 to the juxtaposed host cardiomyocytes after transplantation in an experimentally infarcted rodent heart. More interestingly, we observed that transfer of mir-210 from stem cells (either preconditioned or transfected for miR-210 overexpression) to the host cardiomyocytes occurred via gap junctions and contributed to protection of the cardiomyocytes during myocardial ischemia. The animals treated with preconditioned stem cells or stem cells with transgenic overexpression of miR-210 showed significantly reduced cardiomyocyte TUNEL positivity. The cardioprotective effects were evident from the significantly attenuated infarct size besides the preserved global cardiac function. These *in vivo* data were well supported by an *in vitro* coculture model between rat-derived neonatal cardiomyocytes and stem cells with induced miR-210 expression by either preconditioning or transfection with miR-210 mimic. Similar gap-junctional transfer of miRs has also been reported by Anversa's group as a part of the mircrine activity of the stem cells in the heart.[169]

The concept of repetitive, intermittent exposure to hypoxia/reoxygenation during ischemic preconditioning has also been exploited in preconditioning at organ level. In an interesting study, female Fisher-344 rats were exposed to intermittent hypoxic exposure (380 Torr) for 15 h per day for 28 days.[170] Molecular analyses showed significantly increased renal HIF-1α expression besides an increase in HIF-1α-dependent renal B-cell lymphoma-2 (Bcl-2) protein expression. Subsequently, renal ischemia/reperfusion injury was mitigated in the animals subjected to repetitive hypoxic preconditioning.

C. Pharmacological Preconditioning with Preconditioning Mimetic

As the field progressed, many research groups simulated the cytoprotective effects of ischemic preconditioning using pharmacological agents, known as preconditioning mimetics, with comparable efficacy.[78,171,172] One of the most popular and extensively studied pharmacological agents used as a preconditioning mimetic is diazoxide, which has been employed to confer cardioprotection in experimental animal models as well as in cultured cardiomyocytes *in vitro*.[173,174] Diazoxide is a classical mitochondrial ATP-sensitive K^+ channel opener (mitoK$_{ATP}$) and, similar to brief exposure to ischemia/reoxygenation, opens the mitoK$_{ATP}$ channels and renders the heart more tolerant to subsequent lethal ischemia.[132,175,176] Our group has already reported that cardiac protection from mitoK$_{ATP}$ channels is dependent on Akt translocation from cytosol to mitochondria.[177] The opening of mitoK$_{ATP}$ channels may act quite early during apoptosis by inhibiting cytochrome *c* release and depolarization, thus altering the earlier steps in the apoptotic cascade.

Using skeletal myoblasts obtained from young male Fischer-344 rats as an *in vitro* model, we optimized the preconditioning protocol by treating the cells with 200 μM diazoxide for 30 min.[178] Treatment of skeletal myoblasts with 200 μM diazoxide did not show any toxicity as was observed during morphological assessment of the cells using phase contrast microscopy (Fig. 1A–C). When subjected to oxidant stress, the preconditioned cells had improved survival, which was evident from the significantly reduced leakage of lactate dehydrogenase (LDH; as a marker of cellular injury) and TUNEL (Fig. 1D–F) as well as Annexin-V positivity by flow cytometry. Staining with DePsipher dye (a lipophilic cation (5,5′,6,6′-tetrachloro-1,1′,3,3′-tetraethylbenzimidazolyl carbocyanin iodide), which is used as a mitochondrial activity marker) showed that mitochondrial membrane potential was significantly preserved in the preconditioned cells. These data were also supported by the decreased cytochrome *c* release in the preconditioned cells under oxidant stress. Transplantation of the male skeletal myoblasts into female Fischer-344 rat hearts with permanent coronary artery ligation showed a multiplefold higher survival of the preconditioned cells during the acute phase after transplantation. Long-term experimental animal studies have shown that transplanted, preconditioned skeletal myoblasts were involved in the angiomyogenic repair of the heart. We observed an extensive repopulation of the infarcted myocardium with neofibers. Besides, with neofibers, we observed a concomitant increase in

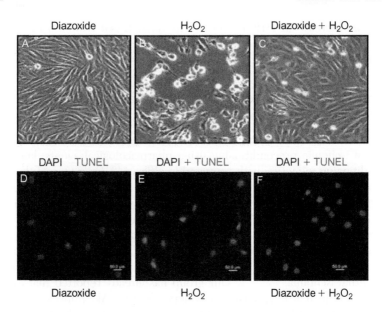

FIG. 1. Pharmacological preconditioning of skeletal myoblasts with diazoxide treatment. Skeletal myoblasts were purified from rat skeletal muscle biopsy samples and used in preconditioning experiments. The cells were seeded 24 h before preconditioning. For pharmacological preconditioning, the cells were treated with 200 µM diazoxide for 30 min. Skeletal myoblasts treated with 200 µM diazoxide did not show any toxic effects of pharmacological treatment. Upon subsequent exposure to 100 µM H_2O_2 treatment, preconditioned skeletal myoblasts showed significant morphological integrity (phase contrast microphotographs; A–C) and reduced TUNEL positivity (fluorescence photomicrographs; D–F) as compared to the nonpreconditioned skeletal myoblasts (cells without diazoxide treatment). (For color version of this figure, the reader is referred to the online version of this chapter.)

blood vessel density in the infarct and peri-infarct regions in the heart. These molecular and anatomical changes led to significantly preserved global cardiac function. More recently, we have also determined the role of miR during treatment of stem cells with diazoxide and reported that miR-146 was critical during preconditioning of skeletal myoblasts.[179] In a recent study, we have also reported that diazoxide treatment of human skeletal myoblast induced interleukin-11, which activated ERK1/2 and Stat3 to improve cell survival upon exposure to oxidant stress.[180]

Besides reports of the use of diazoxide as a pharmacological agent, there are reports of pharmacological agents with inhibitory effects on phosphodiestarse-5a being used successfully to precondition stem cells.[78,181] These studies are part of the endeavor to determine the effectiveness of pharmacological agents for the preconditioning of stem cell safety, which has already been well established in the clinical setting.

D. Preconditioning of Stem Cells by Gene Modification

During the past two decades, both stem cell therapy and genetic therapy have shown great promise as emerging new strategies for the repair of ischemic damage in the heart. However, neither of these approaches is without its pitfalls, and precise treatment plans have yet to be determined. For making the best use of the advantages of both these strategies, it would be prudent to combine them. Stem cells are excellent carriers of transgene/s and therefore can be genetically modified to overexpress growth factors, and survival signaling molecules such as the transgene/s expression products condition the cells for improved survival as well as differentiation fate.[182,183] The efficiency of stem cell therapy may be increased by injecting genetically modified stem cells that overexpress bioactive growth factors and cytokines, which will not only precondition the cells but also allow for the development of the functioning vasculature network for biological bypass of the occluded blood vessel to improve regional blood flow to the ischemic myocardium. Such a strategy will create a microenvironment in the heart that will be more suitable for the survival of the donor stem cell and the newly developing cardiomyocytes.

Activation of PI3K and serine threonine kinase-B (Akt) is central to survival signaling and protects the cells against apoptosis in the wake of any stress stimulus. The cardioprotective role of activated Akt has been attributed to the prevention of cardiomyocyte apoptosis.[184] Mangi and colleagues reported that genetic modification of mesenchymal stem cells from the bone marrow for Akt transgene overexpression enhanced their differentiation potential besides improving posttransplantation cell survival and engraftment rate in an infarcted rat heart.[120] Moreover, cells with transgenic Akt expression significantly attenuated infarct size expansion, which was attributed to the altered paracrine activity of mesenchymal stem cells overexpressing Akt.[185] Since the publication of these encouraging data, other research groups have also focused on genetic modification of stem cells with survival signaling molecules for improved survival and engraftment during stem cell therapy.[119,186] There is mounting evidence supporting Pim-1 as a cardioprotective kinase that prevents cardiomyocyte apoptosis via mitochondrial protection.[187,188] Cardiac progenitor cells genetically engineered to overexpress Pim-1 showed significantly improved survival as well as cardiac reparability as compared to the native cardiac progenitor cells.[189] On the same note, our group has shown a clear advantage of the combinatorial approach wherein overexpression of survival signaling molecule Akt was combined with angio-competent angiopoietin-1 expression.[89] Our multimodal therapeutic approach not only improved the survival fate of the bone marrow stem cells but also helped the cells to undergo angiomyogenic differentiation. Besides survival signaling molecules, stem cells have also been modified genetically to express growth factors. Noticeable among these growth factors are Insulin-like growth factor (IGF-1),[116] hepatocyte growth factor (HGF),[190] stromal cell-derived factor-1α (SDF-1α),[191] VEGF,[192–194] netrin-1,[195] sonic hedgehog,[196] angiopoietin-1,[197] FGF2,[198] heme oxygenase,[199,200] and HIF-1α.[159]

In our latest endeavor to exploit genetic modification of stem cells for preconditioning, we genetically modified syngenic rat skeletal myoblasts with a quartet of plasmids encoding for growth factors SDF-1α, HGF-1, VEGF, and IGF-1. The rationale for the selection of the group of growth factors was based on simultaneous exploitation of multiple growth factor ligand/receptor systems; the aim was to create a favorable concentration gradient of these growth factors in the heart for mobilization of various stem/progenitor cells, including c-met-expressing resident cardiac progenitors in response to HGF, CXCR4-positive cells from bone marrow in response to SDF-1α, and endothelial progenitor cells in the peripheral circulation in response to VEGF. IGF-1 was intended to provide proproliferative, prosurvival, and prodifferentiation stimulus for the mobilized stem cells that homed in to the infarcted myocardium in response to the growth factor gradient developed at that site of the cell graft. IGF-1 also enhanced connexin-43 (Cx43) expression in skeletal myoblasts and supported their coupling with cardiomyocytes (Fig. 2A and B). Moreover, these growth factors primed the transplanted skeletal myoblasts for improved survival after transplantation in the infarcted myocardium. The growth factor-transfected skeletal myoblasts (1.5×10^6 per animal heart at multiple sites) were transplanted in a rat heart model of acute coronary artery ligation by intramyocardial injection. Extensive mobilization of $cMet^+$, $ckit^+$, $ckit^+/GATA^+$, $CXCR4^+$, $CD44^+$, $CD31^+$, and $CD59^+$ cells was observed on day 7 after transplantation of growth factor-expressing skeletal myoblasts. The mobilized cells homed into the infarcted heart ($p < 0.05$ vs. animal hearts transplanted with nontransfected skeletal myoblasts) and participated in the ongoing repair process. We observed extensive neomyogenesis and angiogenesis in hearts transplanted with growth factor-transfected skeletal myoblasts, with the resultant attenuation of infarct size and improved global heart function at 8 weeks of observation (Fig. 2C and D). Our multimodal strategy based on preconditioning of stem cells by genetic modification not only ensured improved survival of the transplanted cells but also made possible the participation of intrinsically available stem/progenitor cells in the myocardial repair and regeneration process.

In continuation of our previous work based on the hypothesis that initiation of embryonic signaling in stem cells may be helpful for their vasculogenic and myogenic differentiation characteristics, we transplanted mesenchymal stem cells overexpressing the sonic hedgehog transgene, a morphogen, during the embryonic development.[196] While analyzing the downstream signaling involved, we observed that netrin-1 expression was significantly increased and was involved in angiogenic response after transplantation of mesenchymal

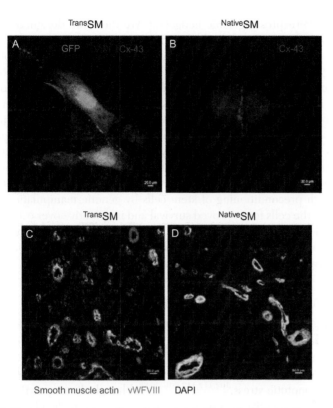

FIG. 2. Transgene expression of a select quartet of growth factors induces connexin-43 expression in skeletal myoblasts. (A and B) Fluorescence immunostaining of skeletal myoblasts (from GFP[+] transgenic rats) transfected with select quartet of growth factors VEGF, SDF-1α, IGF-1, and HGF-1 and subsequently cocultured with unlabeled neonatal cardiomyocytes. Skeletal myoblasts were transfected using polyethyleneimine nanoparticles for each of the growth factor individually. Forty-eight hours after transfection, the transfected skeletal myoblasts were pooled and cultured for 24 h before being cocultured with cardiomyocytes. The cells (skeletal myoblasts and cardiomyocytes) in the coculture were later fixed and immunostained for connexin-43 (Cx43, red fluorescence). The nuclei were visualized with DAPI (blue fluorescence). Monoculture of neonatal cardiomyocytes was used as a control and was immunostained for Cx43 expression (red fluorescence). Skeletal myoblasts with transfection of four growth factors showed significant expression of Cx43 and formed gap junctions with juxtaposed cardiomyocytes (original magnification = 40×). (C and D) Double fluorescence immunostaining of histological sections from rat heart at 4 weeks after transplantation of (C) skeletal myoblast transfected with four growth factors ([Trans]SM) and (D) ative nontransfected skeletal myoblast ([Native]SM). The cells were transplanted intramyocardially in a rat heart model of permanent coronary artery ligation during acute phase after coronary artery ligation. The histological tissue sections were immunostained for smooth muscle actin (red fluorescence) and vWFactor VIII (green fluorescence) antigens to visualize vascular density. DAPI was used to visualize the nuclei (blue fluorescence). The vascular density was significantly higher in [Trans]SM transplanted animal hearts as compared to [Native]SM. (For interpretation of the references to color in this figure legend, the reader is referred to the online version of this chapter.)

stem cells overexpressing sonic hedgehog. We therefore designed a study to simulate the effects of sonic hedgehog transgene overexpression via cell-based netrin-1 transgene delivery to the heart with anti-apoptotic and proangiogenic effects.[195] Sca-1$^+$-like cells were isolated and propagated in vitro and were successfully transduced with adenoviral vector encoding for netrin-1. During in vivo studies, cells overexpressing netrin-1 were transplanted in the experimental heart model of ischemia–reperfusion injury. The cells showed extensive survival and released netrin-1 as a part of their paracrine activity. A significant finding of the study was that netrin-1 decreased apoptosis in the host cardiomyocytes and endothelial cells via activation of Akt besides its proangiogenic activity and reduction in ischemia/reperfusion injury (Fig. 3).

Although preconditioning of stem cells by genetic manipulation successfully primes the cells for improved survival and is operative over extended time durations depending upon the type of the vector used for DNA delivery, the genetic manipulation protocols, especially in cases where viral vectors are used, raises safety concerns for clinical application.

E. Preconditioning of Stem Cells by Growth Factor Treatment

Besides preconditioning by genetic modification, treatment with bioactive growth factors and cytokines effectively induces prosurvival as well as differentiation-related signaling in stem cells and improves their defense against proapoptotic stress.[102,112,113,201–205] Although these growth factors act through their respective receptor systems, mostly the molecular mechanism and signal transduction converges on activation of PI3K/Akt and ERK1/2. Moreover, growth factor treatment also alters the paracrine release of bioactive molecules from the preconditioned cells. A recent study showed that TGF-β treatment of mesenchymal stem cells enhanced VEGF secretion from the preconditioned cells via MAPK activation and improved their survival as well as their ability to protect the cardiomyocytes post transplantation against ischemia–reperfusion injury.[112] Besides prosurvival effects, pretreatment of stem cells with growth factors in culture conditions also improves their differentiation characteristics.[206]

IGF-1 is a small polypeptide protein hormone, which is involved in growth and anabolic activity in the body under the influence of growth hormones. IGF-1 interacts with its transmembrane receptor IGF-1R, which has a wide distribution in the various cell types including the stem cells and critically regulates various cellular functions including cell proliferation, growth, differentiation, and survival.[207,208] Given that bone-marrow-derived Sca-1$^+$ cells have strong expression of IGF-1R, preconditioning of these cells with IGF-1 activated IGF-1R to initiate survival signaling in the growth factor-treated cells

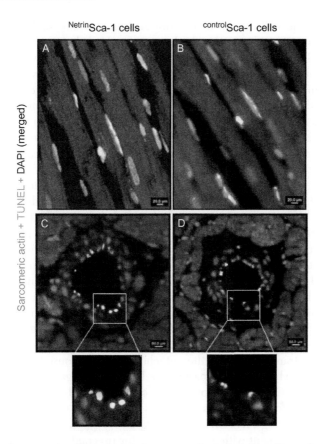

FIG. 3. Cytoprotective effects of netrin-1 overexpression in the infarcted heart. Sca-1[+] like cell population (Sca-1[+] cells) was isolated from young male Wistar rat hearts using EasySep® isolation kit (Stem Cell Technology Inc.) per manufacturer's protocol. Adenoviral vectors encoding for netrin-1 or empty vector were developed and used for transduction of the cells using our optimized protocols. Sca-1 cells overexpressing netrin-1 (netrin-1 transduced Sca-1[+] cells, [Netrin]Sca-1[+] cells) or transduced with empty vector ([Control]Sca-1[+] cells) were transplanted in a rat heart model of ischemia–reperfusion injury. A total of 2×10^6 cells were injected per heart at multiple sites (at least 4–5 sites per heart). The animals ($n = 3$) were euthanized on day 4 after cell transplantation and were used for immunohistological studies. Immunostaining for sarcomeric actin (green fluorescence) combined TUNEL (red fluorescence) showed significantly higher TUNEL positivity in the [Control]Sca-1[+] cell transplanted animal hearts as compared to the [Netrin]Sca-1[+] cells transplanted animal hearts. DAPI (blue fluorescence) was used to visualize the nuclei. A and C are representative images of TUNEL positive cardiomyocytes and endothelial cells in the blood vessels respectively in the peri-infarct region in the [Netrin]Sca-1[+] cell transplanted animal hearts whereas, B and D show representative images from [Control]Sca-1[+] cell transplanted animal hearts. White box in C and D have been magnified for clarity to show endothelium of the lumen with TUNEL positivity. (For interpretation of the references to color in this figure legend, the reader is referred to the online version of this chapter.)

and primed the cells to become more resistant to any subsequent lethal stress.[102] While elucidating the mechanism involved therein, treatment with IGF-1, which is a known activator of Akt activity in the cells,[208] was observed to initiate survival signaling in the preconditioned cells, which was characterized by activation of IGF-1R and its downstream signaling. Moreover, there was a concomitant increase in Cx43 expression in the IGF-1-preconditioned stem cells, a molecular event that was subsequently established as a critical determinant of the cytoprotection when the cells were exposed to 8 h of lethal oxygen–glucose deprivation.[209] Moreover, there was a significant PI3K/Akt-dependent reduction in cytochrome c release and caspase-3 activity in the preconditioned stem cells, which incidentally showed significantly preserved mitochondrial membrane potential.[102] Loss-of-function studies by Cx43-specific RNA interference showed significant abolishment of the prosurvival effects of preconditioning when the cells were subjected to 8 h of oxygen and glucose deprivation. Experimental animal studies in a female rat model of acute coronary artery ligation showed that intramyocardial transplantation of sex-mismatched male preconditioned Sca-1$^+$ showed a 5.5-fold higher survival as compared to the nonpreconditioned stem cells 7 days after transplantation. Confocal imaging after actinin- and Cx43-specific immunostaining demonstrated that the preconditioned Sca-1$^+$ cells showed extensive myogenic differentiation in the infarcted heart. Extensive in-depth mechanistic studies confirmed that Cx43 was translocated into the mitochondria in response to preconditioning with IGF-1 which critically participated in survival signaling during IGF-1 preconditioning. On the same note, we have also reported that preconditioning with IGF-1, as a matter of fact, accentuates the intrinsic survival instinct of the cells by orchestrating PKC/ERK1/2 activation.[210]

Although treatment with growth factors for preconditioning of cells has a very short window of effectiveness and may protect the preconditioned cells for only a limited time duration post transplantation in the ischemic heart, preconditioning strategy using growth factors has multiple advantages. For example, it involves a simple protocol and exploits the cells' inherent survival instinct.[210] Moreover, the approach does not involve genetic manipulation of the cells and, therefore, it is safe from the clinical perspective.

F. A Novel Strategy of "Sub-Cellular" Preconditioning of Stem Cells

A computational analysis during IGF-1 preconditioning studies revealed the presence of Bcl-2 homology domain-3 (BH3) motif in Cx43 and there was a conserved pattern of amino acids similar to Bcl-2 family of prosurvival proteins that regulated cytochrome c release from the mitochondria.[209] Moreover, secondary structure prediction indicated an extended alpha-helix in this region, a known condition for BH3-driven protein–protein interactions.[209] In the light of these computational findings and the critical participation of mitochondrial Cx43 in prosurvival signaling in the preconditioned cells,[211] we are investigating whether Cx43 may be bracketed with the Bcl-2 family of prosurvival proteins.

Based on computational data and the results from our mechanistic studies, we are currently working on the novel concept of subcellular preconditioning, an approach that will emphasize preconditioning of subcellular organelles to simulate the effects of cellular preconditioning. The mitochondria in this regard would be the primary targets because of their significance in the preconditioning approach.[212–215] Moreover, the role of mitochondria in apoptosis has been extensively studied, and mitochondrial participation in the apoptotic cascade is based on a range of mechanisms that involve the release of inter-membrane space proteins, that is, cytochrome c, to promote caspase activation in the cytoplasmic compartment.[216] The release of cytochrome c implies the loss of integrity of the outer mitochondrial membrane under the influence of proapoptotic members of the Bcl-2 family. More recent studies have shown that mitophagy (elimination of the damaged mitochondria) during ischemic preconditioning of the heart constitutes an important mechanism to prevent myocyte apoptosis.[217] Alternatively, we propose that mitochondria-specific targeting of the Cx43 transgene in stem cells would support their protection, thereby enhancing the resistance of the cells upon exposure to subsequent ischemic insult.[209] To determine the role of mitochondrial Cx43, a vector encoding for full-length mouse Cx43 with a mitochondria localization signal was designed, which was subsequently cloned into a shuttle vector (pShuttle-IRES-hrGFP-1) for mitochondrial targeting of Cx43 (mito-Cx43). Sca-1$^+$ cells were successfully transfected for mitochondria-specific overexpression of Cx43.[209] On subsequent exposure to oxygen–glucose deprivation, a lower cytosolic accumulation of cytochrome c and significantly reduced caspase-3 activity was observed in stem cells with overexpression of mitochondria-specific Cx43 compared to the cells with mitochondria-specific GFP overexpression as control. The survival of stem cells with overexpression of mitochondria-specific Cx43 was significantly higher as determined by TUNEL and LDH assays.

Although our strategy of nonviral mitochondrial targeting of Cx43 transgene reproduced the effects of IGF-1 treatment in terms of significantly enhanced survival of stem cells and provided a proof of concept, the strategy suffered from the limitation of poor transfection efficiency with nonviral-vector-based transgene delivery to the mitochondria. We therefore focused on the development of a high-efficiency adenoviral vector system that encoded for the Cx43 transgene with mitochondria localization signal and elucidated the prosurvival effects of mitochondria-specific Cx43 transgene overexpression.[218] The transduction efficiency of the adenoviral vector thus developed was more than 90%, with very negligible nontargeted subcellular expression of Cx43. Double fluorescence immunostaining for cytochrome c and FLAG-tag protein (Fig. 4) supported by Western blotting confirmed the mitochondria-specific expression of the Cx43 transgene. Subsequent oxygen–glucose deprivation led to a significantly improved survival of Cx43-overexpressing cells as compared to the control cells. Again, we observed a significantly reduced caspase-3 activity and a concomitantly reduced cytochrome c release from the mitochondria into the cytoplasm. An important observation during this study was a direct relationship between mitochondrial Cx43 transgene expression and an altered expression of the Bcl-2 family members. Although there are still many unanswered questions regarding the exact mechanism by which the Bcl-2 family of proteins modulates apoptosis, structural studies of these proteins have provided a deeper insight into the understanding of apoptosis on the molecular level.[219] The distribution pattern of the anti- and proapoptotic Bcl-2 family members between the mitochondrial and cytoplasmic compartments is a critical determinant of cell survival and remains integral to the release of cytochrome c from the mitochondria into the cytoplasm during the apoptotic cascade.[220] Initiation of the apoptotic cascade is indicated by the higher release

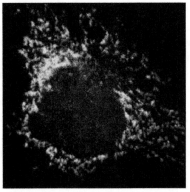

FIG. 4. Bone-marrow-derived Sca-1[+] cells infected with Ad-mito-Cx43-FLAG (adenoviral vector encoding for Cx43 with FLAG-tag protein and mitochondrial localization signal). The cells after transduction with the vector were double immunostained for detection of FLAG-tag protein (red fluorescence) using anti-FLAG primary antibody and detected with Alexa Fluor-546 conjugated secondary antibody and anti-cytochrome c (green fluorescence). DAPI was used to visualize nuclei. Merged image shows punctuate distribution of FLAG (red fluorescence) co-localized with cytochrome c (green fluorescence) indicating its mitochondrial localization. (For interpretation of the references to color in this figure legend, the reader is referred to the online version of this chapter.)

of cytochrome c from the mitochondrial compartment into the cytoplasm.[220] We observed that the mitochondrial fraction of the stem cell protein lysate had a higher mitochondrial Bcl-xL/Bak ratio, which was accompanied by reduced cytochrome c release into the cytoplasmic fraction and abolished caspase-3 activity compared to GFP-overexpressing cells as controls. For *in vivo* extrapolation of the data, sex-mismatched transplantation of 2×10^6 male cells overexpressing mitochondria-specific Cx43 into a female rat model of acute myocardial infarction was carried out. GFP-transfected stem cells and basal DMEM-injected animals were used as controls. Based on *sry*-gene expression by PCR, we observed a multiple fold higher survival of the mitochondria-specific Cx43-expressing stem cells posttransplantation, while DMEM-injected animal hearts served as negative controls. Moreover, animals treated with stem cells having mitochondrial overexpression of Cx43 showed improved LV ejection fraction, LV fractional shortening as well as LV end-diastolic dimension, and reduced infarction size. These results clearly supported our novel concept of "subcellular preconditioning" by targeting the Cx43 transgene for mitochondria-specific overexpression to address the problem of massive cell death of stem cells post transplantation in the infarcted heart.

V. Conclusions

Many of the fundamental issues including massive death of the donor cells need to be resolved before stem cell therapy for the heart can advance to become a clinical reality. In fact, these issues are the limiting factors that are slowing down the successful transition of stem cell therapy from bench to bedside. Despite the adoption of various prosurvival strategies with encouraging results, the preconditioning paradigm, especially based on physical methods of stem cell manipulation prior to transplantation such as ischemic preconditioning[162] as well as preconditioning by heat shock,[221] cold shock,[222] and electrical stimulation[223], provides effective and safe options to strengthen the cellular survival instinct. Similarly, growth factor treatment of stem cells prior to transplantation is equally effective. However, these approaches have a common limitation in that the stimulus has a narrow window of effectivity, which may be insufficient in the clinical perspective where the cells may require extended acclimatization and engraftment. Moreover, with current research focused on the identification of novel mechanisms and signaling pathways involved in stem cell death post transplantation, and considering the multifactorial nature of the problem at hand, it will be prudent to develop a multimodal strategy that will simultaneously address the different aspects of the problem. The use of a cocktail of growth factors with different receptors and downstream signaling cascades might be more effective in this regard than using a single growth factor treatment approach.[224] Similarly, with the emerging role for miRs as critical regulators of cellular functions including their survival,[225] more recent studies have manipulated stem cells with single as well as multiple miR precursors and observed a significant improvement in stem cell survival post transplantation.[226,227] An alternative, or a supportive, approach to prosurvival strategies would be to promote donor cell proliferation to compensate for cell loss.[228] However, such an approach would require controlled proliferative activity of the transplanted cells to avoid any undesired outcome. In conclusion, a combined strategy based on preconditioning and manipulation of stem cells for controlled proproliferative activity may provide a clinically relevant solution to the menace of massive donor stem cell death for optimal prognosis.

ACKNOWLEDGMENTS

This work was supported by National Institutes of Health grants # HL-23597, HL70062, and HL-080686 (to M. A) and HL-087288, HL-089535, and HL106190-01 (to Kh. H. H).

REFERENCES

1. Ravingerova T, Matejikova J, Neckar J, Andelova E, Kolar F. Differential role of PI3K/Akt pathway in the infarct size limitation and antiarrhythmic protection in the rat heart. *Mol Cell Biochem* 2007;**297**:111–20.
2. Mocanu MM, Bell RM, Yellon DM. PI3 kinase and not p42/p44 appears to be implicated in the protection conferred by ischemic preconditioning. *J Mol Cell Cardiol* 2002;**34**:661–8.
3. Miki T, Swafford AN, Cohen MV, Downey JM. Second window of protection against infarction in conscious rabbits: real or artifactual. *J Mol Cell Cardiol* 1999;**31**:809–16.
4. Guo Y, Wu WJ, Qiu Y, Tang XL, Yang Z, Bolli R. Demonstration of an early and a late phase of ischemic preconditioning in mice. *Am J Physiol* 1998;**275**:H1375–87.
5. Yamashita N, Hoshida S, Taniguchi N, Kuzuya T, Hori M. A "second window of protection" occurs 24 h after ischemic preconditioning in the rat heart. *J Mol Cell Cardiol* 1998;**30**: 1181–9.
6. Takano H, Tang XL, Kodani E, Bolli R. Late preconditioning enhances recovery of myocardial function after infarction in conscious rabbits. *Am J Physiol Heart Circ Physiol* 2000;**279**: H2372–81.
7. Vinten-Johansen J. Postconditioning: a mechanical maneuver that triggers biological and molecular cardioprotective responses to reperfusion. *Heart Fail Rev* 2007;**12**:235–44.
8. Penna C, Cappello S, Mancardi D, Raimondo S, Rastaldo R, Gattullo D, et al. Postconditioning reduces infarct size in the isolated rat heart: role of coronary flow and pressure and the nitric oxide/cGMP pathway. *Basic Res Cardiol* 2006;**101**:168–79.
9. Sun H, Guo T, Liu L, Yu Z, Xu W, Chen W, et al. Ischemic postconditioning inhibits apoptosis after acute myocardial infarction in pigs. *Heart Surg Forum* 2010;**13**:E305–10.
10. Hausenloy DJ, Yellon DM. Remote ischaemic preconditioning: underlying mechanisms and clinical application. *Cardiovasc Res* 2008;**79**:377–86.

11. Andreka G, Vertesaljai M, Szantho G, Font G, Piroth Z, Fontos G, et al. Remote ischaemic postconditioning protects the heart during acute myocardial infarction in pigs. *Heart* 2007;**93**:749–52.

12. Schott RJ, Rohmann S, Braun ER, Schaper W. Ischemic preconditioning reduces infarct size in swine myocardium. *Circ Res* 1990;**66**:1133–42.

13. Donato M, D'Annunzio V, Berg G, Gonzalez G, Schreier L, Morales C, et al. Ischemic postconditioning reduces infarct size by activation of A1 receptors and K+(ATP) channels in both normal and hypercholesterolemic rabbits. *J Cardiovasc Pharmacol* 2007;**49**:287–92.

14. Ravingerova T, Matejikova J, Pancza D, Kolar F. Reduced susceptibility to ischemia-induced arrhythmias in the preconditioned rat heart is independent of PI3-kinase/Akt. *Physiol Res* 2009;**58**:443–7.

15. Grund F, Sommerschild HT, Kirkeboen KA, Ilebekk A. Proarrhythmic effects of ischemic preconditioning in anesthetized pigs. *Basic Res Cardiol* 1997;**92**:417–25.

16. Sadat U. Signaling pathways of cardioprotective ischemic preconditioning. *Int J Surg* 2009;**7**:490–8.

17. Otani H. Ischemic preconditioning: from molecular mechanisms to therapeutic opportunities. *Antioxid Redox Signal* 2008;**10**:207–47.

18. Liem DA, Manintveld OC, Schoonderwoerd K, McFalls EO, Heinen A, Verdouw PD, et al. Ischemic preconditioning modulates mitochondrial respiration, irrespective of the employed signal transduction pathway. *Transl Res* 2008;**151**:17–26.

19. Lim SY, Davidson SM, Hausenloy DJ, Yellon DM. Preconditioning and postconditioning: the essential role of the mitochondrial permeability transition pore. *Cardiovasc Res* 2007;**75**: 530–5.

20. Kamota T, Li TS, Morikage N, Murakami M, Ohshima M, Kubo M, et al. Ischemic pre-conditioning enhances the mobilization and recruitment of bone marrow stem cells to protect against ischemia/reperfusion injury in the late phase. *J Am Coll Cardiol* 2009;**53**:1814–22.

21. Kaur S, Jaggi AS, Singh N. Molecular aspects of ischaemic postconditioning. *Fundam Clin Pharmacol* 2009;**23**:521–36.

22. Murry CE, Jennings RB, Reimer KA. Preconditioning with ischemia: a delay of lethal cell injury in ischemic myocardium. *Circulation* 1986;**74**:1124–36.

23. Ugurlucan M, Yerebakan C, Furlani D, Ma N, Steinhoff G. Cell sources for cardiovascular tissue regeneration and engineering. *Thorac Cardiovasc Surg* 2009;**57**:63–73.

24. Singla DK, Hacker TA, Ma L, Douglas PS, Sullivan R, Lyons GE, et al. Transplantation of embryonic stem cells into the infarcted mouse heart: formation of multiple cell types. *J Mol Cell Cardiol* 2006;**40**:195–200.

25. Fatma S, Selby DE, Singla RD, Singla DK. Factors released from embryonic stem cells stimulate c-kit-FLK-1(+ve) progenitor cells and enhance neovascularization. *Antioxid Redox Signal* 2010;**13**:1857–65.

26. Singla DK, McDonald DE. Factors released from embryonic stem cells inhibit apoptosis of H9c2 cells. *Am J Physiol Heart Circ Physiol* 2007;**293**:H1590–5.

27. Crisostomo PR, Abarbanell AM, Wang M, Lahm T, Wang Y, Meldrum DR. Embryonic stem cells attenuate myocardial dysfunction and inflammation after surgical global ischemia via paracrine actions. *Am J Physiol Heart Circ Physiol* 2008;**295**:H1726–35.

28. Fong CY, Gauthaman K, Bongso A. Teratomas from pluripotent stem cells: a clinical hurdle. *J Cell Biochem* 2010;**111**:769–81.

29. Hentze H, Graichen R, Colman A. Cell therapy and the safety of embryonic stem cell-derived grafts. *Trends Biotechnol* 2007;**25**:24–32.

30. Boheler KR, Joodi RN, Qiao H, Juhasz O, Urick AL, Chuppa SL, et al. Embryonic stem cell-derived cardiomyocyte heterogeneity and the isolation of immature and committed cells for cardiac remodeling and regeneration. *Stem Cells Int* 2011;**2011**:214203.

31. Behfar A, Perez-Terzic C, Faustino RS, Arrell DK, Hodgson DM, Yamada S, et al. Cardio-poietic programming of embryonic stem cells for tumor-free heart repair. *J Exp Med* 2007;**204**:405–20.

32. Christoforou N, Miller RA, Hill CM, Jie CC, McCallion AS, Gearhart JD. Mouse ES cell-derived cardiac precursor cells are multipotent and facilitate identification of novel cardiac genes. *J Clin Invest* 2008;**118**:894–903.

33. Dey D, Evans GR. Generation of induced pluripotent stem (iPS) cells by nuclear reprogramming. *Stem Cells Int* 2011;**2011**:619583.

34. Takahashi K, Yamanaka S. Induction of pluripotent stem cells from mouse embryonic and adult fibroblast cultures by defined factors. *Cell* 2006;**126**:663–76.

35. Darabi R, Pan W, Bosnakovski D, Baik J, Kyba M, Perlingeiro RC. Functional myogenic engraftment from mouse iPS cells. *Stem Cell Rev* 2011;**7**:948–57.

36. Singla DK, Long X, Glass C, Singla RD, Yan B. Induced pluripotent stem (iPS) cells repair and regenerate infarcted myocardium. *Mol Pharm* 2011;**8**:1573–81.

37. Nelson TJ, Martinez-Fernandez A, Yamada S, Perez-Terzic C, Ikeda Y, Terzic A. Repair of acute myocardial infarction by human stemness factors induced pluripotent stem cells. *Circulation* 2009;**120**:408–16.

38. Liu SV. iPS cells: a more critical review. *Stem Cells Dev* 2008;**17**:391–7.

39. Cantz T, Martin U. Induced pluripotent stem cells: characteristics and perspectives. *Adv Biochem Eng Biotechnol* 2010;**123**:107–26.

40. Lee H, Park J, Forget BG, Gaines P. Induced pluripotent stem cells in regenerative medicine: an argument for continued research on human embryonic stem cells. *Regen Med* 2009;**4**:759–69.

41. Yoshida Y, Yamanaka S. iPS cells: a source of cardiac regeneration. *J Mol Cell Cardiol* 2011;**50**:327–32.

42. Ahmed RP, Ashraf M, Buccini S, Shujia J, Haider H. Cardiac tumorigenic potential of induced pluripotent stem cells in an immunocompetent host with myocardial infarction. *Regen Med* 2011;**6**:171–8.

43. Okita K, Hong H, Takahashi K, Yamanaka S. Generation of mouse-induced pluripotent stem cells with plasmid vectors. *Nat Protoc* 2010;**5**:418–28.

44. Zhao R, Daley GQ. From fibroblasts to iPS cells: induced pluripotency by defined factors. *J Cell Biochem* 2008;**105**:949–55.

45. Nakagawa M, Koyanagi M, Tanabe K, Takahashi K, Ichisaka T, Aoi T, et al. Generation of induced pluripotent stem cells without Myc from mouse and human fibroblasts. *Nat Biotechnol* 2008;**26**:101–6.

46. Park IH, Lerou PH, Zhao R, Huo H, Daley GQ. Generation of human-induced pluripotent stem cells. *Nat Protoc* 2008;**3**:1180–6.

47. Hamilton B, Feng Q, Ye M, Welstead GG. Generation of induced pluripotent stem cells by reprogramming mouse embryonic fibroblasts with a four transcription factor, doxycycline inducible lentiviral transduction system. *J Vis Exp* 2009;**13**:1447.

48. Pasha Z, Haider H, Ashraf M. Efficient non-viral reprogramming of myoblasts to stemness with a single small molecule to generate cardiac progenitor cells. *PLoS One* 2011;**6**:e23667.

49. Ahmed RP, Haider HK, Buccini S, Li L, Jiang S, Ashraf M. Reprogramming of skeletal myoblasts for induction of pluripotency for tumor-free cardiomyogenesis in the infarcted heart. *Circ Res* 2011;**109**:60–70.

50. Nelson TJ, Faustino RS, Chiriac A, Crespo-Diaz R, Behfar A, Terzic A. CXCR4+/FLK-1+ biomarkers select a cardiopoietic lineage from embryonic stem cells. *Stem Cells* 2008;**26**:1464–73.

51. Anversa P, Leri A, Kajstura J. Cardiac regeneration. *J Am Coll Cardiol* 2006;**47**:1769–76.

52. Beltrami AP, Barlucchi L, Torella D, Baker M, Limana F, Chimenti S, et al. Adult cardiac stem cells are multipotent and support myocardial regeneration. *Cell* 2003;**114**:763–76.

53. Brasselet C, Morichetti MC, Messas E, Carrion C, Bissery A, Bruneval P, et al. Skeletal myoblast transplantation through a catheter-based coronary sinus approach: an effective means of improving function of infarcted myocardium. *Eur Heart J* 2005;**26**:1551–6.

54. Ott HC, Kroess R, Bonaros N, Marksteiner R, Margreiter E, Schachner T, et al. Intramyocardial microdepot injection increases the efficacy of skeletal myoblast transplantation. *Eur J Cardiothorac Surg* 2005;**27**:1017–21.

55. He KL, Yi GH, Sherman W, Zhou H, Zhang GP, Gu A, et al. Autologous skeletal myoblast transplantation improved hemodynamics and left ventricular function in chronic heart failure dogs. *J Heart Lung Transplant* 2005;**24**:1940–9.

56. Imanishi Y, Miyagawa S, Saito A, Kitagawa-Sakakida S, Sawa Y. Allogenic skeletal myoblast transplantation in acute myocardial infarction model rats. *Transplantation* 2011;**91**:425–31.

57. Giraud MN, Flueckiger R, Cook S, Ayuni E, Siepe M, Carrel T, et al. Long-term evaluation of myoblast seeded patches implanted on infarcted rat hearts. *Artif Organs* 2011;**34**:E184–92.

58. Oshima H, Payne TR, Urish KL, Sakai T, Ling Y, Gharaibeh B, et al. Differential myocardial infarct repair with muscle stem cells compared to myoblasts. *Mol Ther* 2005;**12**: 1130–41.

59. Loffredo FS, Steinhauser ML, Gannon J, Lee RT. Bone marrow-derived cell therapy stimulates endogenous cardiomyocyte progenitors and promotes cardiac repair. *Cell Stem Cell* 2011;**8**:389–98.

60. See F, Seki T, Psaltis PJ, Sondermeijer HP, Gronthos S, Zannettino AC, et al. Therapeutic effects of human STRO-3-selected mesenchymal precursor cells and their soluble factors in experimental myocardial ischemia. *J Cell Mol Med* 2011;**15**:2117–29.

61. Dixon JA, Gorman RC, Stroud RE, Bouges S, Hirotsugu H, Gorman 3rd JH, et al. Mesenchymal cell transplantation and myocardial remodeling after myocardial infarction. *Circulation* 2009;**120**:S220–9.

62. Chugh AR, Zuba-Surma EK, Dawn B. Bone marrow-derived mesenchymal stems cells and cardiac repair. *Minerva Cardioangiol* 2009;**57**:185–202.

63. Orlic D, Kajstura J, Chimenti S, Jakoniuk I, Anderson SM, Li B, et al. Bone marrow cells regenerate infarcted myocardium. *Nature* 2001;**410**:701–5.

64. Cui XJ, Xie H, Wang HJ, Guo HD, Zhang JK, Wang C, et al. Transplantation of mesenchymal stem cells with self-assembling polypeptide scaffolds is conducive to treating myocardial infarction in rats. *Tohoku J Exp Med* 2010;**222**:281–9.

65. Durrani S, Konoplyannikov M, Ashraf M, Haider KH. Skeletal myoblasts for cardiac repair. *Regen Med* 2011;**5**:919–32.

66. Haider H, Ashraf M. Bone marrow stem cell transplantation for cardiac repair. *Am J Physiol Heart Circ Physiol* 2005;**288**:H2557–67.

67. Gepstein L, Ding C, Rahmutula D, Wilson EE, Yankelson L, Caspi O, et al. In vivo assessment of the electrophysiological integration and arrhythmogenic risk of myocardial cell transplantation strategies. *Stem Cells* 2010;**28**:2151–61.

68. Bel A, Messas E, Agbulut O, Richard P, Samuel JL, Bruneval P, et al. Transplantation of autologous fresh bone marrow into infarcted myocardium: a word of caution. *Circulation* 2003;**108**(Suppl. 1):II247–52.

69. Murry CE, Soonpaa MH, Reinecke H, Nakajima H, Nakajima HO, Rubart M, et al. Haematopoietic stem cells do not transdifferentiate into cardiac myocytes in myocardial infarcts. *Nature* 2004;**428**:664–8.

70. Haider H, Ashraf M. Bone marrow cell transplantation in clinical perspective. *J Mol Cell Cardiol* 2005;**38**:225–35.

71. Haider H, Tan AC, Aziz S, Chachques JC, Sim EK. Myoblast transplantation for cardiac repair: a clinical perspective. *Mol Ther* 2004;**9**:14–23.

72. Dib N, Michler RE, Pagani FD, Wright S, Kereiakes DJ, Lengerich R, et al. Safety and feasibility of autologous myoblast transplantation in patients with ischemic cardiomyopathy: four-year follow-up. *Circulation* 2005;**112**:1748–55.

73. Gavira JJ, Herreros J, Perez A, Garcia-Velloso MJ, Barba J, Martin-Herrero F, et al. Autologous skeletal myoblast transplantation in patients with nonacute myocardial infarction: 1-year follow-up. *J Thorac Cardiovasc Surg* 2006;**131**:799–804.

74. Mansour S, Roy DC, Bouchard V, Nguyen BK, Stevens LM, Gobeil F, et al. COMPARE-AMI trial: comparison of intracoronary injection of CD133+ bone marrow stem cells to placebo in patients after acute myocardial infarction and left ventricular dysfunction: study rationale and design. *J Cardiovasc Transl Res* 2011;**3**:153–9.

75. Wollert KC, Meyer GP, Lotz J, Ringes-Lichtenberg S, Lippolt P, Breidenbach C, et al. Intracoronary autologous bone-marrow cell transfer after myocardial infarction: the BOOST randomised controlled clinical trial. *Lancet* 2004;**364**:141–8.

76. Baldi A, Abbate A, Bussani R, Patti G, Melfi R, Angelini A, et al. Apoptosis and post-infarction left ventricular remodeling. *J Mol Cell Cardiol* 2002;**34**:165–74.

77. Bai X, Alt E. Myocardial regeneration potential of adipose tissue-derived stem cells. *Biochem Biophys Res Commun* 2011;**401**:321–6.

78. Hoke NN, Salloum FN, Kass DA, Das A, Kukreja RC. Preconditioning by phosphodiesterase-5 inhibition improves therapeutic efficacy of adipose derived stem cells following myocardial infarction in mice. *Stem Cells* 2012;**30**:326–35. http://dx.doi.org/10.1002/stem.789.

79. Shi CZ, Zhang XP, Lv ZW, Zhang HL, Xu JZ, Yin ZF, et al. Adipose tissue-derived stem cells embedded with eNOS restore cardiac function in acute myocardial infarction model. *Int J Cardiol* 2011;**154**:2–8.

80. Hu CH, Li ZM, Du ZM, Zhang AX, Yang DY, Wu GF. Human umbilical cord-derived endothelial progenitor cells promote growth cytokines-mediated neorevascularization in rat myocardial infarction. *Chin Med J (Engl)* 2009;**122**:548–55.

81. Wu KH, Mo XM, Zhou B, Lu SH, Yang SG, Liu YL, et al. Cardiac potential of stem cells from whole human umbilical cord tissue. *J Cell Biochem* 2009;**107**:926–32.

82. Copeland N, Harris D, Gaballa MA. Human umbilical cord blood stem cells, myocardial infarction and stroke. *Clin Med* 2009;**9**:342–5.

83. Wojakowski W, Tendera M, Kucia M, Zuba-Surma E, Paczkowska E, Ciosek J, et al. Mobilization of bone marrow-derived Oct-4+ SSEA-4+ very small embryonic-like stem cells in patients with acute myocardial infarction. *J Am Coll Cardiol* 2009;**53**:1–9.

84. Sovalat H, Scrofani M, Eidenschenk A, Pasquet S, Rimelen V, Henon P. Identification and isolation from either adult human bone marrow or G-CSF-mobilized peripheral blood of CD34(+)/CD133(+)/CXCR4(+)/ Lin(-)CD45(-) cells, featuring morphological, molecular, and phenotypic characteristics of very small embryonic-like (VSEL) stem cells. *Exp Hematol* 2011;**39**:495–505.

85. Zuba-Surma EK, Guo Y, Taher H, Sanganalmath SK, Hunt G, Vincent RJ, et al. Transplantation of expanded bone marrow-derived very small embryonic-like stem cells (VSEL-SCs) improves left ventricular function and remodelling after myocardial infarction. *J Cell Mol Med* 2011;**15**:1319–28.

86. Muller-Ehmsen J, Kedes LH, Schwinger RH, Kloner RA. Cellular cardiomyoplasty—a novel approach to treat heart disease. *Congest Heart Fail* 2002;**8**:220–7.

87. Marelli D, Desrosiers C, el-Alfy M, Kao RL, Chiu RC. Cell transplantation for myocardial repair: an experimental approach. *Cell Transplant* 1992;**1**:383–90.

88. Takamiya M, Haider KH, Ashraf M. Identification and characterization of a novel multipotent sub-population of Sca-1 cardiac progenitor cells for myocardial regeneration. *PLoS One* 2011;**6**:e25265.

89. Jiang S, Haider H, Idris NM, Salim A, Ashraf M. Supportive interaction between cell survival signaling and angiocompetent factors enhances donor cell survival and promotes angiomyogenesis for cardiac repair. *Circ Res* 2006;**99**:776–84.

90. Kocher AA, Schuster MD, Szabolcs MJ, Takuma S, Burkhoff D, Wang J, et al. Neovascularization of ischemic myocardium by human bone-marrow-derived angioblasts prevents cardiomyocyte apoptosis, reduces remodeling and improves cardiac function. *Nat Med* 2001;**7**:430–6.

91. Kamihata H, Matsubara H, Nishiue T, Fujiyama S, Tsutsumi Y, Ozono R, et al. Implantation of bone marrow mononuclear cells into ischemic myocardium enhances collateral perfusion and regional function via side supply of angioblasts, angiogenic ligands, and cytokines. *Circulation* 2001;**104**:1046–52.

92. Pagani FD, DerSimonian H, Zawadzka A, Wetzel K, Edge AS, Jacoby DB, et al. Autologous skeletal myoblasts transplanted to ischemia-damaged myocardium in humans. Histological analysis of cell survival and differentiation. *J Am Coll Cardiol* 2003;**41**:879–88.

93. Menasche P, Hagege AA, Scorsin M, Pouzet B, Desnos M, Duboc D, et al. Myoblast transplantation for heart failure. *Lancet* 2001;**357**:279–80.

94. Drexler H, Meyer GP, Wollert KC. Bone-marrow-derived cell transfer after ST-elevation myocardial infarction: lessons from the BOOST trial. *Nat Clin Pract Cardiovasc Med* 2006;**3**(Suppl 1):S65–8.

95. Siminiak T, Grygielska B, Jerzykowska O, Fiszer D, Kalmucki P, Rzezniczak J, et al. Autologous bone marrow stem cell transplantation in acute myocardial infarction-report of two cases. *Kardiol Pol* 2003;**59**:502–10.

96. Muller-Ehmsen J, Whittaker P, Kloner RA, Dow JS, Sakoda T, Long TI, et al. Survival and development of neonatal rat cardiomyocytes transplanted into adult myocardium. *J Mol Cell Cardiol* 2002;**34**:107–16.

97. Qu Z, Balkir L, van Deutekom JC, Robbins PD, Pruchnic R, Huard J. Development of approaches to improve cell survival in myoblast transfer therapy. *J Cell Biol* 1998;**142**:1257–67.

98. Fan Y, Maley M, Beilharz M, Grounds M. Rapid death of injected myoblasts in myoblast transfer therapy. *Muscle Nerve* 1996;**19**:853–60.

99. Toma C, Pittenger MF, Cahill KS, Byrne BJ, Kessler PD. Human mesenchymal stem cells differentiate to a cardiomyocyte phenotype in the adult murine heart. *Circulation* 2002;**105**:93–8.

100. Skuk D, Goulet M, Roy B, Tremblay JP. Efficacy of myoblast transplantation in nonhuman primates following simple intramuscular cell injections: toward defining strategies applicable to humans. *Exp Neurol* 2002;**175**:112–26.

101. Haider H, Jiang SJ, Ye L, Aziz S, Law PK, Sim EK. Effectiveness of transient immunosuppression using cyclosporine for xenomyoblast transplantation for cardiac repair. *Transplant Proc* 2004;**36**:232–5.

102. Lu G, Haider HK, Jiang S, Ashraf M. Sca-1+ stem cell survival and engraftment in the infarcted heart: dual role for preconditioning-induced connexin-43. *Circulation* 2009;**119**:2587–96.

103. Bosio E, Lee-Pullen TF, Fragall CT, Beilharz MW, Bennett AL, Grounds MD, et al. A comparison between real-time quantitative PCR and DNA hybridization for quantitation of male DNA following myoblast transplantation. *Cell Transplant* 2004;**13**:817–21.

104. Hodgetts SI, Beilharz MW, Scalzo AA, Grounds MD. Why do cultured transplanted myoblasts die in vivo? DNA quantification shows enhanced survival of donor male myoblasts in host mice depleted of CD4+ and CD8+ cells or Nk1.1+ cells. *Cell Transplant* 2000;**9**:489–502.

105. Pouzet B, Vilquin JT, Hagege AA, Scorsin M, Messas E, Fiszman M, et al. Intramyocardial transplantation of autologous myoblasts: can tissue processing be optimized? *Circulation* 2000;**102**:III210–5.

106. Pouzet B, Vilquin JT, Hagege AA, Scorsin M, Messas E, Fiszman M, et al. Factors affecting functional outcome after autologous skeletal myoblast transplantation. *Ann Thorac Surg* 2001;**71**:844–50.

107. Skuk D, Caron NJ, Goulet M, Roy B, Tremblay JP. Resetting the problem of cell death following muscle-derived cell transplantation: detection, dynamics and mechanisms. *J Neuropathol Exp Neurol* 2003;**62**:951–67.

108. Geng YJ. Molecular mechanisms for cardiovascular stem cell apoptosis and growth in the hearts with atherosclerotic coronary disease and ischemic heart failure. *Ann N Y Acad Sci* 2003;**1010**:687–97.

109. Camirand G, Caron NJ, Turgeon NA, Rossini AA, Tremblay JP. Treatment with anti-CD154 antibody and donor-specific transfusion prevents acute rejection of myoblast transplantation. *Transplantation* 2002;**73**:453–61.

110. Hodgetts SI, Grounds MD. Complement and myoblast transfer therapy: donor myoblast survival is enhanced following depletion of host complement C3 using cobra venom factor, but not in the absence of C5. *Immunol Cell Biol* 2001;**79**:231–9.

111. Singla DK, Singla RD, Lamm S, Glass C. TGF-beta2 treatment enhances cytoprotective factors released from embryonic stem cells and inhibits apoptosis in infarcted myocardium. *Am J Physiol Heart Circ Physiol* 2011;**300**:H1442–50.

112. Herrmann JL, Wang Y, Abarbanell AM, Weil BR, Tan J, Meldrum DR. Preconditioning mesenchymal stem cells with transforming growth factor-alpha improves mesenchymal stem cell-mediated cardioprotection. *Shock* 2011;**33**:24–30.

113. Kinoshita I, Vilquin JT, Tremblay JP. Pretreatment of myoblast cultures with basic fibroblast growth factor increases the efficacy of their transplantation in mdx mice. *Muscle Nerve* 1995;**18**:834–41.

114. Cizkova D, Rosocha J, Vanicky I, Radonak J, Galik J, Cizek M. Induction of mesenchymal stem cells leads to HSP72 synthesis and higher resistance to oxidative stress. *Neurochem Res* 2006;**31**:1011–20.

115. Tang YL, Tang Y, Zhang YC, Qian K, Shen L, Phillips MI. Improved graft mesenchymal stem cell survival in ischemic heart with a hypoxia-regulated heme oxygenase-1 vector. *J Am Coll Cardiol* 2005;**46**:1339–50.

116. Haider H, Jiang S, Idris NM, Ashraf M. IGF-1-overexpressing mesenchymal stem cells accelerate bone marrow stem cell mobilization via paracrine activation of SDF-1alpha/CXCR4 signaling to promote myocardial repair. *Circ Res* 2008;**103**:1300–8.

117. Liu TB, Fedak PW, Weisel RD, Yasuda T, Kiani G, Mickle DA, et al. Enhanced IGF-1 expression improves smooth muscle cell engraftment after cell transplantation. *Am J Physiol Heart Circ Physiol* 2004;**287**:H2840–9.

118. Kutschka I, Kofidis T, Chen IY, von Degenfeld G, Zwierzchoniewska M, Hoyt G, et al. Adenoviral human BCL-2 transgene expression attenuates early donor cell death after cardiomyoblast transplantation into ischemic rat hearts. *Circulation* 2006;**114**:I174–80.

119. Li W, Ma N, Ong LL, Nesselmann C, Klopsch C, Ladilov Y, et al. Bcl-2 engineered MSCs inhibited apoptosis and improved heart function. *Stem Cells* 2007;**25**:2118–27.

120. Mangi AA, Noiseux N, Kong D, He H, Rezvani M, Ingwall JS, et al. Mesenchymal stem cells modified with Akt prevent remodeling and restore performance of infarcted hearts. *Nat Med* 2003;**9**:1195–201.

121. Gross GJ, Auchampach JA. Blockade of ATP-sensitive potassium channels prevents myocardial preconditioning in dogs. *Circ Res* 1992;**70**:223–33.

122. Miyazaki T, Zipes DP. Protection against autonomic denervation following acute myocardial infarction by preconditioning ischemia. *Circ Res* 1989;**64**:437–48.

123. Li GC, Vasquez JA, Gallagher KP, Lucchesi BR. Myocardial protection with preconditioning. *Circulation* 1990;**82**:609–19.

124. Schultz JE, Hsu AK, Gross GJ. Morphine mimics the cardioprotective effect of ischemic preconditioning via a glibenclamide-sensitive mechanism in the rat heart. *Circ Res* 1996;**78**:1100–4.

125. Liu GS, Thornton J, Van Winkle DM, Stanley AW, Olsson RA, Downey JM. Protection against infarction afforded by preconditioning is mediated by A1 adenosine receptors in rabbit heart. *Circulation* 1991;**84**:350–6.

126. Lankford AR, Yang JN, Rose'Meyer R, French BA, Matherne GP, Fredholm BB, et al. Effect of modulating cardiac A1 adenosine receptor expression on protection with ischemic preconditioning. *Am J Physiol Heart Circ Physiol* 2006;**290**:H1469–73.

127. Peart JN, Gross ER, Gross GJ. Opioid-induced preconditioning: recent advances and future perspectives. *Vascul Pharmacol* 2005;**42**:211–8.

128. Saraiva J, Oliveira SM, Rocha-Sousa A, Leite-Moreira A. Opioid receptors and preconditioning of the heart. *Rev Port Cardiol* 2004;**23**:1317–33.

129. Tong H, Rockman HA, Koch WJ, Steenbergen C, Murphy E. G protein-coupled receptor internalization signaling is required for cardioprotection in ischemic preconditioning. *Circ Res* 2004;**94**:1133–41.

130. Hausenloy DJ, Yellon DM. Survival kinases in ischemic preconditioning and postconditioning. *Cardiovasc Res* 2006;**70**:240–53.

131. Wang Y, Takashi E, Xu M, Ayub A, Ashraf M. Downregulation of protein kinase C inhibits activation of mitochondrial K(ATP) channels by diazoxide. *Circulation* 2001;**104**:85–90.

132. Wang Y, Ashraf M. Role of protein kinase C in mitochondrial KATP channel-mediated protection against Ca2 + overload injury in rat myocardium. *Circ Res* 1999;**84**:1156–65.

133. Semenza GL. O2 sensing: only skin deep? *Cell* 2008;**133**:206–8.

134. Lahiri S, Roy A, Baby SM, Hoshi T, Semenza GL, Prabhakar NR. Oxygen sensing in the body. *Prog Biophys Mol Biol* 2006;**91**:249–86.

135. Davy P, Allsopp R. Hypoxia: are stem cells in it for the long run? *Cell Cycle* 2011;**10**:206–11.

136. Jaderstad J, Brismar H, Herlenius E. Hypoxic preconditioning increases gap-junctional graft and host communication. *Neuroreport* 2010;**21**:1126–32.

137. Rasmussen JG, Frobert O, Pilgaard L, Kastrup J, Simonsen U, Zachar V, et al. Prolonged hypoxic culture and trypsinization increase the pro-angiogenic potential of human adipose tissue-derived stem cells. *Cytotherapy* 2011;**13**:318–28.

138. Mendez O, Zavadil J, Esencay M, Lukyanov Y, Santovasi D, Wang SC, et al. Knock down of HIF-1alpha in glioma cells reduces migration in vitro and invasion in vivo and impairs their ability to form tumor spheres. *Mol Cancer* 2011;**9**:133.

139. Santilli G, Lamorte G, Carlessi L, Ferrari D, Rota Nodari L, Binda E, et al. Mild hypoxia enhances proliferation and multipotency of human neural stem cells. *PLoS One* 2011;**5**:e8575.

140. Liu L, Yu Q, Lin J, Lai X, Cao W, Du K, et al. Hypoxia-inducible factor-1alpha is essential for hypoxia-induced mesenchymal stem cell mobilization into the peripheral blood. *Stem Cells Dev* 2011;**20**:1961–71.

141. Semenza GL. Regulation of metabolism by hypoxia-inducible factor 1. *Cold Spring Harb Symp Quant Biol* 2011; (Published ahead of print).

142. Semenza GL. Regulation of vascularization by hypoxia-inducible factor 1. *Ann N Y Acad Sci* 2009;**1177**:2–8.

143. Semenza GL. Oxygen homeostasis. *Wiley Interdiscip Rev Syst Biol Med* 2010;**2**:336–61.

144. Semenza G. Signal transduction to hypoxia-inducible factor 1. *Biochem Pharmacol* 2002;**64**:993–8.

145. Li JH, Zhang N, Wang JA. Improved anti-apoptotic and anti-remodeling potency of bone marrow mesenchymal stem cells by anoxic pre-conditioning in diabetic cardiomyopathy. *J Endocrinol Invest* 2008;**31**:103–10.

146. Theus MH, Wei L, Cui L, Francis K, Hu X, Keogh C, et al. In vitro hypoxic preconditioning of embryonic stem cells as a strategy of promoting cell survival and functional benefits after transplantation into the ischemic rat brain. *Exp Neurol* 2008;**210**:656–70.

147. Chacko SM, Ahmed S, Selvendiran K, Kuppusamy ML, Khan M, Kuppusamy P. Hypoxic preconditioning induces the expression of prosurvival and proangiogenic markers in mesenchymal stem cells. *Am J Physiol Cell Physiol* 2010;**299**:C1562–70.

148. Hu X, Yu SP, Fraser JL, Lu Z, Ogle ME, Wang JA, et al. Transplantation of hypoxia-preconditioned mesenchymal stem cells improves infarcted heart function via enhanced survival of implanted cells and angiogenesis. *J Thorac Cardiovasc Surg* 2008;**135**:799–808.

149. Wang JA, Chen TL, Jiang J, Shi H, Gui C, Luo RH, et al. Hypoxic preconditioning attenuates hypoxia/reoxygenation-induced apoptosis in mesenchymal stem cells. *Acta Pharmacol Sin* 2008;**29**:74–82.

150. Stubbs SL, Hsiao ST, Peshavariya H, Lim SY, Dusting GJ, Dilley RJ. Hypoxic preconditioning enhances survival of human adipose-derived stem cells and conditions endothelial cells in vitro. *Stem Cells Dev* 2011; (Published ahead of print).

151. Rosova I, Dao M, Capoccia B, Link D, Nolta JA. Hypoxic preconditioning results in increased motility and improved therapeutic potential of human mesenchymal stem cells. *Stem Cells* 2008;**26**:2173–82.

152. Tang YL, Zhu W, Cheng M, Chen L, Zhang J, Sun T, et al. Hypoxic preconditioning enhances the benefit of cardiac progenitor cell therapy for treatment of myocardial infarction by inducing CXCR4 expression. *Circ Res* 2009;**104**:1209–16.

153. Semenza GL. HIF-1: mediator of physiological and pathophysiological responses to hypoxia. *J Appl Physiol* 2000;**88**:1474–80.

154. Pedersen M, Lofstedt T, Sun J, Holmquist-Mengelbier L, Pahlman S, Ronnstrand L. Stem cell factor induces HIF-1alpha at normoxia in hematopoietic cells. *Biochem Biophys Res Commun* 2008;**377**:98–103.

155. Gorlach A. Regulation of HIF-1alpha at the transcriptional level. *Curr Pharm Des* 2009;**15**:3844–52.

156. Milosevic J, Adler I, Manaenko A, Schwarz SC, Walkinshaw G, Arend M, et al. Non-hypoxic stabilization of hypoxia-inducible factor alpha (HIF-alpha): relevance in neural progenitor/stem cells. *Neurotox Res* 2009;**15**:367–80.

157. Wang Y, Feng C, Xue J, Sun A, Li J, Wu J. Adenovirus-mediated hypoxia-inducible factor 1alpha double-mutant promotes differentiation of bone marrow stem cells to cardiomyocytes. *J Physiol Sci* 2009;**59**:413–20.

158. Azarnoush K, Maurel A, Sebbah L, Carrion C, Bissery A, Mandet C, et al. Enhancement of the functional benefits of skeletal myoblast transplantation by means of coadministration of hypoxia-inducible factor 1alpha. *J Thorac Cardiovasc Surg* 2005;**130**:173–9.

159. Wu W, Chen X, Hu C, Li J, Yu Z, Cai W. Transplantation of neural stem cells expressing hypoxia-inducible factor-1alpha (HIF-1alpha) improves behavioral recovery in a rat stroke model. *J Clin Neurosci* 2011;**17**:92–5.

160. Keranen MA, Nykanen AI, Krebs R, Pajusola K, Tuuminen R, Alitalo K, et al. Cardiomyocyte-targeted HIF-1alpha gene therapy inhibits cardiomyocyte apoptosis and cardiac allograft vasculopathy in the rat. *J Heart Lung Transplant* 2011;**29**:1058–66.

161. McDunn JE, Cobb JP. That which does not kill you makes you stronger: a molecular mechanism for preconditioning. *Sci STKE* 2005;**2005**:pe34.

162. Kim HW, Haider HK, Jiang S, Ashraf M. Ischemic preconditioning augments survival of stem cells via MIR-210 expression by targeting caspase-8 associated protein 2. *J Biol Chem* 2009;**284**:33161–7.

163. Wurdinger T, Costa FF. Molecular therapy in the microRNA era. *Pharmacogenomics J* 2007;**7**:297–304.

164. Kulshreshtha R, Ferracin M, Negrini M, Calin GA, Davuluri RV, Ivan M. Regulation of microRNA expression: the hypoxic component. *Cell Cycle* 2007;**6**:1426–31.
165. Kulshreshtha R, Ferracin M, Wojcik SE, Garzon R, Alder H, Agosto-Perez FJ, et al. A microRNA signature of hypoxia. *Mol Cell Biol* 2007;**27**:1859–67.
166. Chan SY, Loscalzo J. MicroRNA-210: a unique and pleiotropic hypoxamir. *Cell Cycle* 2011;**9**:1072–83.
167. Kulshreshtha R, Davuluri RV, Calin GA, Ivan M. A microRNA component of the hypoxic response. *Cell Death Differ* 2008;**15**:667–71.
168. Mutharasan RK, Nagpal V, Ichikawa Y, Ardehali H. microRNA-210 is upregulated in hypoxic cardiomyocytes through Akt- and p53-dependent pathways and exerts cytoprotective effects. *Am J Physiol Heart Circ Physiol* 2011;**301**:H1519–30.
169. Hosoda T, Zheng H, Cabral-da-Silva M, Sanada F, Ide-Iwata N, Ogorek B, et al. Human cardiac stem cell differentiation is regulated by a mircrine mechanism. *Circulation* 2011;**123**:1287–96.
170. Yang CC, Lin LC, Wu MS, Chien CT, Lai MK. Repetitive hypoxic preconditioning attenuates renal ischemia/reperfusion induced oxidative injury via upregulating HIF-1 alpha-dependent bcl-2 signaling. *Transplantation* 2009;**88**:1251–60.
171. Das S, Cordis GA, Maulik N, Das DK. Pharmacological preconditioning with resveratrol: role of CREB-dependent Bcl-2 signaling via adenosine A3 receptor activation. *Am J Physiol Heart Circ Physiol* 2005;**288**:H328–35.
172. Caparrelli DJ, Cattaneo 2nd SM, Bethea BT, Shake JG, Eberhart C, Blue ME, et al. Pharmacological preconditioning ameliorates neurological injury in a model of spinal cord ischemia. *Ann Thorac Surg* 2002;**74**:838–44.
173. Garlid KD, Paucek P, Yarov-Yarovoy V, Murray HN, Darbenzio RB, D'Alonzo AJ, et al. Cardioprotective effect of diazoxide and its interaction with mitochondrial ATP-sensitive K + channels. Possible mechanism of cardioprotection. *Circ Res* 1997;**81**:1072–82.
174. Ghosh S, Standen NB, Galinanes M. Evidence for mitochondrial K ATP channels as effectors of human myocardial preconditioning. *Cardiovasc Res* 2000;**45**:934–40.
175. Takashi E, Wang Y, Ashraf M. Activation of mitochondrial K(ATP) channel elicits late preconditioning against myocardial infarction via protein kinase C signaling pathway. *Circ Res* 1999;**85**:1146–53.
176. Oldenburg O, Cohen MV, Downey JM. Mitochondrial K(ATP) channels in preconditioning. *J Mol Cell Cardiol* 2003;**35**:569–75.
177. Ahmad N, Wang Y, Haider KH, Wang B, Pasha Z, Uzun O, et al. Cardiac protection by mitoKATP channels is dependent on Akt translocation from cytosol to mitochondria during late preconditioning. *Am J Physiol Heart Circ Physiol* 2006;**290**:H2402–8.
178. Niagara MI, Haider H, Jiang S, Ashraf M. Pharmacologically preconditioned skeletal myoblasts are resistant to oxidative stress and promote angiomyogenesis via release of paracrine factors in the infarcted heart. *Circ Res* 2007;**100**:545–55.
179. Suzuki Y, Kim HW, Ashraf M, Haider H. Diazoxide potentiates mesenchymal stem cell survival via NF-kappaB-dependent miR-146a expression by targeting Fas. *Am J Physiol Heart Circ Physiol* 2010;**299**:H1077–82.
180. Idris NM, Ashraf M, Ahmed RP, Shujia J, Haider KH. Activation of IL-11/STAT3 pathway in preconditioned human skeletal myoblasts blocks apoptotic cascade under oxidant stress. *Regen Med* 2011;**7**:47–57.
181. Haider H, Lee YJ, Jiang S, Ahmed RP, Ryon M, Ashraf M. Phosphodiesterase inhibition with tadalafil provides longer and sustained protection of stem cells. *Am J Physiol Heart Circ Physiol* 2010;**299**:H1395–404.
182. Phillips MI, Tang YL. Genetic modification of stem cells for transplantation. *Adv Drug Deliv Rev* 2008;**60**:160–72.

183. Wang X, Zhao T, Huang W, Wang T, Qian J, Xu M, et al. Hsp20-engineered mesenchymal stem cells are resistant to oxidative stress via enhanced activation of Akt and increased secretion of growth factors. *Stem Cells* 2009;**27**:3021–31.

184. Matsui T, Tao J, del Monte F, Lee KH, Li L, Picard M, et al. Akt activation preserves cardiac function and prevents injury after transient cardiac ischemia in vivo. *Circulation* 2001;**104**:330–5.

185. Gnecchi M, He H, Liang OD, Melo LG, Morello F, Mu H, et al. Paracrine action accounts for marked protection of ischemic heart by Akt-modified mesenchymal stem cells. *Nat Med* 2005;**11**:367–8.

186. Fan L, Lin C, Zhuo S, Chen L, Liu N, Luo Y, et al. Transplantation with survivin-engineered mesenchymal stem cells results in better prognosis in a rat model of myocardial infarction. *Eur J Heart Fail* 2009;**11**:1023–30.

187. Muraski JA, Rota M, Misao Y, Fransioli J, Cottage C, Gude N, et al. Pim-1 regulates cardiomyocyte survival downstream of Akt. *Nat Med* 2007;**13**:1467–75.

188. Borillo GA, Mason M, Quijada P, Volkers M, Cottage C, McGregor M, et al. Pim-1 kinase protects mitochondrial integrity in cardiomyocytes. *Circ Res* 2011;**106**:1265–74.

189. Fischer KM, Cottage CT, Wu W, Din S, Gude NA, Avitabile D, et al. Enhancement of myocardial regeneration through genetic engineering of cardiac progenitor cells expressing Pim-1 kinase. *Circulation* 2009;**120**:2077–87.

190. Duan HF, Wu CT, Wu DL, Lu Y, Liu HJ, Ha XQ, et al. Treatment of myocardial ischemia with bone marrow-derived mesenchymal stem cells overexpressing hepatocyte growth factor. *Mol Ther* 2003;**8**:467–74.

191. Elmadbouh I, Haider H, Jiang S, Idris NM, Lu G, Ashraf M. Ex vivo delivered stromal cell-derived factor-1alpha promotes stem cell homing and induces angiomyogenesis in the infarcted myocardium. *J Mol Cell Cardiol* 2007;**42**:792–803.

192. Ye L, Haider H, Guo C, Sim EK. Cell-based VEGF delivery prevents donor cell apoptosis after transplantation. *Ann Thorac Surg* 2007;**83**:1233–4.

193. Ye L, Haider H, Jiang S, Ling LH, Ge R, Law PK, et al. Reversal of myocardial injury using genetically modulated human skeletal myoblasts in a rodent cryoinjured heart model. *Eur J Heart Fail* 2005;**7**:945–52.

194. Yau TM, Kim C, Ng D, Li G, Zhang Y, Weisel RD, et al. Increasing transplanted cell survival with cell-based angiogenic gene therapy. *Ann Thorac Surg* 2005;**80**:1779–86.

195. Durrani S, Haider KH, Ahmed RP, Jiang S, Ashraf M. Cytoprotective and proangiogenic activity of ex-vivo netrin-1 transgene overexpression protects the heart against ischemia/reperfusion injury. *Stem Cells Dev* 2011. Published ahead of print.

196. Ahmed RP, Haider KH, Shujia J, Afzal MR, Ashraf M. Sonic Hedgehog gene delivery to the rodent heart promotes angiogenesis via iNOS/netrin-1/PKC pathway. *PLoS One* 2010;**5**:e8576.

197. Ye L, Haider H, Jiang S, Tan RS, Toh WC, Ge R, et al. Angiopoietin-1 for myocardial angiogenesis: a comparison between delivery strategies. *Eur J Heart Fail* 2007;**9**:458–65.

198. Song H, Kwon K, Lim S, Kang SM, Ko YG, Xu Z, et al. Transfection of mesenchymal stem cells with the FGF-2 gene improves their survival under hypoxic conditions. *Mol Cells* 2005;**19**:402–7.

199. Zeng B, Lin G, Ren X, Zhang Y, Chen H. Over-expression of HO-1 on mesenchymal stem cells promotes angiogenesis and improves myocardial function in infarcted myocardium. *J Biomed Sci* 2011;**17**:80.

200. Shu T, Zeng B, Ren X, Li Y. HO-1 modified mesenchymal stem cells modulate MMPs/TIMPs system and adverse remodeling in infarcted myocardium. *Tissue Cell* 2011;**42**:217–22.

201. Kofidis T, de Bruin JL, Yamane T, Balsam LB, Lebl DR, Swijnenburg RJ, et al. Insulin-like growth factor promotes engraftment, differentiation, and functional improvement after transfer of embryonic stem cells for myocardial restoration. *Stem Cells* 2004;**22**:1239–45.

202. Bartunek J, Croissant JD, Wijns W, Gofflot S, de Lavareille A, Vanderheyden M, et al. Pretreatment of adult bone marrow mesenchymal stem cells with cardiomyogenic growth factors and repair of the chronically infarcted myocardium. *Am J Physiol Heart Circ Physiol* 2007;**292**:H1095–104.

203. Mahboubi K, Biedermann BC, Carroll JM, Pober JS. IL-11 activates human endothelial cells to resist immune-mediated injury. *J Immunol* 2000;**164**:3837–46.

204. Mias C, Trouche E, Seguelas MH, Calcagno F, Dignat-George F, Sabatier F, et al. Ex vivo pretreatment with melatonin improves survival, proangiogenic/mitogenic activity, and efficiency of mesenchymal stem cells injected into ischemic kidney. *Stem Cells* 2008;**26**:1749–57.

205. Khan M, Akhtar S, Mohsin S, N Khan S, Riazuddin S. Growth factor preconditioning increases the function of diabetes-impaired mesenchymal stem cells. *Stem Cells Dev* 2011;**20**:67–75.

206. Joo HJ, Kim H, Park SW, Cho HJ, Kim HS, Lim DS, et al. Angiopoietin-1 promotes endothelial differentiation from embryonic stem cells and induced pluripotent stem cells. *Blood* 2010;**118**:2094–104.

207. Fujio Y, Nguyen T, Wencker D, Kitsis RN, Walsh K. Akt promotes survival of cardiomyocytes in vitro and protects against ischemia-reperfusion injury in mouse heart. *Circulation* 2000;**101**:660–7.

208. Kadowaki T, Tobe K, Honda-Yamamoto R, Tamemoto H, Kaburagi Y, Momomura K, et al. Signal transduction mechanism of insulin and insulin-like growth factor-1. *Endocr J* 1996;**43** (Suppl):S33–41.

209. Lu G, Haider H, Porollo A, Ashraf M. Mitochondria-specific transgenic overexpression of connexin-43 simulates preconditioning-induced cytoprotection of stem cells. *Cardiovasc Res* 2010;**88**:277–86.

210. Lu G, Ashraf M, Haider KH. Insulin-like growth factor-1 preconditioning accentuates intrinsic survival mechanism in stem cells to resist ischemic injury by orchestrating protein kinase calpha-erk1/2 activation. *Antioxid Redox Signal* 2012;**16**:217–27.

211. Schulz R, Boengler K, Totzeck A, Luo Y, Garcia-Dorado D, Heusch G. Connexin 43 in ischemic pre- and postconditioning. *Heart Fail Rev* 2007;**12**:261–6.

212. Han JS, Wang HS, Yan DM, Wang ZW, Han HG, Zhu HY, et al. Myocardial ischaemic and diazoxide preconditioning both increase PGC-1alpha and reduce mitochondrial damage. *Acta Cardiol* 2011;**65**:639–44.

213. Zu L, Zheng X, Wang B, Parajuli N, Steenbergen C, Becker LC, et al. Ischemic preconditioning attenuates mitochondrial localization of PTEN induced by ischemia-reperfusion. *Am J Physiol Heart Circ Physiol* 2011;**300**:H2177–86.

214. Quarrie R, Cramer BM, Lee DS, Steinbaugh GE, Erdahl W, Pfeiffer DR, et al. Ischemic preconditioning decreases mitochondrial proton leak and reactive oxygen species production in the postischemic heart. *J Surg Res* 2011;**165**:5–14.

215. Gonzalez-Loyola A, Barba I. Mitochondrial metabolism revisited: a route to cardioprotection. *Cardiovasc Res* 2011;**88**:209–10.

216. Martinou JC, Youle RJ. Mitochondria in apoptosis: Bcl-2 family members and mitochondrial dynamics. *Dev Cell* 2011;**21**:92–101.

217. Huang C, Andres AM, Ratliff EP, Hernandez G, Lee P, Gottlieb RA. Preconditioning involves selective mitophagy mediated by Parkin and p62/SQSTM1. *PLoS One* 2011;**6**:e20975.

218. Lu G, Ashraf M, Shujia J, Haider HK. Mitochondrial transgenic expression of connexin 43 confers cytoprotection to the stem cells and cardiomyocytes in ischemic myocardium. *Circulation* 2011; (Abstract). Nov. (suppl).

219. Petros AM, Olejniczak ET, Fesik SW. Structural biology of the Bcl-2 family of proteins. *Biochim Biophys Acta* 2004;**1644**:83–94.

220. Yang J, Liu X, Bhalla K, Kim CN, Ibrado AM, Cai J, et al. Prevention of apoptosis by Bcl-2: release of cytochrome c from mitochondria blocked. *Science* 1997;**275**:1129–32.
221. Feng Y, Haider KH, Shujia S, Ashraf M. Pre-induction of Hsp70 is associated with stem cell resistance to ischemic stress via HSF1-miR34a-Hsp70 interaction. *Circulation* 2011; (Abstract). Nov. (Suppl).
222. Ferry AL, Vanderklish PW, Dupont-Versteegden EE. Enhanced survival of skeletal muscle myoblasts in response to overexpression of cold shock protein RBM3. *Am J Physiol Cell Physiol* 2011;**301**:C392–402.
223. Gary DS, Malone M, Capestany P, Houdayer T, McDonald JW. Electrical stimulation promotes the survival of oligodendrocytes in mixed cortical cultures. *J Neurosci Res* 2011;**90**:72–83.
224. Laflamme MA, Chen KY, Naumova AV, Muskheli V, Fugate JA, Dupras SK, et al. Cardiomyocytes derived from human embryonic stem cells in pro-survival factors enhance function of infarcted rat hearts. *Nat Biotechnol* 2007;**25**:1015–24.
225. Wang Z. MicroRNA: a matter of life or death. *World J Biol Chem* 2011;**1**:41–54.
226. Kim H, Ashraf M, Haider K. Direct transfer of miR-210 from preconditioned stem cells to the host cardiomyocytes via gap junctions. *Circulation* 2011; (Abstract). Nov. (Suppl).
227. Hu S, Huang M, Nguyen PK, Gong Y, Li Z, Jia F, et al. Novel microRNA prosurvival cocktail for improving engraftment and function of cardiac progenitor cell transplantation. *Circulation* 2010;**124**:S27–34.
228. Lai VK, Ashraf M, Haider KH. MicroRNA-143 is critical regulator of cell cycle activity in stem cells with co-expression of Akt and angiopoietin-1 transgenes via transcriptional regulation of Cyclin D1. *Circulation* 2011; (Abstract). Nov. (Suppl).

Index

Note: Page numbers followed by "*f*" indicate figures, and "*t*" indicate tables.

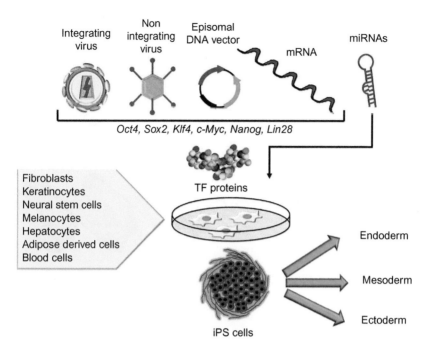

SOHN *ET AL.*, FIG. 1.

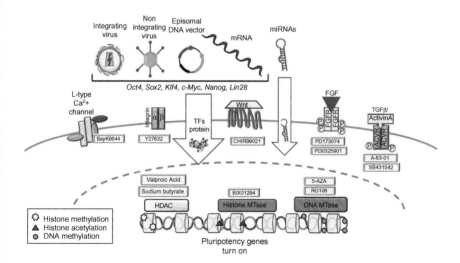

SOHN *ET AL.*, FIG. 2.

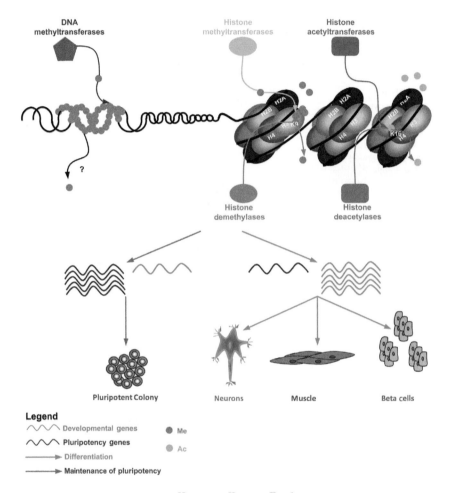

DNA
methyltransferases

Histone
methyltransferases

Histone
acetyltransferases

Histone
demethylases

Histone
deacetylases

?

Pluripotent Colony

Neurons

Muscle

Beta cells

Legend

〰〰 Developmental genes

〰〰 Pluripotency genes

⟶ Differentiation

⟶ Maintenance of pluripotency

● Me

● Ac

Hoxha and Kishore, Fig. 1.

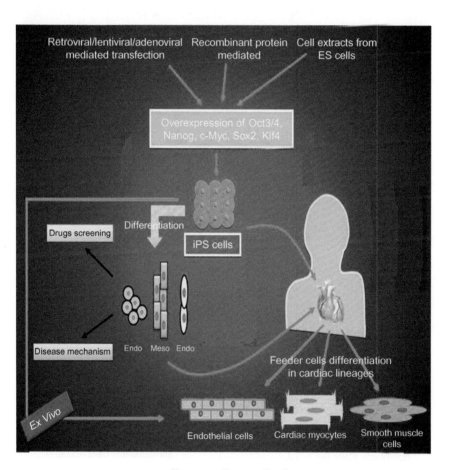

Hoxha and Kishore, Fig. 2.

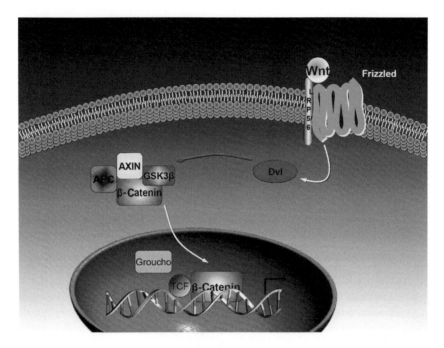

CHIMENTI *ET AL.*, FIG. 2.

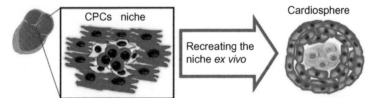

- CSCs + supporting cells (myocytes, fibroblasts, Cajal-like cells).

- Junctions (cadherins, connexins, integrins).

- Structural organization of cell types.

- Physiologic ECM.

- CPCs (c-kit⁺) + supporting fibroblast-like cells (CD90+).

- Espression of connexin 43, cadherin 1, and integrin β2.

- BrdU+/c-kit⁺ cell in the core; differentiation gradient.

- Production of ECM (collagens, fibronectin, laminins).

CHIMENTI *ET AL.*, FIG. 3.

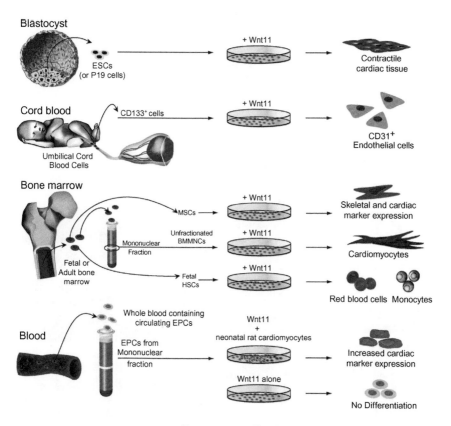

Blastocyst

ESCs
(or P19 cells)

+ Wnt11

Contractile
cardiac tissue

Cord blood

CD133⁺ cells

Umbilical Cord
Blood Cells

+ Wnt11

CD31⁺
Endothelial cells

Bone marrow

Fetal or
Adult bone
marrow

Mononuclear
Fraction

MSCs

+ Wnt11

Skeletal and cardiac
marker expression

Unfractionated
BMMNCs

+ Wnt11

Cardiomyocytes

Fetal
HSCs

+ Wnt11

Red blood cells Monocytes

Blood

Whole blood containing
circulating EPCs

EPCs from
Mononuclear
fraction

Wnt11
+
neonatal rat cardiomyocytes

Increased cardiac
marker expression

Wnt11 alone

No Differentiation

FLAHERTY *ET AL.*, FIG. 4.

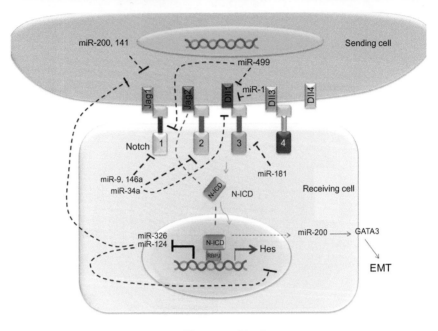

Tang *et al.*, Fig. 2.

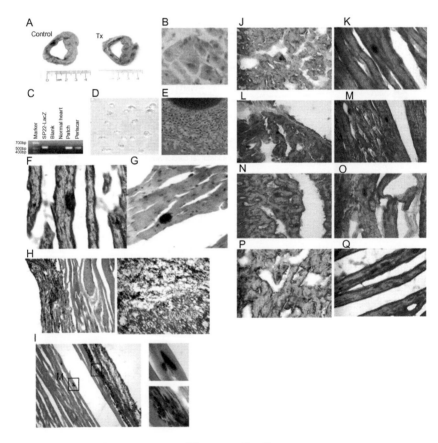

Wang *et al.*, Fig. 5.

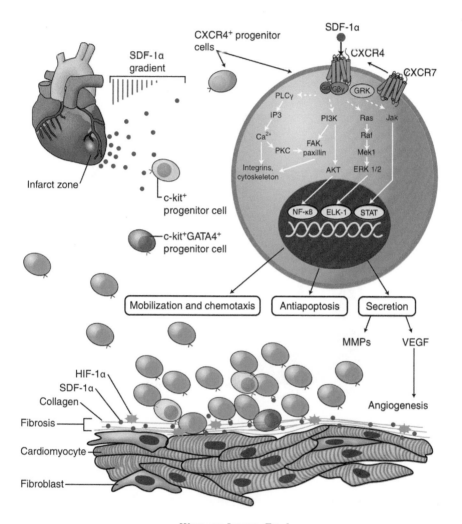

WANG AND LUTHER, FIG. 1.

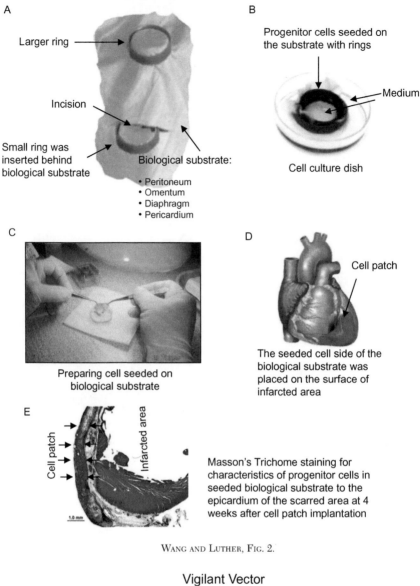

WANG AND LUTHER, FIG. 2.

Vigilant Vector

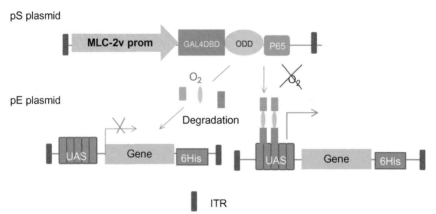

PHILLIPS AND TANG, FIG. 1.

Printed and bound by CPI Group (UK) Ltd, Croydon, CR0 4YY

08/05/2025

01864957-0002